Ecology and Behaviour of the African Buffalo

Chapman & Hall Wildlife Ecology and Behaviour Series

Series editor: Dr R.J. Putman, Department of Biology, University of Southampton, UK

This series brings together a number of authoritative monographs on ecology and behaviour written by experts in their field. They are concisely presented, scholarly works at research level, but written in such a way as to be readable and affordable for the interested layperson.

Each book presents an up-to-date and authoritative review of the ecology and behaviour of a particular animal species or group, and in addition uses detailed case studies to explore a number of more general conceptual issues or principles in behavioural ecology. Developed from the original research of the author, each account offers a definitive review of the species or group 'under the spotlight' with that unmistakable 'immediacy' of style gained from being written by someone directly involved with ongoing research on the animals and issues concerned. In addition, through this detailed case study approach, they also offer unique insights into a number of more general principles of behavioural ecology and can often provide a critical test of accepted theory.

Published

Ecology and Behaviour of the African Buffalo: Social inequality and decision making
H.H.T. Prins

Behavioural Ecology of Siberian and European Roe Deer
A. Danilkin (in association with *A.J.M. Hewison*)

In preparation or being planned

Comparative Behavioural Ecology of Madagascan Lemurs
J. Ganzhorn, P. Kappeler and S. O'Connor

Behavioural Ecology of Black Bears: An analysis of factors determining home range behaviour
R.A. Powell

Fallow Deer: Behavioural plasticity in contrasting environments
J. Langbein and S. Thirgood

Behavioural Ecology of Muntjac Deer
N. Chapman and S. Harris

Competition and Resource Partitioning in Ungulate Assemblies
R.J. Putman

Ecology and Behaviour of the African Buffalo

Social inequality and decision making

H.H.T. Prins
Professor of Tropical Nature Conservation
Wageningen Agricultural University
Wageningen, The Netherlands

CHAPMAN & HALL
London · Glasgow · Weinheim · New York · Tokyo · Melbourne · Madras

Published by Chapman & Hall, 2–6 Boundary Row, London SE1 8HN, UK

Chapman & Hall, 2–6 Boundary Row, London SE1 8HN, UK

Blackie Academic & Professional, Wester Cleddens Road, Bishopbriggs, Glasgow G64 2NZ, UK

Chapman & Hall GmbH, Pappelallee 3, 69469 Weinheim, Germany

Chapman & Hall USA, 115 Fifth Avenue, New York, NY 10003, USA

Chapman & Hall Japan, ITP-Japan, Kyowa Building, 3F, 2-2-1 Hirakawacho, Chiyoda-ku, Tokyo 102, Japan

Chapman & Hall Australia, 102 Dodds Street, South Melbourne, Victoria 3205, Australia

Chapman & Hall India, R. Seshadri, 32 Second Main Road, CIT East, Madras 600 035, India

First edition 1996

Typeset in Palatino 10/12pt by Saxon Graphics Ltd

ISBN-13: 978-0-412-72520-3 e-ISBN-13: 978-94-009-1527-5
DOI: 10.1007/978-94-009-1527-5

A catalogue record for this book is available from the British Library

Library of Congress Catalog Card Number: 95-71090

Printed on permanent acid-free text paper, manufactured in accordance with ANSI/NISO Z39.48-1992 and ANSI/NISO Z39.48-1984 (Permanence of Paper).

For my mother, Pietertje Anna Catharina (Ita) Prins neé Poorter

Contents

Series editor's foreword

Over the past 30 years or so, research effort in behaviour and ecology has progressed from simple documentation of the habits or habitats of different species to asking more searching questions about the adaptiveness of the patterns of behaviour observed; moved from documenting simply **what** occurs, to trying to understand **why**. Increasingly, studies of behaviour or ecology explore the function of particular responses or patterns of behaviour in individuals or populations – looking for the adaptiveness that has led to the adoption of such patterns either at a proximate level (what environmental circumstances have favoured the adoption of some particular strategy or response from within the animal's repertoire at that specific time) or at an evolutionary level (speculating upon what pressures have led to the inclusion of a particular pattern of behaviour within the repertoire in the first place).

Many common principles have been established – common to a wide diversity of animal groups, yet showing some precise relationship between a given aspect of behaviour or population dynamics and some particular ecological factor. In particular, tremendous advances have been made in understanding the foraging behaviour of animals – and the 'decision rules' by which they seek and select from the various resources on offer – and patterns of social organization and behaviour: the adaptiveness of different social structures, group sizes or reproductive tactics. Equal progress has been made in understanding the relationship of many population processes – rates of increase or mortality, relative stability of population size and structure – to environmental character.

Such principles or **explanations** of patterns of behaviour or ecology are now widely recognized and find treatment in almost any undergraduate text. Yet, important though it is to recognize such generalities of pattern, in the very derivation of such general principles often much subtlety and elegance apparent in an individual case is lost. And it may be that very subtlety which is of critical importance in fine-tuning adaptation. Indeed, it has become increasingly apparent over more recent years that there is a tremendous flexibility of behaviour and ecology even within a species.

While earlier research had concentrated on exploring more general patterns, had sought to explain interspecific differences in behaviour or individual or population ecology in relation to observed environmental differences, in practice almost equal variation is apparent between different populations of one and the same species.

This new series tries to develop a focus on that subtlety of adaptation by exploring in detail the behaviour and ecology of particular species or closely related groups and trying to show how particular aspects of the behaviour or ecology described may be adaptive. In focusing on the subtlety of adaptation shown within a single species or even single population however, they may in addition add new perspective, a new depth to our understanding even of those more general principles. Many of the earlier breakthroughs in developing the broader conceptual framework in behavioural ecology derived from comparative studies of closely related species – or by investigation of the wealth of increasingly recognized intraspecific variation in behaviour and ecology – and such comparative studies often reveal important new insight into the shaping of relationships between behaviour, ecology and environment.

This series then brings together a sequence of authoritative monographs on behaviour and ecology of particular species or species groups. They are concisely presented, scholarly works at research level, written by experts in their field, but also written in such a way as to be readable and accessible also by the interested layperson. Each book in the series presents an up-to-date and comprehensive review of the behaviour and ecology of some particular animal species or group, but in addition uses these detailed case studies to explore a number of more general conceptual issues or principles in behavioural ecology or population ecology.

In each case, developed upon the original research of the author, each account thus offers a definitive review of the species or group 'under the spotlight', with that unmistakable immediacy of style and understanding of having been written by someone directly involved with ongoing research on the animal/issues concerned. But in addition, through this detailed case-study approach, each also offers unique insight into a number of more general principles of behaviour or ecology and can often provide a critical test of accepted theory.

For the first volume in this new series, Professor Herbert Prins offers a personal perspective on the ecology and behaviour of African buffalo – reviewing the results of many years of original and innovative research, particularly at Lake Manyara in northern Tanzania. The primary focus of his research, throughout, has been on the central question: 'What are the benefits that individual animals gain from living within a social group?'. While a number of general advantages are now widely recognized, in terms of a reduced risk of predation, increased access to potential mates or in increased efficiency of foraging on certain types of resources, such explanations are based upon considerations of a group of 'average' individuals and take little account of variation in behaviour, status or condition of individuals within that group. Prins focuses on this very variation between individuals to consider the position and role of different individuals of different status within the group as a whole. Recognizing the group as an assemblage of varied individuals, he also addresses the possibility of an additional advantage of group living in the transfer of information between individuals. While the idea of information transfer within the group is widely accepted in invertebrate societies (particularly among social insects), until recently far less has been published in relation to vertebrates, on the extent to which the individual may be able to make use of information gathered by conspecifics, adjusting its own behaviour in response.

Professor Prins uses his detailed studies of the African buffalo as a model through which to explore some of these fascinating questions of interaction between individuals within a social species and the position of the individual within the population. As anticipated for the series as a whole in its original conception therefore, this book is much more than simply a definitive monograph, an up-to-date review of current knowledge on the behaviour and ecology of the African buffalo itself; it has a 'hidden agenda' in exploring, through this example, more general principles about why individuals within a population behave as they do and the ecological implications of 'the individual within the population'.

R.J. Putman
University of Southampton, UK

If we want to appreciate the wonders of the natural world, we increasingly have to go to National Parks and other protected areas. These areas are protected against Man. This has become necessary as with the ever-increasing human population the really pristine areas are disappearing at a tremendous rate. This book on the African buffalo is about research that takes place in such an area.

Only forty years ago when I came to the shores of Lake Manyara in the Great Rift of Tanzania, the area truly could be called a wild place. In 1958, when it was gazetted a Game Reserve, there was a teeming black rhinoceros population and the highest density of African elephant and African buffalo occurred here as compared to any other area in Africa. However, outside the park more and more people settled and, how difficult to imagine at present, an official programme aimed at eradicating black rhino and buffalo was executed just outside the park's boundaries at that time to give way for irrigated cultivation. In 1962 the area received an increased protection status and Manyara became a National Park. We should be grateful for the foresight of Governments at that time that the protection of this and other areas took place before all the land was to be allotted to use by humans only.

Even though the human population increased, one still could experience Lake Manyara National Park as forming a part of the true African wilderness. Professor Prins has been lucky enough to be allowed to work in this unique area before disastrous events took place, but he has been working there also after these events, and witnessed their impact. I am referring to those bleak years at the end of the 1980s, when the black rhino was poached to extinction in Manyara and its unique elephant population was reduced by over eighty per cent.

Simultaneously, the park became a small island in the middle of cultivation. Increasingly, events play havoc with the small populations of mammals living in the park: it becomes more and more difficult for individuals of these mammals to recolonize the park, and random extinctions slowly but surely remove more and more species from the original faunal assemblage.

Eland, hartebeest, oribi and wild dog have disappeared from Manyara over the last thirty years and the status of reedbuck and klipspringer is precarious. The same applies to Manyara's bird species, with, for example, Narina's trogon extinct and the Giant Kingfisher disappearing. As an ecologist Professor Prins registered those events so that through his study of these 'natural experiments' we can learn about cause and effect in these tropical savanna ecosystems. Ultimately, it is only through such studies that nature protection and nature management will be on a firm footing instead of being based on mere guesses and good intentions. Yet, I know that he, at the same time, is deeply moved by the havoc played by Man, often unknowingly, in these natural ecosystems.

This book is mainly about the social organization of African buffalo in Manyara. His study shows that besides being an animal that can ferociously protect itself against its enemies, the buffalo's herd life is a wonder of adaptation to finding food and raising offspring. Far from being just a mob of animals, a buffalo herd has a clear organization closely akin to the sociality one normally associates with primates. The study even hints at the possibility of communal decision making in buffalo, even though many of us would think, unjustified, this to be the prerogative of Man alone.

B

In the present time and day, a study like this can only be executed in a protected area. Only there, animals such as buffalo, can be observed and studied long enough that one can learn to recognize individuals, which is necessary for unravelling their behavioural and especially their social organization. Only here wild animals trust people enough to show their natural behaviour because here they can lay down the fear accumulated over the generations since hunting and harassment are absent. This animal trust should not be broken since otherwise not only the animals will come to grieve but without it we will live in a poorer world too. I have been working many years to help to save wildlife and so have hundreds of rangers and scouts, National Park officials, and Governments all over Africa, not the least those of Tanzania. The scientific facts presented in this book show that Lake Manyara National Park is too small to ensure the continued survival of its wild animals, and it is very gratifying to know that the Government of Tanzania through its custodian of wildlife, that is Tanzania National Parks, has been working hard to extend Manyara's boundaries. Already twice they have succeeded, first in 1972 and again in 1991.

Now the inclusion of the large Marang Forest can proceed any time, and further extension to the south of the park is investigated. Only through the close cooperation of Government institutions, scientists, nature protection organizations, and the public, both in Africa and in the rich countries, can Nature survive so that our children and grandchildren can learn from it and enjoy it. I sincerely hope that the African buffalo will be amongst the survivors.

Bernhard

Prince of the Netherlands

Introduction

Only two hundred years ago, the African buffalo was hardly known to Europeans, and for a long time descriptions of the animal were surprisingly vague. For instance, in a dictionary of 1849 it could be learnt that 'Two or three species of Buffalo inhabit the woods and marshy grounds [of Africa], but little is known of their form or habits... he is a savage and dangerous animal' (Goodrich 1849, pp. 821, 822). Yet, with the progress of exploration of the African interior a new edition of this dictionary gave a more adequate description: 'The African buffalo... is of a rough, shaggy, wild appearance, and in a state of nature is altogether a savage and formidable brute... They live in large herds, and though they generally fly from man, when wounded are extremely dangerous. Sometimes, also, they will make sudden and fierce attacks upon hunters and travellers they chance to meet...' (Goodrich 1875, p. 489). Fleeing was often the best way to deal with a wounded buffalo, especially in days when repeating guns had not yet been invented:

> The Cape buffalo is, at all times, a dangerous animal to hunt; as when wounded or closely pressed, he will not unfrequently turn and run down his pursuer, whose only chance of escape, if he be a colonist or European, is in the swiftness of his steed... When the Cape buffalo is wounded, he ploughs up the ground with his horns, scattering the turf and stones on all sides, and rushes through the thickets and up the rugged sides of the steep ascent with astonishing energy. Often, when urged by revenge, instead of escaping, he returns suddenly to the attack, and appears in the middle of his assailants, one or more of whom often fall victim to his rage: he gores them, tramples upon them, and presses upon them his whole weight, as if to glut his revenge...
>
> (Williams 1859, p. 108)

By the end of the nineteenth century, the African buffalo had gained a notorious reputation of ferocity and cunning. Stanley, the great explorer, was one in a series who canonized the buffalo as one of the few worthy adversaries to be faced by White Man in his struggle to conquer the Dark Continent. A typical example of his writing serves to highlight the human's bravery, coolness and superiority but, at the same time, reinforces the buffalo's reputation:

> This morning... Saat Tato the hunter set out to shoot game for the camp, accompanied by a few young fellows anxious to participate in the sport. Two buffalo fell victims to the hunter's unerring aim, but a third one, wounded only in the leg, according to the cunning instinct of the beast, rushed away, and making a circle hid himself in some branchy acacias to await his opponent. Mabruki, the son of Kassim, thought he knew the art of buffalo hunting, and set out on the tracks of the wounded animal. The buffalo on the alert no sooner discovered his enemy, than uttering a hoarse bellow charged and tossed him, one of his horns entering the thigh of the unhappy man. While thus prostrate, he was pounded with the head, gored in the side, arms, and ripped in the body, until Saat Tato, hearing the screams, rushed to rescue when almost too late, and planting a shot in the buffalo's head, rolled him over, dead.
>
> (Stanley 1890, p. 385)

Big game hunters have continued to elaborate on the cunning, aggression and almost supernatural strength of the African buffalo for the last hundred years:

> Buff are highly inconsiderate once their dander is up, regularly shrugging off major holes in their exteriors that would stall an armoured personnel carrier. There is no rationing on dander when it comes to Inyati [buffalo], probably because their irritability with fumble-fingered hunters is expressed in pure adrenaline. This is translated into the following formula: B + LS = F. For the uninitiated, this means Buff + Lousy Shot = Funeral. Excellent historical precedent indicates that it will be yours.
>
> (Capstick 1983, p. 24)

As well as their formidable strength, buffalo have many other facets. I am the last to deny that they are dangerous but, whether they are more or less dangerous than, say, elephant or lion, is an ideal subject for a camp fire talk. During my years in Africa I have witnessed accidents with buffalo and heard many hunting stories, some of them quite unbelievable, some intended to show the buffalo's meanness:

> In one hunt, a British officer who was pursued by a wounded buffalo took refuge amidst the branches of a low, stunted tree. The infuri-

> ated animal, though unable to reach him with its horns, effectually used its tongue as a weapon of offence, with whose rough, grating surface – by licking the legs and the thighs of the unfortunate sufferer – it so completely denuded them of flesh, that although at last rescued from so dreadful a situation by some Hottentot attendants, who shot his tormentor, the colonel only lingered on for a few days, when he died in the most excruciating agonies.
>
> (Williams 1859, p. 112)

My research on buffalo, however, follows a different line of tradition, namely the one which was started by naturalists at the turn of the nineteenth century into the twentieth. This tradition culminated in the careful studies by today's behavioural scientists. One of those early naturalists was Radclyffe Dugmore, one of the very first wildlife photographers in Africa. In his books (1910, 1925) he painted a different picture of the African buffalo:

> Judging from the numerous stories I had read of the terrible ferocity of these powerful beasts, I had fully expected to have a lot of trouble, but from what I have since learned it seems that buffalo will not, as a rule, attack when in herds unless they have been fired at, and even then, unless a cow is killed or wounded, they will more often than not run away, just as most animals do if they have the chance.
>
> (Radclyffe Dugmore 1910, p. 42)

Radclyffe Dugmore, it will be no surprise to learn, was also a man who called for the conservation of African wildlife. Another famous photographer at that time, Johnson, summarized the buffalo's behaviour as follows:

> The African wild buffalo... is exactly like a man who is always being stampeded by the latest rumour. He is continually and unnecessarily on the alert; not sensibly... but in a silly, apprehensive way that must take much of the joy out of his life. He will graze for a few moments in a natural manner; then up goes the head and he feels the wind. There may be no enemy within many miles, and his herd may have been unmolested by man or beast for months. It makes no difference, he is always nervous and ready to enjoy a panic at the slightest provocation... Perhaps it is the same nervousness that makes the buffaloes such aggressive animals.
>
> (Johnson 1928, p. 99, 100)

Although Radclyffe Dugmore and Johnson were not scientists, MacLatchy (1932) was. He discovered that buffalo are not mere cattle but have a clear social organization. He observed, for example, that 'in the rear of a group, some metres behind, there is often an old male, which the local hunters call '*l'homme pour le palabre*; they think he prefers the duty of a rear-guard to defend the animals which move slowly... these are often the largest and the oldest. This would be against their instinct because they prefer to be alone'. MacLatchy also observed so-called bachelor bulls, and wrote 'Some authors claim that young and vigorous bulls chase [these] other bulls out, but how can this be the case if they are so large? Is it perhaps hypochondria that causes these large males to become solitary?'. The wording is not that of a modern behavioural ecologist, but the questions and observations were in the same vein as of today. The result was that buffalo became to be seen as individuals in their own right, and not only as specimens to be bagged by a hunter.

A few researchers have paid particular attention to the behaviour and the ecology of the African buffalo. I refer especially to the study of Sinclair (1977), which focused on its ecology in the Serengeti of northern Tanzania, and that of Grimsdell (1969), which concentrated on the ecology and behaviour of buffalo in Uganda. From these studies a clear picture of social organization and its relation to the buffalo's ecology emerged. This laid the foundation for my own work on buffalo in northern Tanzania. However, since the science of behavioural ecology lays more and more emphasis on individuals instead of on populations, I increasingly tried to envisage buffalo as individuals within a society. This means that observed behaviour was interpreted by me in terms of individual gains and losses. Because of this emphasis on individuals, and thus on individual recognition, I chose Lake Manyara National Park as the area where I would concentrate my efforts. This national park has three major advantages: the buffalo density is extremely high, the park is small, and it is narrow. Individual recognition had quite a high success rate in this park.

At the end of my study on the buffalo I attempted to answer one of the major questions that has coloured much of my research on geese in temperate and arctic areas (for example, Prins *et al.* 1980, Ydenberg *et al.* 1983, Ydenberg and Prins 1984), as well as on buffalo, namely 'What are the benefits that individuals gain from living socially?'. This is the reason why the present book finishes with communal decision-making by buffalo. At present my thinking on animals is geared more and more towards what is now called cognitive ethology. Biologists now dare to ask questions from which we have refrained too long, such as whether animals use abstract concepts and have motives, beliefs, and desires or whether they can count or have numerical competence. The question of numerical competence is particularly important for understanding the buffalo's approach to the exploitation of its food resources. Hence, the present research finishes with the buffalo's cognitive abilities (in juxtaposition to adaptive behaviour shaped through natural selection in the course of evolution). For those readers who shy away from the idea that animals could have cognition, I present the definition as used by Cheney and Seyfarth (1992), namely 'cognition [is] the ability to relate different unconnected pieces of information in new ways and to apply the results in an adaptive manner'. Domestic cattle and water buffalo have a reputation of stupidity because too many people confuse stupidity with docility. However, they can be trained easily – just as horses and dogs. David Livingstone, as most other travellers at the time, used trained riding cattle, and not horses, for exploring southern Africa. Even today the Brazilian police on the swampy island of Marajo in the Amazon estuary use riding buffalo. Williams (1859, p. 23) described how the Hottentots in South Africa had trained their cattle to herd sheep and goats instead of using dogs:

> The ox... seems little inferior to that most sagacious of all quadrupeds, the dog. Among the Hottentots, these animals are their domestics, and the companions of their pleasures and fatigues. While the sheep are grazing, the faithful *backeley*, as this kind of oxen is called, stand and grazes besides them. Still attentive, however, to the looks of its master, the *backeley* flies round the field, obliges the flocks of sheep that are straying, to keep within proper limits, and shows no mercy to robbers, who attempt to plunder, nor even to strangers. It is not the plunderers of the flock alone but even the enemies of the nation that these oxen are taught to combat. Every army of Hottentots was furnished with a proper herd of these creatures, which were let loose against the foe. Thus sent forward, they overturned all before them, and often procured their masters, even before they had struck a blow, an easy victory.
>
> (Williams 1859, p. 23)

Whether or not one accepts Williams' description of these oxen is irrelevant. Anybody who has observed the close rapport between water buffalo and little boys in South-East Asia or between Masai and their cattle will, undoubtedly I think, accept that animals such as buffalo are not just *automata* but are gifted with specific cognitive abilities.

The African buffalo used to be a highly successful animal in ecological terms. It occurred over vast areas of Africa, hardly surpassed by any other mammal save the leopard and the African elephant. It was found in the near-deserts of West Africa, the rain forests of the Congo Basin, the savannahs of eastern and southern Africa, and even at very high altitudes. When McGregor Ross (1911) described the find of a frozen carcass of a buffalo near a glacier (4800 m above sea level) on Mount Kenya, it inspired Brian Brooke to compose a poem (quoted by Beard 1988):

Under the mists of the Kenya snow
When the game has passed from the realm he held,
And the waterless plains are fenced and welled,
When the forests are ploughed and the trees are felled,
As a monument to the Bush and Plain,
Where in ages past he was wont to reign,
There the Buffalo Bull will still remain,
Safe, in the mists of the Kenya snow.

Indeed, perhaps the carcass of this buffalo was to remain safe; the living buffalo is not. Although it recovered from the great rinderpest plague of the 1890s, many buffalo populations are at present under severe threat. The plight of the African elephant is well publicized, that of the African buffalo is not. Even as recently as 1988 buffalo populations were wiped out by rinderpest in the Waza and Benoué National Parks of Cameroon without anybody outside the area paying attention. However, the present-day drive for more food production for the ever-growing human population has caused a great reduction in the area left for wild animals: fencing and welling, ploughing and felling take place all over the continent. This has caused a drastic decline in the buffalo's range, and thus its numbers. In addition, there is an increased demand for meat, from which the buffalo also suffers.

> Commercial meat poaching is dangerously affecting northern Serengeti buffalo herds, whose numbers have dropped from over 11,000 in 1986 to 6,142 in 1992, a decline rate of about ten percent per year.
>
> [Daily News (Tanzania), 4 November 1992]

Around Tarangire National Park this decline is at the moment even about 30% per year. The vast buffalo herds of Mozambique were wiped out over the last two decades because people needed meat. Even if the buffalo herds were to survive these onslaughts on their numbers, development agencies will ensure their extinction through opening up vast areas of interior Africa by tsetse control operations: cattle will move into areas where they could not survive before and ultimately there will be no place left for wild herbivores such as buffalo (Prins 1992). It will be a great shame if the African buffalo become extinct at the hands of humans, in the same way as so many other creatures that live and lived in Africa. The great African writer, Laurens van der Post (1965, pp. 125–126), when he described the African buffalo, also described what we will lose if the buffalo should become extinct:

> Twelve yards away, his muzzle wet and shining with light and health, grey scimitar of powerful horn thrown back, expert nostrils

> searching the air, stood our first buffalo bull, the grass wherein he stood clasped around his throat and lying like a cape of gold against his purple flanks. It is for me always a solemn moment when I first see buffalo again after a long time. The buffalo's powerful head darkening the yellow grass, like the lion's imperative roar and the elephant's long, somnambulistic stride, has more quintessential Africa in it for me than any other manifestation of all the scores of animals that I love and know. It's as if in the buffalo the very stones and earth of Africa, all its vast, ponderable, withdrawn and solified matter have turned magically into a living design of life; as if all the tranced magnetic power of the land is released in their drumming hooves and in the easy lift of the weight of their head. In their passionate defence of themselves, in their unpredictable reaction to intrusion, indifferent or benign, in their readiness to give way, their determination to avenge an injury with indefatigable cunning, in all these characteristics it's as if they're a living expression of the inarticulate character of the African earth.

My writing, although not as evocative and beautiful as Van der Post's description of the African buffalo, provides, I hope, additional arguments for saving this fellow-species from extinction. Its social organization and decision-making procedures are as fascinating as our own.

Acknowledgements

It is a pleasure to acknowledge the support of so many people during my research. To mention them all is almost impossible. Yet, my gratitude for their help in shaping my ideas and for their active participation is sincere. I would like to mention especially my colleagues Jan Beekman, Henrich Bruggemann, Floris Deodatus, Glenn Iason, Henk van der Jeugd and Paul Loth. We spent so much time together collecting data in Manyara, and Paul and I particularly shared many adventures all over Africa. In the same class are Fred de Boer, Iain Douglas-Hamilton, David Scarce and Rick Weyerhaeuser, who helped to analyse the data.

The study would have been impossible without permission to work in Lake Manyara National Park. Permits were granted by the Serengeti Wildlife Research Institute (SWRI), the Tanzanian National Commission for Science and Technology and the Board of Trustees and Directors of Tanzania National Parks (TaNaPa). I would like particularly to thank Dr C. Mollel, Professor Dr K. Hirji, Dr G. Sabuni (SWRI) and Messrs T.M.L. ole Konchella, D.S. Babu, J. Magombe and E. ole Kapelondo (TaNaPa). The successive ambassadors and staff of the Royal Netherlands Embassy in Dar-es-Salaam were also always very helpful.

My time in Ndala Research Camp in the middle of Manyara was always a pleasure because of the many rangers of Tanzania National Parks. Very special thanks are due to Mhoja Burengo and Renatus Alois, who accompanied me in the field, often on foot, and who have saved my life when elephants or rhinos thought it fit to attack. They also were a permanent source of information. The wazee Shaban Swalleh, Tseama Pissa, Lulu Thomas, Taataa and Louisi Mollel made life sweet in Ndala as members of my household. They provided me with company, food and general support but also with endless detailed information on people, plants and animals. I am especially grateful to Rudi Drent and Jaap Kruijt (University of Groningen), Tim Clutton-Brock (Cambridge University) and the late Dick Thalen, who I still miss.

Many thanks are due to Per and Margaret Kullander, Jim and Jacky Marshal, Genetta Hanby, David Bygott, the late Bupath Keshwala, Mike and Gisela Leach, John Scherlis, Alan and Sally Johnston, Aadje Geertsema, Tim and Monique Caro and Suleh, Giteh, Rajinder and Paul Mandh: without you life would have been much more difficult, especially in the 1980s, and definitely less fun!

I would especially like to thank my parents. My father kindled my interest in East Africa, where he conducted much research himself as an anthropologist. My mother deserves a special *dank U wel*: she kept me informed about home and family, and was an intermediary on my behalf with the bureaucracy at large. Without her, I am sure, it would have been impossible for me to be in the field so much.

The main funders of the research have been, successively, The Netherlands Foundation for the Advancement of Tropical Research, the University of Groningen, the Royal Netherlands Academy of Sciences and Wageningen Agricultural University. I wish to acknowledge my indebtedness to the governing boards of these institutions. Other organizations that funded parts of the research are the A.P. Fokkerfonds, Bachienne Stichting, Cambridge Philosophical Society, FONA Fonds, Noorthey Fonds, Universiteitsfonds of the Rijksuniversiteit Groningen, Treub Maatschappij and World Wide Fund for Nature (The Netherlands). I am happy to acknowledge their support, and that of HRH Prince Bernhard. Ecological research somewhere in the middle of Africa is expensive, and I am grateful that so many organizations have been willing to spend so much.

Permission to use previously published material was granted by the following organizations and/or journals: E.J. Brill/*Behaviour* (material contained in Chapters 4, 5 and 7), Blackwell Science/*Journal of Biogeography* (material contained in Chapters 1 and 2), Blackwell Science/*African Journal of Ecology* (material contained in Chapters 1, 2 and 6), International Institute for Aerospace Survey and Earth Sciences (ITC)/*ITC Journal* (material contained in Chapter 1), *Oikos* (material contained in Chapters 5 and 6), Paul Parey/*Ethology* (formerly *Zeitschrift für Tierpsychologie*) (material contained in Chapters 3 and 5), and Springer Verlag/*Oecologia* (material contained in Chapters 5 and 6).

Max Rietkerk and Marjolijn de Brabander did a tremendous job on checking and correcting the references. Dick Visser is in a league apart, since he prepared all figures and often understood better than myself what I wanted to convey. Finally, I am very grateful for the patience of the editor of this series, Rory Putman, and for that of Clem Earle and Bob Carling (Senior Editors Life Sciences, Chapman & Hall). Without their patience and support I doubt whether I would have finished this book.

1

Distribution of resources in time and space

1.1 GENERAL BACKGROUND

Lake Manyara National Park is nearly unsurpassed in its extremely high density of large mammals. In terms of biomass, this constitutes about 17 000 kg km^{-2}. Its high density of buffalo, about 20 individuals km^{-2}, is also rare, and, until about 1987, its density of African elephant (6 km^{-2}) was exceptional (for details see Chapter 3). Manyara lies in the rift valley of northern Tanzania. In this so-called Eastern Rift or Gregory Rift a number of soda lakes lie like a string from north to south. Lake Manyara is one of those lakes. It is situated at latitude 3°30′ south and longitude 35°45′ east. On the oldest map of the area by Krapf (1849; facsimile in Beard 1988, p. 86) the Manyara area is named Olmaniara, along the caravan route of the Swahili from the coast to Lake Victoria. The name Manyara is a Masai description not for the lake, but in general for a lake shore region (Harris 1951). The park occupies an area of 325 km^2, of which 225 km^2 is lake. The park proper lies between the lake and the steeply rising escarpment of the rift valley. This lake is situated approximately 960 m above sea level and the escarpment rises to a height of about 1200–1300 m in the northern part and to more than 1700 m at the southern end. The high Mbulu Plateau (called after 'Mbulu', a derogatory term of the Masai for the Iraqw tribe living there) above the northwestern part of the park is composed mainly of lava and layers of volcanic ash, overlying the basement rocks. Farther south, ancient crystalline rocks are exposed. Whereas the great rift valley at most places is delineated by both western and eastern rift walls, in Manyara only the western side is bordered by an escarpment. At the eastern side the Masai Steppe gently dips to the west, forming a depression at the foot of the fault scarp where the lake lies. At the western side the escarpment rises steeply and many perennial freshwater springs feed the abundant growth of vegetation and present permanent sources of water to the wild animals of Manyara.

In the oral history of the Iraqw it is remembered that once they lived in Manyara before the advance of the Masai. On the basis of the oral history of the Masai, H.A. Fosbrooke (personal communication) put this at the end of the eighteenth century, which accords with clan leader succession recitals of Iraqw if one generation is taken to be 25 years. According the Iraqw, their ancestors husbanded especially goats but also cattle in what is now Manyara. Two hundred years ago Manyara became unoccupied tribal boundary land, and apparently at that time bush encroachment with *Acacia tortilis* took place; in 1991 very old *Acacia* trees were still standing that, apparently, date back to this event (Prins and Van der Jeugd 1993). I believe that the density of large mammals increased in the Manyara area when it became a tribal boundary wilderness but that later the great rinderpest epidemic of 1889 (Chapter 5) reduced this density again. Kjekshus (1977, p. 75) thought that in Manyara there was no game left at all, but in my opinion he relied too much on Baumann, who crossed Manyara in 1892, and did not note any game observations in his diary. This is because Baumann was incapacitated by malaria in the 3 days he took to pass through Manyara (Baumann 1894).

It is likely that Wandorobo started hunting and collecting honey in this tribal boundary land once it was vacated by the Iraqw (S. Swalleh, personal communication). Their presence was noted by Arbuthnot (1954), who hunted the area in 1927. The members of this hunting party, led by Boers, were the first Europeans ever to hunt in Manyara. Manyara itself stayed tribal boundary land and was not occupied by Masai, probably because of tsetse infestation. When they arrived in the late eighteenth century, they

> for the next hundred years... remained in sole occupation of the northern shore of the lake, where they found abundant pasture and... perennial water supply for their herds. Only in recent years has their hold gradually weakened not from human agency but by the insidious spread of the tsetse fly... this malevolent insect now occupies land which was formerly the home of thousands of cattle, and today [1940] only their Masai bomas remain.
>
> (Watermeyer and Elliott 1943)

The hunting Wandorobo are still remembered by present-day Iraqw (S. Swalleh and A. Seif in interview, 1991): Wandorobo hunted in Manyara with poisoned arrows; their bands were small and ranged between six and ten (including children), and they only killed for meat.

From the 1930s to the 1950s, Manyara was especially used by Europeans to hunt black rhinoceros, for which the area was famous, but they also hunted impala, zebra, buffalo and elephant. In particular, Boers from Eldoret in Kenya annually hunted buffalo in Manyara to obtain leather bands for their trek oxen. It is estimated that they shot about 500 buffalo per year and that the total buffalo population in Manyara and on the plateau was about 5000 individuals during the 1940s and 1950s (D.J. Ruysenaar, personal communication). Neither Masai nor Mbugwe nor Iraqw hunted buffalo (the Mbugwe preferred zebra) and mixed herds of buffalo and cattle could still be observed until the 1960s on the rift valley floor outside the present park (M. Burengo, personal communication). Iraqw rarely hunted buffalo or other game since they were agriculturalists (S. Swalleh, personal communication), and if they hunted it was not in Manyara because it was too far from Mbulu, the core area of their land, but also 'because they did not know the value of rhino horn'. It was only in the 1950s that they started killing rhinos for their horn (A. Seif, personal communication). Because of too much hunting by Europeans, and because of the beginning of cultivation on European estates to the south of Manyara, the area was declared a game reserve in 1958 and was elevated to national park status when Tanzania became independent: at that time there were about 2200 buffalo in Manyara.

Until the First World War there was no permanent settlement around the present-day park. Now, however, the park is hemmed in by cultivation. Immediately northeast of the park there is maize cultivation and there are extensive banana groves. These plantations were established from about 1925 onwards on cleared groundwater forest; the last large parts of this forest outside the reserve were cut down in the 1950s (Warden's Report 1961). In the 1940s these banana plantations were not yet very important, as the area around the village of Mto-wa-Mbu was then mainly cultivated for sugar cane, maize, peppers and sorghum but cotton was the main crop (Watermeyer and Elliott 1943). The alkaline grasslands north of the lake are used by large numbers of wild herbivores, especially wildebeest, zebra and Thompson's gazelle; however, their numbers are declining notably while new settlements accrue. To the west, on the volcanic soils of the plateau, are frequently burned grasslands and scattered Iraqw homesteads. Towards the south, the plateau becomes hillier, with woodlands used for cattle grazing and some cultivated fields. Over the years the plateau has changed dramatically. When the first European explorers traversed this land one hundred years ago, they found lush glades, many rivulets and forests, as did Radclyffe Dugmore (1925) still in 1921. Now it is a rather arid land to a large extent overgrazed or cultivated since 'bush clearing' took place between 1947 and 1962.

During the Second World War the clearing started further north to grow wheat for the so-called War Effort (McDonald 1946). The first Iraqw families settled here in 1947 and definite settlement started in 1957 (S. Swalleh, personal communication). As result of the conversion of the natural forest into agricultural land, a continuous decrease in annual rainfall since about 1945 has taken place on the Mbulu Plateau south of the Ngorongoro Crater (that is in the Karatu subdistrict) (Prins and Loth 1988). Also, because of increased hunting outside the park, the number of buffalo has decreased dramatically since 1991: in 1994 there were about 700 individual buffalo left out of some 2300 in the 1960s to 1980s.

1.2 ECOLOGY OF THE LANDSCAPE

Three principal factors account for the high diversity and density of animal life, both mammals and birds, in Manyara. These are water, soil fertility and the diversity of the landscape. I will discuss them in turn. The following description is based on Loth and Prins (1986), in which publication a detailed vegetation map (scale 1:50 000) can be found.

From the plateau above the escarpment a number of freshwater rivers drain into the park. These form waterfalls in deep gorges, which are lined, at least in the north, with evergreen **gallery forest**, dominated by *Ficus wakefieldii (vasta)*. This type of forest is most noticeable along the Kirurumi River and the Mchanga River (Figure 1.1).

Also at the northern-most end of the park many freshwater springs emerge at the foot of the escarpment, most of them along the road just to the south of the entrance gate. These springs are fed by rainfall over the Ngorongoro Forest through underground seepage. This results in freshwater rivulets at intervals of only a few hundred metres, which caused an evergreen **groundwater forest** to grow up in the middle of the arid steppe-like vegetation of the rift valley. This groundwater forest is dominated by *Trichilia emetica* trees, but other tree species also occur, such as *Croton macrostachyus* and *Cordia africana*. Besides the many tree species, the most common shrubs are *Rauvolfia caffra* and *Cordia goetzei*. Fresh water also comes from the Marang Forest block in the south, which catches considerable rain because of its high elevation, resulting in permanent rivers at the most southern part of the park, again lined by gallery forest, as can be seen along the Yambi River.

Where these rivers flow more widely apart through the lower area of the park, gallery forest did not develop but instead **riverine bushland** came to dominate. Many of the species in that type of bushland are evergreen, but here again, in contrast to the gallery forest and the groundwater forest, many perennial grasses also grow. These bushlands are most noticeable along the Rivers Chem chem, Ndala and Bagayo but also Endabash. Typical bushes here are *Maerua triphylla*, *Cadaba farinosa*, *Capparis tomentosa*, *Justicia cordata* and *Acalypha fruticosa*. The many rivers

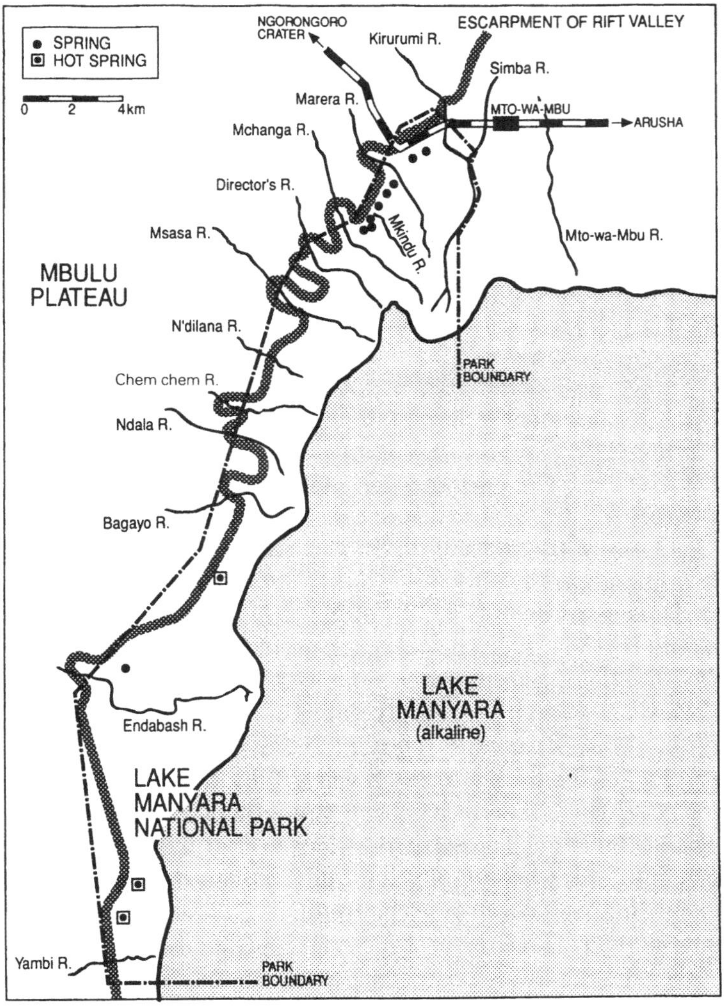

Figure 1.1 Major features of Lake Manyara National Park.

form a series of repeating landscape elements dominated by fresh water, but there is, of course, more to the topography of the landscape. From the west to the east this landscape is dominated by the fault of the rift along which formed the **escarpment**. This escarpment is steep to very steep, and is dominated by *Euphorbia nyikae*, *Commiphora* spp. and *Terminalia brownii* trees and the shrubs *Grewia tembensis* and *Euphorbia scheffleri*. At the northern end it is about 350 m above the rift valley floor. Further south it becomes less steep and less high (about 200 m) between the Msasa and Ndala Rivers (Figure 1.1), and in the past animals such as zebra, eland and hartebeest (the last two now extinct in Manyara) could climb the escarpment here to graze the fertile Chem chem plains (now cultivated) on the **plateau**, formerly dominated by perennial *Themeda triandra* and *Pennisetum mezianum* grass. Here also lay the old caravan route connecting Arusha with Oldeani and beyond to Lake Victoria. This plateau becomes hillier between Ndala and Endabash Rivers, and to the south of the latter it rises even higher to the Marang elevated block, which is capped by the Marang Forest. This forest, which is very infrequently visited by ecologists is often shrouded in mist and has an elevation of 700 m above the rift valley floor. Here the escarpment is so steep that trees cannot grow and bare cliffs are visible. The Marang Forest is poorly known and dominated by *Olea* spp., *Albizia* spp. and *Podocarpus* spp. trees. It is floristically akin to the Ngorongoro Forest.

Soil fertility also plays an important role in explaining the animal abundance of Manyara. The northern part of the plateau, from the Ngorongoro area towards the Ndala River, is of volcanic origin, and weathered basalt forms the basis of the red soil there. This **volcanic plateau** has a high soil fertility and had, until cultivation started here in the 1950s, a high animal density. However, the area to the south of Ndala River, that is including the Marang Forest, is formed from gneisses of the pre-Cambrian **basement complex**. Weathering of this rock type leads to brownish, sandy, infertile soils. Because of these two radically different soil types on the plateau and because of transport by the rivers of sediment into the rift valley, the northern part of the park has a high soil fertility while the southern end is nutrient poor. This gradient has been strengthened by deposition of volcanic ash in the rift valley at the northern end of the park but not at the southern end. This gradient of soil fertility explains the much higher density of game in the north than in the south.

When the rivers leave the groundwater forest at the eastern side, that is at the lakeside, the effect of the lake itself becomes noticeable. This lake has a high alkalinity and even some salinity because all water discharged into it ultimately evaporates and salt and carbonates stay behind. During a wet cycle (Figure 1.4) the lake level rises, and salts infiltrate into the soil; even when the lake recedes it is difficult for most forest species to establish themselves on this soil. Hence, grassy **forest glades** dominated by the grass *Cynodon dactylon* with *Phoenix reclinata* palms and yellow fever trees *Acacia xanthophloea* dominate the landscape where the groundwater table is high. Further to the south, discharge by the rivers is less and an *Acacia tortilis*-dominated landscape has developed in areas where the soil fertility is still high. In this area many herbs are found, but also the grass *Cynodon plectostachyus*. Wherever the salinity is high, the evergreen shrub *Salvadora persica* and the perennial grass *Sporobolus consimilis* grow. These

Acacia woodlands give way south of Bagayo River (Figure 1.1); this is especially noticeable in the Endabash area. This huge alluvial fan of the Endabash River is dominated by *Balanites aegyptiaca* and *Acacia sieberiana* trees, interspersed with shrubs such as *Justicia cordata, Ocimum suave* and *Vernonia cinerascens*. Grass cover is high, particularly with *Urochloa pullulans* but also with *Cynodon dactylon*. Closer to Endabash River conditions become more humid and a dense wooded bush is dominated by *Pluchea dioscoridis* and *Hippocratea paniculata* shrubs and interspersed with the grass *Panicum maximum*. Closer to the river a gallery forest with *Trichilia emetica* and *Ficus wakefieldii* trees is found.

Along the lake shore a narrow beach wall has developed with *Tephrosia* and *Indigofera* woody herbs. Behind this the *Sporobolus spicatus* alkaline grass forms a sward a few centimetres high. In some areas of the park fresh water wells up at the foot of the riverine deltas, and here grows the evergreen perennial sedge *Cyperus laevigatus*. This sedge forms the mainstay during the dry season for many herbivores. Finally, at both the northern and southern ends of the lake swamps with a surface area of several square kilometres have developed. In these swamps *Cyperus laevigatus* is interspersed with stands of the tall bulrush *Typha*. Because many of the migratory animals, such as wildebeest and zebra, return during the dry season to this swamp and its direct surroundings, Lake Manyara National Park forms a cornerstone of the Masai Ecosystem, 35 000 km^2 in size (Prins 1987). Another cornerstone is the Tarangire River (Figure 9.5), where, again, fresh water can be found in the dry season.

The landscape of Manyara that I have described here is the home of the Manyara buffalo. In this landscape they have to find water and food, and shelter against the heat of the sun and against their major predator, the lion. Since water is a daily need, I will first discuss the distribution of water in Manyara.

1.3 WATER REQUIREMENTS AND DISTRIBUTION OF WATER

African buffalo are never found very far from water. Yet they can graze very dry standing hay and roughage nearly devoid of water. The daily water requirements of buffalo can be calculated on the basis of information collated by Taylor (1968). At 22°C their water demand (expressed on a metabolic weight basis) is 12.7 l 100 $kg^{0.75}$ day^{-1}; hence, an adult bull of 700 kg needs about 55 l day^{-1} and an adult cow of about 450 kg requires 39 l day^{-1}. These water requirements have to come from different sources, that is from water in the food, from drinking and from metabolic water through catabolizing body reserves.

Calculations on the water content of the food (Table 1.1) and the total food intake of buffalo (Chapter 2) make it possible to calculate the water intake with the food.

Table 1.1 Moisture contents of green leaves of grasses

Date/season	*Moisture contents as percentage of fresh leaf weight*					
	14 August 1983 Dry season	*16 November 1983 Short rain*	*10 December 1983 Short rain*	*26 January 1984 Short rain*	*3 March 1984 Long rain*	*15 April 1984 Long rain*
Annual grasses						
Group I						
Aristida adscensionis	–	–	–	–	–	63.9
Cymbosetaria sagittifolia	–	–	–	–	–	75.3
Digitaria velutina	–	–	–	–	–	80.0
Group II						
Brachiaria deflexa	–	–	–	78.9	55.2	88.3
Chloris virgata	–	–	–	74.1	65.2	69.2
Dactyloctenium aegyptium	–	–	–	83.5	69.1	73.0
Sporobolus cordofanus	–	–	–	67.2	36.3*	71.7
Tragus berteronianus	–	–	–	69.7	53.3	59.5
Perennial grasses						
Group I						
Eragrostis superba	–	–	77.3	76.1	32.4*	71.3
Panicum maximum	46.9 *	–	72.1	72.4	37.7*	77.2
Sporobolus pyramidalis	41.9 *	–	78.2	65.2	49.1	64.5
Group II						
Cenchrus ciliaris	38.1	–	77.9	73.5	41.3	85.8
Chloris roxburghiana	38.1	36.4	76.8	73.3	40.5	72.3
Cynodon plectostachyus	42.5	–	75.5	81.3	45.4	78.1
Enneapogon cenchroides	37.8	–	72.8	60.6	44.7	67.1
Enteropogon macrostachyus	37.8	–	62.8	67.3	48.3	54.6

Heteropogon contortus	n.c.	–	52.7	68.9	62.4	65.7
Urochloa pullulans	45.0	–	75.0	71.9	73.8	83.0
Group III						
Chloris gayana	55.3	50.0	61.5	63.4	65.8	64.2
Cynodon dactylon	52.6	47.2	60.6	64.8	51.8	65.8
Sporobolus consimilis	57.0	36.3	58.4	57.0	65.7	66.4
Sporobolus spicatus	44.3	47.1	58.1	56.1	46.0	57.7
Perennial sedge						
Cyperus laevigatus	81.3	72.8	81.4	80.8	77.3	74.2

Wilting is indicated by an asterisk. When green leaves were not available, moisture contents could not be determined (–); n.c., not collected. The groups refer to ecological groups with an identical phenological pattern (Prins 1988).

Comparing these amounts with the total water requirements shows that grazing the most succulent food, that is the sedge *Cyperus laevigatus*, can just cover the buffalo's water requirements. However, the other grasses have too low a water content to satisfy the buffalo's water needs. To obtain its water requirements from food, the buffalo would have to ingest large quantities of fresh food. For example, if the water content were 80%, and if the dry matter intake were to cover the energy requirements, then an adult cow would have to ingest between 34 and 61 kg of fresh food and an adult bull between 48 and 100 kg, depending on the exact dry matter requirements. It is doubtful whether buffalo can ingest so much fresh food during a day. Hence, it appears safe to assume that buffalo nearly always have to drink water in Manyara. Some African bovids have the ability to produce dry to very dry faeces to save water (Delany and Happold 1979, p. 279 ff.). Buffalo do not fall into this category. I investigated this by measuring the water content of buffalo dung. The water content of the faeces over the year (Figure 1.2; 153 samples) shows that buffalo dung in Manyara always has a very high water content, slightly over 80%, much higher than the water content of most of its food sources (on average about 60%).

Thus, as the buffalo do not appear to reduce water loss by producing dry faeces, I conclude that water was never limiting for them during the study period. For this reason, the Manyara buffalo can only obtain their water requirements from food if they graze *Cyperus laevigatus* only; this sedge, however, cannot meet their protein requirements (Chapter 2). Since buffalo do not save water by producing dry faeces, they have to

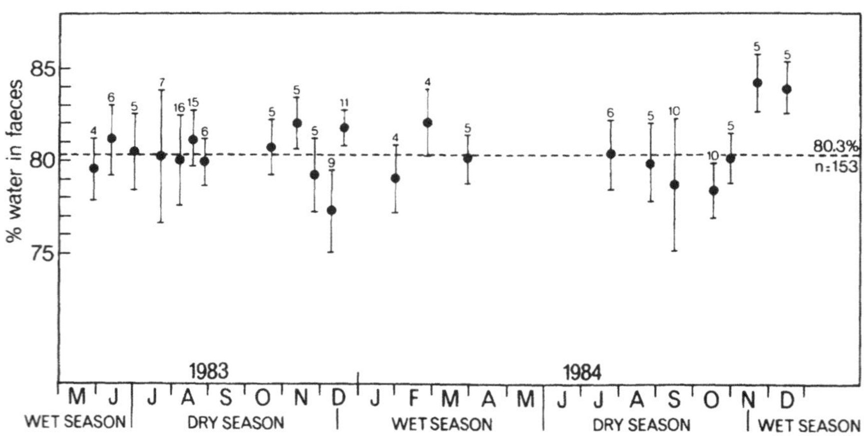

Figure 1.2 Water content of fresh buffalo dung over the year: means and 95% confidence limits. The numbers denote sample size.

drink. This is borne out by the observations that they drink nearly every day in Manyara, as was also observed by Mloszewski (1983).

In Manyara buffalo can find water in many places but this situation is deteriorating slowly. The springs at the base of the escarpment in the north produce large quantities of excellent drinking water. In addition, many of the rivers (Figure 1.1) contain fresh water all year round; this is true of the Rivers Simba, Kirurumi, Marera, Mkindu, Director's, Msasa (all in the north), Bagayo (in the middle of the park) and Yambi (in the south). The Rivers Ndala, Endabash and possibly Chem chem used to be perennial in the 1940s (A. Seif, personal communication), and apparently even still in the 1960s (I. Douglas-Hamilton, personal communication). In the early 1980s these rivers still flowed for about 300 days a year and had standing water in pools during the remaining period. However, at the beginning of the 1990s overgrazing by domestic stock on the plateau above the escarpment, about 10 000 head, affected these rivers badly; even Msasa River stopped flowing for a number of weeks. The pools at the foot of the escarpment in Ndala and Chem chem no longer had any standing water during the dry season.

Before the loss of these pools, the average shortest linear distance between permanent rivers was 3.6 km. Even after the loss of permanent water in Ndala, Chem chem and Endabash Rivers in the 1980s, the average shortest distance between permanent rivers was still only 4.9 km, although the longest distance increased from 7.5 km (between Yambi and Endabash) to 15 km (between Yambi and Bagayo; Figure 1.1). Since buffalo, just like cattle, have to drink at least every 2 days and easily cover a linear distance of 20 km day^{-1}, the distribution of water in the park implies that water was always within easy access for the Manyara buffalo. Choice of grazing grounds or of food plants is, thus, most likely not governed by the water distribution in the park.

1.4 WAXING AND WANING OF FOOD RESOURCES

Manyara lies in the semiarid region of north Tanzania. The mean annual temperature is approximately 22°C and the mean monthly temperatures do not deviate more than 3°C from the yearly mean. Generally it is overcast during many months of the dry season while during the period of the long rains the sky can be very clear resulting in a high insolation and high animal skin temperatures. Because Manyara has a semiarid climate (Figure 1.3), the annual cycle of wet and dry seasons has a profound influence on the food resources of the herbivores.

The mean annual rainfall (1958–90) is only 647 mm. Rainfall is divided over a short rainy season (November to February) and a long rainy season (February to May), with a prolonged dry period from June to November. Rain in October and November is erratic, sometimes even absent, and the

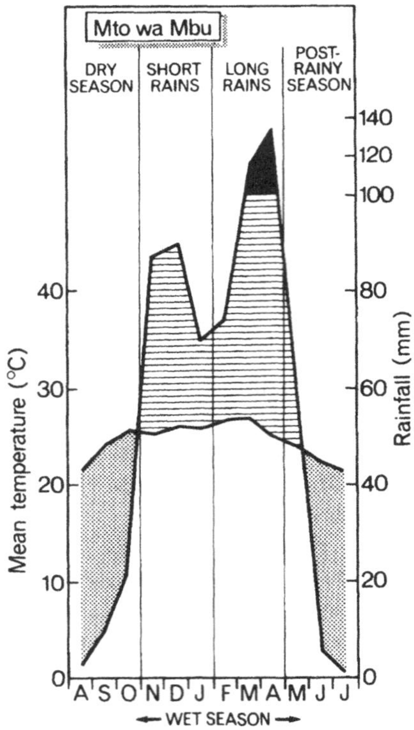

Figure 1.3 Climate diagram for Lake Manyara National Park headquarters in the village of Mto-wa-Mbu. Black indicates a semihumid period, the hatched lines a humid period, and the stippling a dry period.

onset of the short rains may be delayed. The variation in rainfall patterns (Figure 1.4) means that some years vegetation is extremely lush, while in other years the rains nearly totally fail (Table 1.2).

If the rainfall has been excessive, the lake level can rise to such an extent that most of the grasslands are flooded. Grazing herbivores may die because of lack of food or because they are trapped between the lake and the escarpment, where they may be killed by lions; in dry years, diseases such as anthrax can almost wipe out populations of herbivores (Chapter 5).

The effect of the rainfall patterns on food resources is strongly dependent on soil type. In sandy areas lack of rain has an effect on plant growth very quickly. On clayey soils the effects are much less severe and fluctuation in rainfall has a much less visible effect. Near-total buffering of the

Table 1.2 Annual rainfall in Lake Manyara National Park headquarters (in Mto-wa-Mbu)

Year	*Annual rainfall (mm)*	*Number of dry months (with rain < 5 mm)*	*Number of wet months (with rain > 50 mm)*	*Drought or wet year?*
1958–59	382	5	3	Drought
1959–60	575	4	3	–
1960–61	214	4	2	Drought
1961–62	1126	3	6	Wet
1962–63	828	4	7	–
1963–64	996	3	6	Wet
1964–65	594	4	4	–
1965–66	555	4	5	–
1966–67	812	3	5	–
1967–68	1275	2	7	Wet
1968–69	500	5	3	–
1969–70	706	4	4	–
1970–71	632	5	4	–
1971–72	475	4	5	–
1972–73	496	5	5	–
1973–74	288	6	2	Drought
1974–75	546	5	3	–
1975–76	497	6	3	–
1976–77	783	4	5	–
1977–78	456	5	3	–
1978–79	928	3	9	Wet
1979–80	677	5	4	–
1980–81	643	7	4	–
1981–82	509	4	5	–
1982–83	954	3	5	Wet
1983–84	624	6	4	–
1984–85	860	4	6	–
1985–86	634	3	5	–
1986–87	702	4	7	–
1987–88	357	4	3	Drought
1988–89	554	4	5	–
1989–90	528	4	4	–
Mean standard deviation	647±236	4.3	4.6	

A drought year is defined as a year in which the total rainfall was less than the mean annual rainfall minus the standard deviation, while a wet year is defined as a year with more than the annual rainfall plus the standard deviation.

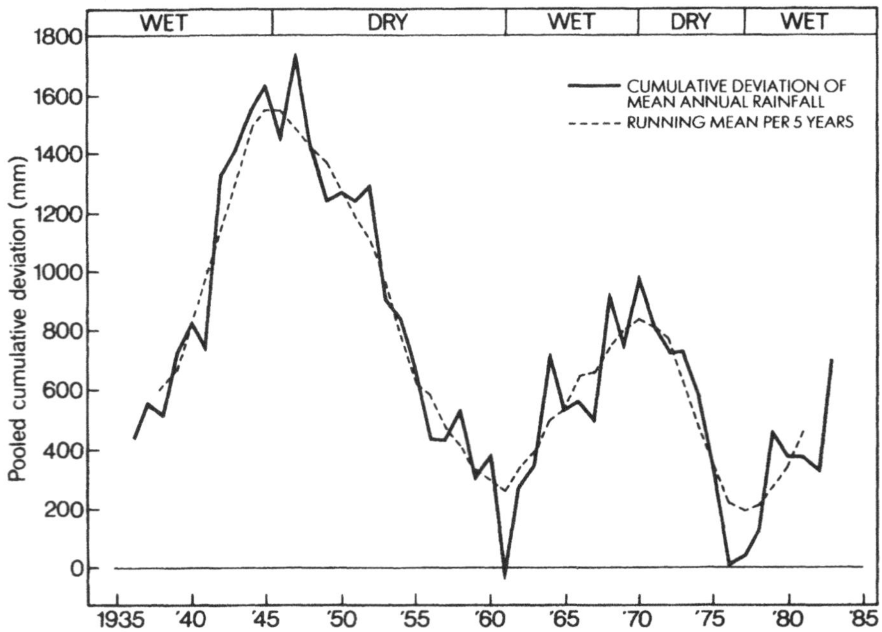

Figure 1.4 Pooled cumulative deviation of mean annual rainfall of nine rainfall stations around Manyara (Prins and Loth 1988). Wet and dry periods of several years' length alternate.

effects of rainfall is evident in the groundwater forest at the northern end of the park. Most ungulates in Manyara are grazers and depend for their food supply on the primary production of grasses and to a lesser extent sedges. This production is especially dependent on an adequate supply of water, mainly from rainfall. Because of this, a positive relation exists between rainfall and production of grass in the semiarid tropics up to an annual total of about 1000 mm (Walter 1962, Phillipson 1975, Coe *et al.* 1976, Breman 1982, Cornet 1984). The amount of rain that will fall at a given place is unpredictable, and over the years rainfall adopts a random pattern in the Masai ecosystems that includes Manyara (Prins and Loth 1988). This means that neither animals nor plants can rely on the onset of the rains (see also Tyrell and Coe 1974, Pratt and Gwynne 1977), and that the annual primary production is not constant (Breman and Krul 1982). In many areas the seasonal variation in rainfall has been well documented, but not so where forage is concerned. Hence I paid particular attention to the cycle of food production in relation to rainfall.

Much of the primary production in Manyara comes from perennial grass species. However, in some vegetation types, especially in the woodlands, annual species may also contribute substantially to the productivity. With the possible exception of elephant, all grazers in these areas prefer to ingest green leaves and avoid stalks and dead material. Green leaves have the highest protein content and the highest digestibility (Van Soest 1982; Figure 1.7). Hence, investigation of the dynamics of the green leaf proportion in the sward is important to understand habitat selection of the large grazers (Drent and Prins 1987). Annuals and perennials show different reactions to rainfall and, thus, vegetations dominated by annuals or by perennials have different sward dynamics. However, browsers as well as grazers make use of plant food, and study of the phenology of browse species, and in particular of the foliage of trees and shrubs, is likewise of interest because the Manyara buffalo browse to some extent.

In 1983 and 1984 I collected data on the growth of grass species and in 1984 and 1985 data on browse species in Manyara (for nomenclature and identification see Prins 1988). All the genera to which the grass species mentioned belong are capable of C_4 photosynthesis (Watson *et al.* 1986), which means that these species are well adapted to warm and dry environments (Tieszen *et al.* 1979, Osmond *et al.* 1982, Prins 1988). Proportions of green leaf in the total standing crop of the different grass species and of green material in the total standing crop of *Cyperus laevigatus* are given in Table 1.3. The green material of the sedge cannot be separated into leaf or stalk because of its growth form. The data indicate that, although rainfall is very seasonal in Manyara, only a few species reflect frank seasonality, namely annual grasses, the perennial grasses of groups I and II and the deciduous shrubs. Details on the grouping of perennial grasses based on their phenological reaction to rainfall can be found in Prins (1988). The data imply that water is in truly short supply for many plant species during the dry season on the higher grounds of the park. However, a large number of species are not, or are only weakly, seasonal. These are the perennial grasses of group III, *Cyperus laevigatus*, the evergreen shrubs and a number of tree species (Prins 1988). Thus for many other plant species water is not in short supply. This has important implications for the herbivores in Manyara, as these groups of plants can act as a food buffer. The presence of a food buffer of plants, especially group III perennial grasses and sedges, that remains green during the dry season might at least partly explain the observed high herbivore biomass supported by the Manyara part of the Masai ecosystem. The observed herbivore biomass (17 ton km^{-2} in the 1960s to 1980s) is much higher than would be expected from the relationship between annual rainfall and herbivore biomass as determined by Coe *et al.* 1976 (cf. Delany and Happold 1979; see also Drent and Prins

1987). The existence of this buffer might explain the striking stability in animal numbers over that period (Chapter 5). Although large fluctuations in annual rainfall are common, and groups of wet or dry years arise in a more or less random fashion (Table 1.2), herbivores in the Manyara area are relatively well protected from starvation as a result of the forage buffer provided by 'aseasonal' plant species. Of the study area of 5400 ha, 24% is made up of 'aseasonal' perennial grasses and sedges swards (with a ground cover of about 80%), while 44% covers terrain with 'seasonal' perennials and annuals (ground cover of about 15%).

The effects of alternations between dry and wet seasons on the quality of the food for herbivores such as buffalo are manifold. The quantities of food on offer vary between different species groups and, hence, habitats (Drent and Prins 1987; Figure 1.5). The proportion of live grass leaves in the sward, the preferred type of food, also changes (Table 1.3).

1.5 BUFFALO DIET COMPOSITION

The diet composition of the Manyara buffalo was determined through microscopic analyses of dung. Buffalo were observed for some time and if possible classified (sex, age, herd). When they were seen defecating, the location of the dung pat was noted and dung was retrieved shortly after the buffalo vacated the area. Additional samples were collected of which only herd identity was known. Buffalo show the hallmarks of herbivory as revealed so clearly in the extensive studies in cattle reviewed by Kalmbacher *et al.* (1984). First, despite a huge choice of plants (hundreds of browse species and 92 Gramineae and Cyperaceae species in Manyara; Greenway and Vesey-FitzGerald 1969, 1972, Loth and Prins 1986) only a few species make up the major portion of the diet. Secondly, on a year-round basis, perennial grasses form the mainstay of the diet. Thirdly, the diet changes during the year (Table 1.4). These features undoubtedly hold for the majority of grass-dependent herbivores.

Analysis of buffalo dung shows that *Cynodon dactylon* is the most important food species of Manyara buffalo, but in the long rainy season *Cynodon plectostachyus* becomes equally important. During the short rains, the woodland species (especially *Leptocarydion vulpiastrum, Panicum maximum* and *Urochloa pullulans*) and escarpment species (*Enneapogon cenchroides, Enteropogon macrostachyus* and *Eragrostis superba*) are selected. These species react strongly and quickly to rainfall and have a very favourable sward structure at this time of year (Table 1.3) and are likely to have a high crude protein concentration. *Chloris gayana* is consumed throughout the year in small quantities, but in exceptionally dry years, such as 1994, it is completely consumed. During the rainy seasons, when

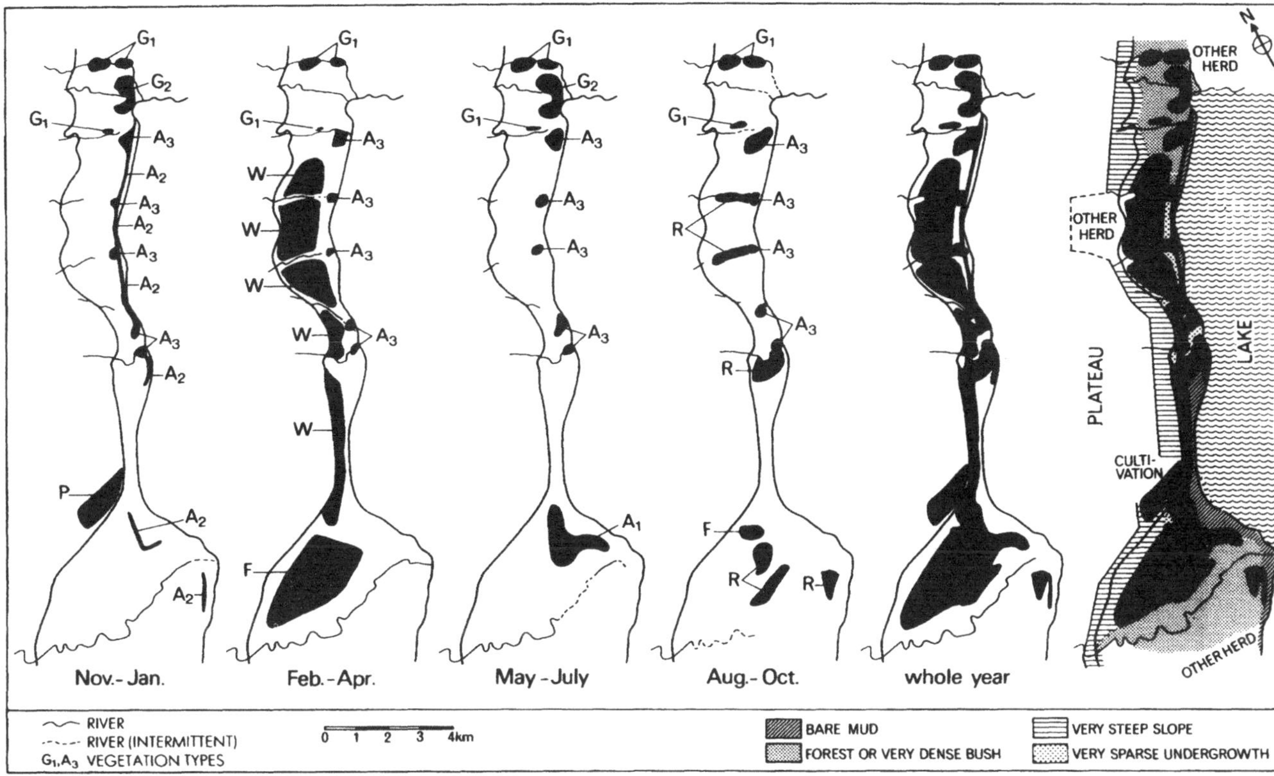

Figure 1.5 Seasonal occupancy of grazing grounds by buffalo of the Central Herd of Manyara. Landscape ecological units: A1, areas with upwelling groundwater with the sedge *Cyperus laevigatus*; A2, lake shore grasslands dominated by the grass *Sporobolus spicatus*; A3, riverine grasslands dominated by the grass *Cynodon dactylon*; F, terrace grasslands dominated by the grass *Sporobolus pyramidalis*; G1 and G2, different types of glades dominated by, respectively, *C. dactylon* and *Chloris gayana* grass; P, plateau grasslands dominated by *Sporobolus pyramidalis*; R, riverine bushlands; W, *Acacia tortilis* woodlands. The figure farthest to the right shows that over the year the whole home range is utilized.

Table 1.3 Green leaf as dry weight percentage of the standing crop

Date/season		*5 April 1983 Long rain*	*14 August 1983 Dry season*	*16 November 1983 Short rain*	*10 December 1983 Short rain*	*26 January 1984 Short rain*	*3 March 1985 Long rain*	*15 April 1984 Long rain*	
Annual grasses									
Group I (n=3)	Mean	66.5[a]	0.0[a]	0.0[a]	0.0[a]	0.0[a]	0.0[a]	59.3[ab]	$F_{5,12}=67.30$
	SE	4.5	0.0	0.0	0.0	0.0	0.0	10.4	$P<0.01$
Group II (n=5)	Mean	62.2[a]	0.0[a]	0.0[a]	0.0[a]	63.7[b]	44.8[b]	49.5[a]	$F_{5,24}=43.18$
	SE	5.3	0.0	0.0	0.0	6.7	4.0	5.4	$P<0.01$
Perennial grasses									
Group I (n=3)	Mean	37.7[a]	5.6[ab]	0.0[a]	96.7[b]	74.9[b]	33.7[b]	51.3[a]	$F_{5,12}=34.03$
	SE	8.3	3.6	0.0	3.3	8.6	8.5	10.9	$P<0.01$
Group II (n=7)	Mean	54.9[a]	16.8[bc]	0.2[a]	63.3[c]	57.4[b]	37.1[b]	69.8[b]	$F_{5,36}=90.82$
	SE	5.6	1.6	0.2	5.5	4.8	3.7	3.7	$P<0.01$
Group III (n=4)	Mean	47.8[a]	39.3[c]	19.5[b]	37.4[c]	48.6[b]	44.2[b]	56.3[ab]	$F_{5,18}=1.12$
	SE	6.7	8.7	10.4	13.4	12.4	11.1	9.5	NS
Perennial sedge									
Cyperus laevigatus		77.6	67.5	64.1	93.1	93.9	97.5		

n, Number of species in a group; NS, not significant at 5%.

During the course of the year grasses show different proportions of green leaf in the above-ground standing crop. In a column mean proportions with the same suffix are not significantly different at the 5% level (*t*-test after arcsine transformation). The effect of seasonality on the data from August 1983 to April 1984 was tested by means of a one-way ANOVA after an arcsine transformation. For species names see Table 1.1.

Table 1.4 The proportion of different food species in faeces of buffalo

Species	*Short rains (Nov–Dec 1983)*	*Long rains (Apr–May 1983)*	*Dry season (Sep 1984)*
Sporobolus spicatus	8.8	3.3	10.5
Cynodon dactylon	36.4^{-*}	34.1	64.2^{+***}
Cynodon plectostachyus	18.1	37.1^{+***}	0.0^{-***}
Leptocarydion vulpiastrum *Panicum maximum* *Urochloa pullulans*	13.6^{+**}	4.6	0.0^{-***}
Enneapogon cenchroides *Enteropogon macrostachyus* *Eragrostis superba*	6.6^{+**}	2.3	0.0^{-**}
Cyperus laevigatus	5.3	1.3^{-**}	16.5^{+**}
Chloris gayana	6.7	5.2	3.5
Browse species	3.4	1.9	3.4
Sample size	7×100	9×100	5×100

The grouping in the table is based on species growing in the same habitat. The asterisks indicate the level of significance of deviation within the seasons (as calculated with Habermann's adjusted residuals after chi-squared analysis on arcsine transformed data) ($^{*}P < 0.05$, $^{**}P < 0.01$, $^{***}P < 0.001$). The superscripted signs indicate whether a deviation is proportionally more (+) or less (–) than expected. All species except *Chloris gayana* and browse species show a significant seasonal variation by an ANOVA test.

there is plenty of food, *Cyperus laevigatus* is more or less ignored by the buffalo, but during the dry season this species becomes increasingly important. Approximately 60% of the rumen contents of buffalo sampled at the end of the dry seasons of 1982 and 1983 consisted of this sedge. It is noteworthy that browse species do not appear to be important for Manyara buffalo (Table 1.4), although quite large quantities of *Justicia cordata* are sometimes eaten during some days at the beginning of the dry season (section 1.8). However, during the dry season of drought years, as in September 1973, it has been observed that the Manyara buffalo switch to browse to a large extent; it is noteworthy that in 1973 they did not do this for long as in October of that year the buffalo left their usual home ranges and settled in the Marang and Firoto Forests (Warden's Reports 1973).

1.6 GRASS AND SEDGE PRODUCTION

On the basis of utilization intensity by the major grazers in the Manyara system I selected vegetation units for measuring production and consumption. For buffalo and other grazers the most important land units were:

- *Cynodon dactylon*-dominated river deltas;
- *Cynodon dactylon*-dominated forest glades;
- *Cyperus laevigatus*-dominated upwelling zones and swamps;
- *Sporobolus spicatus*-dominated alkaline grasslands;
- *Chloris gayana*-dominated wooded (bush) grasslands;
- Woodlands with *Cynodon plectostachyus*, *Sporobolus helvolus*, *S. cordofanus* and other grasses.

This importance was underscored by the faecal analyses of buffalo dung and by time budget measurements of buffalo, zebra, wildebeest and elephant (Chapters 2 and 5). In all grassland types except the last, I selected monospecies stands of *Cynodon dactylon*, *Cyperus laevigatus*, *Sporobolus spicatus* and *Chloris gayana*. In the woodlands the food species occurred in mixed stands, and I considered these species together. In these vegetations I placed eight elephant- and buffalo-proof exclosures with a ground area of 2 × 2 m. These exclosures did not necessarily exclude small herbivores such as rodents and insects. The exclosures were shifted within the stand every 10 weeks. In a few cases the exclosures were damaged by buffalo or elephant. In this event they were replaced and the measurement series started again at that site. Every time I shifted the exclosures, herbage was clipped to ground level both inside and outside the exclosure. The samples were then hand-sorted into green leaf, dead leaf, green stalk and dead stalk and subsequently dried. I determined the standing crop of green leaf at approximately 2-weekly intervals using a comparison of incoming radiation and spectral reflectance (at 840 nm, 660 nm and 550 nm) of the vegetation (for a description see Van Kasteren 1982). Each measurement with the 'biomass meter' consisted of the average of 10 readings within a few minutes (Van Wirdum 1983). The instrument was calibrated using clipped samples; the ratio of infrared (840 nm) to red (660 nm) reflectance gave a satisfactory estimate of green leaf standing crop (R^2 ranging from 0.62 to 0.91) but not of the standing crop of other sward components (R^2 ranging from 0.02 to 0.13) (see Esselink and Van Gils 1985 for a full description of procedures).

I estimated the production of leaf material in the ungrazed situation (gross production) from the sum of leaf accumulation inside the exclosure (LAI) plus the accumulation of dead material inside the exclosure (DAI). I estimated the production in the grazed situation as the sum of leaf accumulation outside the exclosure (LAO) plus consumption (C) plus the accumulation of dead material outside the exclosure (DAO). In the calculations, I used the empirical results that 0.5 g dry weight dead material equals 1.0 g dry weight green material, except for *Sporobolus spicatus*, *Cyperus laevigatus* and *Chloris gayana*, for which I found that 0.8 g of dry weight dead material equals 1.0 g of dry weight green material. The estimate of the consumption is then given by (LAI – LAO) + (DAI – DAO), but because I observed frequently that grazers are reluctant to eat dead material DAI and DAO have been taken to be equal. This actually means that I have not made a distinction between ingestion or trampling followed by subsequent removal by wind and rain. Strictly speaking I use the term 'consumption' in the sense of utilization by herbivores ('t Mannetje 1978). The figures are minimum estimates since consumption by small herbivores is ignored and I assumed that under the prevailing grazing regime growth stimulation (Chapter 7) could be ignored too. Furthermore, it is assumed that there was no growth reduction in the exclosures because I shifted the exclosures frequently. I do not think that consumption by small vertebrates is very important in Manyara; during the years that I have been working in this park during the day and at night I have only rarely observed voles or mice. Sometimes there was visible termite infestation in or around the exclosures. In April and May 1983 there was a severe army worm outbreak in some patches of *Cynodon dactylon* in the northern part of the park. Army worms are the caterpillars of the moth *Spodoptera exempta*, which can be as devastating as locusts. Infested exclosures were replaced and I did not carry out measurements in them.

The production of the different grasses and the sedge (Table 1.5) generally reflects the rainfall pattern in Manyara, that is production is high during the rainy seasons and low during the dry season (Figure 1.6).

This is especially true for *Sporobolus spicatus*, which grows on sandy soils in which rainfall penetrates easily but where no water remains within augering depth (1.5 m). As was outlined above, *Cynodon dactylon* grows in riverine areas on more loamy and clayey soils that are less easily penetrable than are sandy soil but which have a higher water retention capacity. Hence, the production peak of this species is later but longer than that of *S. spicatus*. The sedge *Cyperus laevigatus* grows in zones with upwelling groundwater, which is presumably replenished after the short rains, and hence the production period is very extended. All species show a higher production peak during the short rains than during the long rains. In part, this response to rainfall is probably due to mineralization in the soil at the end of the long dry season with a smaller nitrogen flux at the end of the short dry season before the long rains (Dradu and Harrington 1972), and perhaps most nitrogen is already 'captured' in the sward during the second rains owing to the vigorous growth in the first rains. The production pattern of *Chloris gayana* differs: although the first

Table 1.5 Production and consumption of green leaves of some monocotyledons and standing crop measurement of green leaves over the year

Green leaves only	*Nov–Dec 1983*	*Dec–Feb 1984*		*Feb–Mar 1984*	*Mar–Jul 1984*	*Jul–Aug 1984*	*Aug–Oct 1984*	*Oct 1984*	*Oct–Nov 1984*	*Year*
Production ($g\ m^{-2}\ day^{-1}$)										
Sporobolus spicatus	3.6	0.0		7.8	0.0	0.0	0.0	0.0	12.0	1068
Cynodon dactylon	6.2	4.2		1.5	2.2	0.0	0.0	1.3	4.6	912
'Woodland species'	3.1	1.0		2.1	0.7	0.0	0.1	0.0	0.7	351
Chloris gayana	5.4	———	1.3	———	6.7	5.5	0.7	4.8	9.2	1752
Cyperus laevigatus	3.7	———	2.6	———	3.5	1.2	0.1	0.1	5.3	860
Consumption ($g\ m^{-2}\ day^{-1}$)										
Sporobolus spicatus	2.7	0.0		7.2	0.0	0.0	0.0	0.0	9.2	871
Cynodon dactylon	5.7	5.3		1.3	1.0	0.3	0.0	0.8	3.8	830
'Woodland species'	0.8	0.9		2.5	0.0	0.0	0.1	0.0	0.1	201
Chloris gayana	0.0	———	0.0	———	5.4	4.4	4.8	5.8	1.4	1137
Cyperus laevigatus	2.8	———	0.8	———	3.2	1.3	1.2	1.1	4.8	793
Standing crop ($g\ m^{-2}$)	15 Nov	15 Dec	23 Feb	25 Mar	19 Jul	21 Aug	1 Oct	25 Oct	1 Dec	31 Dec
Sporobolus spicatus	20.8	138.9	133.1	119.2	87.0	61.6	48.8	50.0	126.6	119.3
(cv)	(112%)	(43%)	(77%)	(15%)	(40%)	(29%)	(51%)	(78%)	(26%)	(54%)
Cynodon dactylon	35.3	84.4	74.0	100.2	69.0	43.2	44.6	47.5	135.3	80.3
(cv)	(49%)	(56%)	(29%)	(11%)	(63%)	(44%)	(37%)	(51%)	(45%)	(56%)
'Woodland species'	0.0	44.9	51.2	46.8	16.3	2.3	0.0	0.0	69.9	32.6
(cv)	(–)	(34%)	(120%)	(35%)	(69%)	(287%)	(–)	(–)	(52%)	(22%)
Cyperus laevigatus	148.6	239.4	296.9	354.4	184.5	176.7	245.2	243.4	442.6	225.9
(cv)	(73%)	(–)	(55%)	(7%)	(88%)	(55%)	(27%)	(32%)	(85%)	(38%)

cv, Coefficient of variation.

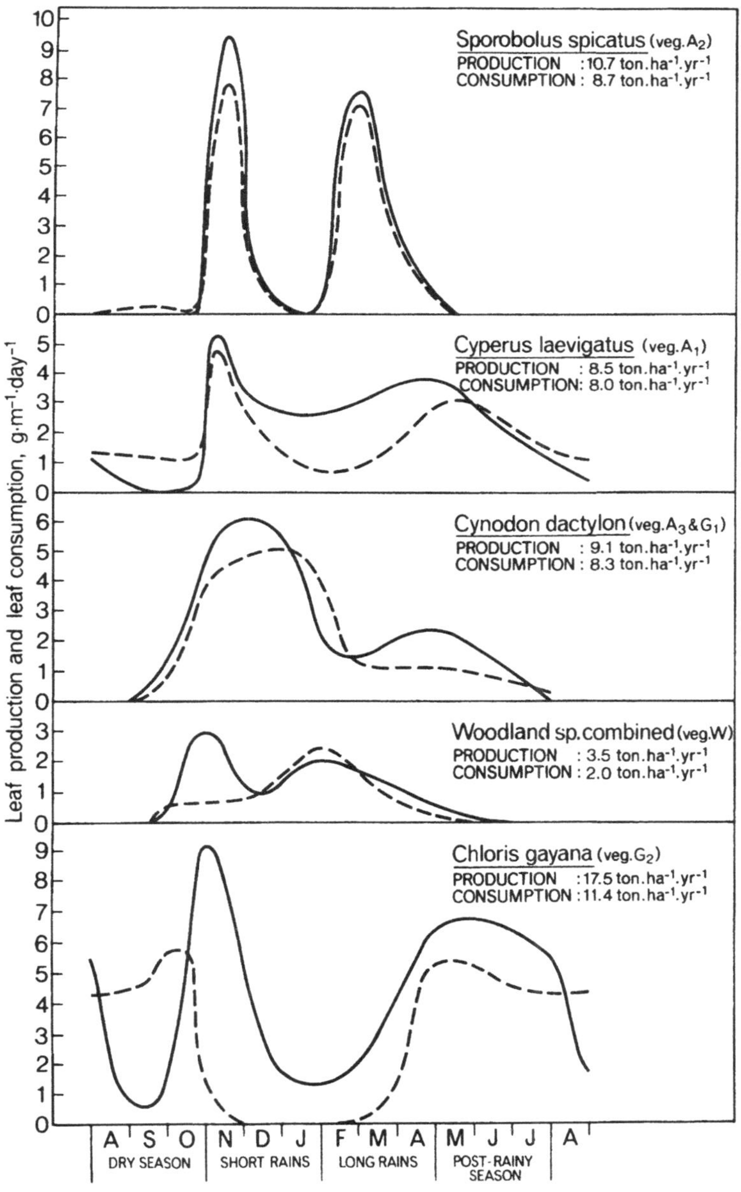

Figure 1.6. Leaf production (solid line) and leaf consumption (broken line) over the year for the monocotyledons that are most important for the Manyara buffalo.

production peak occurs immediately during the short rains, the second peak occurs at the end of the long rains. It is not easy to explain this pattern, but a major difference between this grass and the other species is that *C. gayana* forms very dense, tall swards (100% ground cover in several layers and a sward height of on average 110 cm) while the other species rarely attain a ground cover of 100% and are normally single-layered, with a sward height of at most 40 cm but normally less (Table 1.6).

Moreover, the late production peak in *C. gayana* might be related to waterlogging of this species' growth sites during the long rains as this species has only a limited tolerance to waterlogging (Humphreys 1981, p. 107 ff.). Lastly, *C. gayana* appears to be ignored by the grazers during the wet season, and is grazed only during the dry season (especially by zebra and elephant) (Figure 1.6). It is possible that herbage removal during the dry season stimulates the production of new leaves in this species.

1.7 GRASS AND SEDGE CONSUMPTION

In Table 1.7 the population sizes of the different large herbivores heavier than 50 kg are given.

These are based on aerial censuses (total counts), on road sampling, and for a few species on estimates. Methodology and data reliability are dealt with in Prins and Douglas-Hamilton (1990). The figures are believed to be correct for the resident species, but are less certain for the migratory species (especially wildebeest) because just outside the northern boundary of the park large concentrations of migratory game (up to 10 000 head of wildebeest and several hundred zebra and Thomson's gazelle) can be found in the dry season. The census figures for the park are daytime estimates, and I consider night-time grazing by migratory game as unlikely (Beekman and Prins 1989). Hence I use the daytime census values. Biomass was calculated as the product of the number of individuals of a species and their average body weight.

I have used two different formulas to calculate grass consumption, namely consumption (kg dry matter per day) equals $0.025 \times W$ (Van Wijngaarden 1985) and consumption (kg dry matter per day) equals $0.09 \times W^{0.75}$ (Sinclair 1975), where W is the animal's live weight in kilograms (Chapter 2). Applying the two formulas results in approximately similar proportions consumed by the different grass-eating herbivores in Manyara, but the estimate of the total grass consumption is different (Table 1.7). This is mainly because of the difference in the estimated consumption by elephant. However, because the elephant is a hindgut fermenter with a higher throughput rate, and consequently a higher consumption, than a ruminant of the same size, and because Sinclair's (1975) formula appears to be devised mainly for ruminants, I believe that his formula underestimates the consumption of non-ruminants such as

Table 1.6 A number of parameters to describe the vegetation for some monocotyledons all considered important to predict the intake of herbivores (see text)

	15 Nov	*15 Dec*	*23 Feb*	*25 Mar*	*19 Jul*	*21 Aug*	*1 Oct*	*25 Oct*	*1 Dec*	*31 Dec*
Sporobolus spicatus										
Leaf/stem ratio	5.70	7.43	5.32	9.46	12.7	5.01	4.90	6.12	12.87	9.81
Green leaf bulk density	0.77	1.45	1.37	0.92	0.99	0.67	0.63	0.74	1.47	1.21
Sward bulk density	0.91	1.65	1.63	1.02	1.07	0.80	0.76	0.86	1.58	1.33
Vegetation height	2.7	9.6	9.7	13.0	8.8	9.2	7.7	6.8	8.6	9.9
Bite size index	2.41	4.90	4.38	5.01	4.95	3.33	3.06	3.29	5.59	5.05
Cynodon dactylon										
Leaf/stem ratio	0.48	1.89	1.00	1.81	1.55	0.52	0.48	0.70	1.04	1.17
Green leaf bulk density	0.26	0.47	0.42	0.96	0.25	0.19	0.31	0.37	0.94	0.61
Sward bulk density	0.80	0.72	0.84	1.49	0.41	0.56	0.96	0.90	1.84	1.13
Vegetation height	13.4	17.8	17.7	10.4	27.4	22.3	14.3	12.8	14.4	13.2
Bite size index	1.73	2.89	2.46	3.04	2.59	1.87	1.87	2.01	3.73	2.40
'Woodland species'										
Leaf/stem ratio	0.00	7.98	0.94	2.22	1.21	(0.50)	0.00	0.00	9.25	2.63
Green leaf bulk density	0.00	0.83	0.68	0.62	0.15	0.03	0.00	0.00	1.34	0.48
Sward bulk density	0.00	0.93	1.40	0.90	0.27	0.09	(0.04)	(0.01)	1.48	0.66
Vegetation height	–	5.4	7.5	7.5	10.7	7.0	7.2	5.0	5.2	4.8
Bite size index	0.00	3.41	2.15	2.47	1.52	0.68	0.00	0.00	4.14	2.35
Cyperus laevigatus										
Sward bulk density	2.56	4.79	5.94	7.09	3.08	2.72	3.77	3.33	4.52	2.93
Vegetation height	5.8	5.0	5.0	5.0	10.0	9.5	6.5	7.3	9.8	7.7
Bite size index	3.85	4.89	5.45	5.95	5.55	5.08	4.95	4.93	6.66	4.75

The green leaf/green stem ratio is on a dry matter basis, the sward bulk density is the standing crop of green biomass per m^2 divided by the sward height (cm). The bite size index is explained in the text.

Table 1.7 Identification of the major grass consumers in the Manyara system

Species	*Body weight (kg)*	*Population size* (n)	*Biomass (ton)*	*Proportion grass in diet*	*Total grass consumption per year (ton)*		*Percentage of grass consumption*	
					A	*B*	*A*	*B*
Elephant	1700	373	634.1	0.70	4050	2271	28.9	21.4
Hippopotamus	1000	179	179.0	1.00	1633	1046	11.6	9.9
Buffalo	325	2097	681.5	1.00	6219	5273	44.3	49.7
Zebra	250	255	63.8	1.00	582	527	4.2	5.0
Wildebeest	225	675	151.9	1.00	1386	1288	9.9	12.1
Waterbuck	200	17	3.4	0.80	25	24	0.2	0.1
Bushbuck	60	50	3.0	0.10	3	4	<0.1	<0.1
Impala	50	150	7.5	0.95	65	88	0.5	0.8
Warthog	50	95	4.8	0.90	39	53	0.3	0.2
Reedbuck	40	10	0.4	1.00	4	5	<0.1	<0.1
Baboon	20	500	10.0	0.20	18	31	0.1	0.3
Total		4401	1739.4		14 024	10 610	100.0	100.0

The consumption of the different herbivores is based on their body weight, diet and numbers (Chapter 5) and on estimated dry matter consumption per day, using two different formulas: A = consumption/day as 2.5% W, and B = consumption/day as 0.09 $W^{0.75}$, in which W = body weight in kg. The African buffalo is the major grass consumer.

the elephant (McKay 1973). Most likely the same applies to the estimates derived from his formula for the hippopotamus, another non-ruminant exhibiting a wide difference when the two formulas are compared. Hence, I prefer to use the formula consumption = 0.025 W. This preference is enforced by the findings that stomach capacity is proportional to $W^{1.0}$ and that stomach capacity is for different species constant at 5% of body mass (Garland 1983).

From Table 1.7 it is clear that most grass consumption in Manyara is by buffalo. According to rumen analyses of the species mentioned in this table (Hofmann and Stewart 1972, Hofmann 1973), the bulk of the grass consumption in Manyara will be green leaves. These leaves form the most nutritious part of the sward (for example, Stobbs 1973a,b, Sinclair 1977, Stanley Price 1978, Van Soest 1982). Even in the dry season, the quality of the green leaves of *Cynodon dactylon* is high enough to meet the nutritional need of the buffalo because they offer the most favourable ratio between crude protein concentration and neutral detergent fibre (NDF) within the sward (Figure 1.7), that is the cell wall fraction is not so high that it depresses food intake while the crude protein content is high enough to guarantee enough digestible crude protein (see Chapter 2 for details).

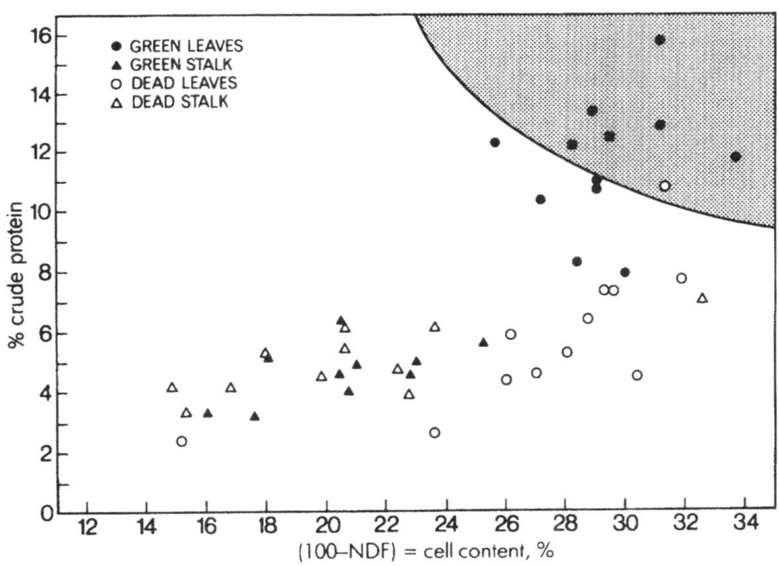

Figure 1.7 Relation between cell contents and crude protein concentration in different fractions of the grass *Cynodon dactylon* on 31 August 1984. Most live leaves can provide buffalo with sufficient digestible crude protein (shaded part), but other sward components cannot [see Appendix for details; based on the assumption that buffalo dry matter voluntary intake is according Moen (1973, p. 169): 110.4 – 1716/(100 – NDF%].

Table 1.8 Estimates of annual production and consumption of green leaves of monocotyledons, and of total annual above-ground production based on the production of green leaves and on the average leaf/stem ratio (from Table 1.6)

	Production of green leaves (ton ha^{-1} year^{-1})	*Consumption of green leaves (ton ha^{-1} year^{-1})*	*Consumption as a percentage of production (green leaves only)*	*Average leaf/stem ratio*	*Estimate of total annual production (ton ha^{-1} year^{-1})*
Sporobolus spicatus	10.7	8.7	82	7.94	12.0
Cynodon dactylon	9.1	8.3	91	1.06	17.7
'Woodland species'	3.5	2.0	57	2.47	4.9
Chloris gayana	17.5	11.4	65	1.00[e]	35.0
Cyperus laevigatus	8.6	7.9	92	All leaf	8.6

Most green leafy material appears to be harvested by herbivores. [e]Estimated.

In Manyara, large herbivores remove the greater part of the production of green leaves (Table 1.8 and Figure 1.6), so most leaves in this grazing system will be consumed by herbivores instead of senescing.

On the basis of the production estimates of green leaves (Table 1.5) and the leaf/stem ratio (Table 1.6), the annual production of above-ground biomass of the Manyara grasslands is estimated to be between 5 and 35 tons ha^{-1} $year^{-1}$ (Table 1.8). This is well within the range given for semiarid regions in the tropics with a reasonable water supply (Snaydon 1981). The grass production in the woodlands is very low, possibly as a result of competition with trees and shrubs (Walker *et al.* 1986).

1.8 FOOD QUALITY

The crude protein concentration of the green leaves of the four main grassland species is shown in Figure 1.8.

The sedge *Cyperus laevigatus*, growing in substrates with upwelling groundwater, does not show significant deviations from its mean annual crude protein content of 6.8%. This species has a very low relative growth rate, being maximally 0.02 g g^{-1} day^{-1} (Prins and Beekman 1989), which results in a high mean age of the green leaves. The perennial grasses, which depend on rainfall for their water supply, reflect in their growth reaction to the bimodal rainfall pattern the intra-annual crude protein content pattern. Although they react in the same manner, these species show different levels of crude protein.

All grasses show their highest levels of growth during the short rains (Figure 1.6), which means that at that time of the year the mean age of the green sward components is at its lowest. During the short rains season the relative growth rate of the green leaves is close to the relative growth rate under optimal conditions as measured for a number of tropical grasses (namely 0.10–0.14 g g^{-1} day^{-1}; Prins and Beekman 1989), which indicates that there is no nutrient limitation at that time of the year.

This is confirmed by nutrient limitation experiments on *C. dactylon* and *Sporobolus spicatus*; at the end of the wet season there is nutrient limitation for these food plants, especially of nitrogen and to a lesser extent of phosphorus and potassium (unpublished results). *S. spicatus* and *Cynodon plectostachyus*, owing to the sandy substrate in which they grow, react to the short dry season with a growth reduction, resulting in an increase in the mean age of the leaves in the sward, and this causes a depression of the crude protein concentration (Figure 1.8). *Cynodon dactylon*, as was outlined before, lacks water only during the long dry season. This means that the sward ages only slowly, and that the species is not affected by the short dry season, and hence the quality of its leaves shows a steady but slow decrease.

In all species at all times of the year, the quality of the green parts is

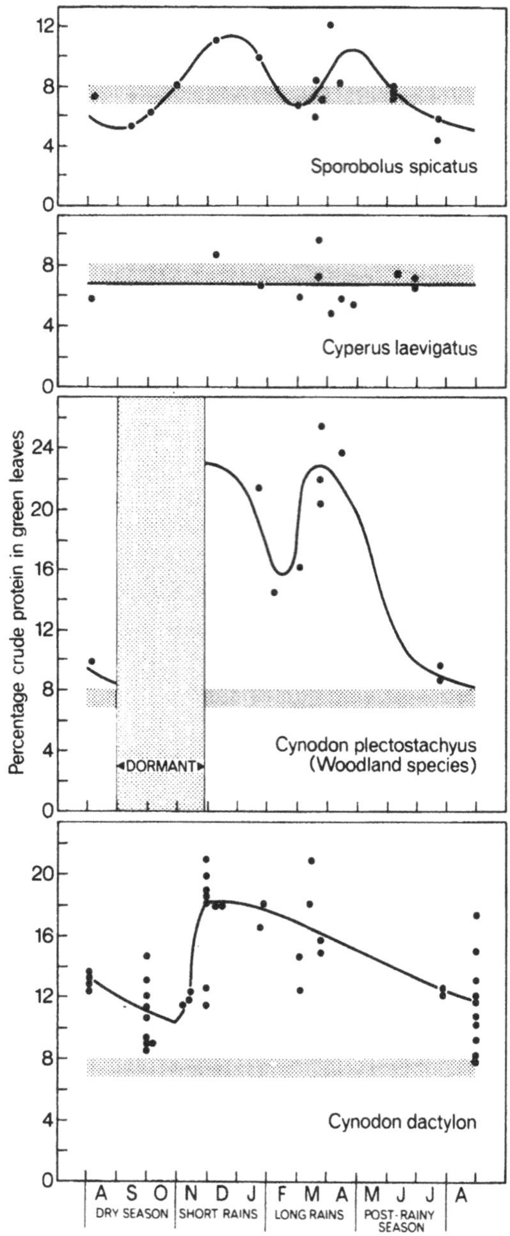

Figure 1.8. Crude protein concentration in live leaves of the most important food species of buffalo. The shaded horizontal bar indicates the buffalo's maintenance requirements (Appendix). The short rains started on 16 November.

higher than that of the dead parts, and leaves always have a higher crude protein concentration than stalks. The crude protein concentration of the dead parts within the sward is (nearly) always lower than the level required for maintenance, that is 7% (Table 2.1 and Chapter 2). The quality of the grasses is also reflected in the ratio between green leaf and green stem (Table 1.6). A high ratio means that herbivores can easily select for the green leaves within the sward (for example Stobbs 1973a,b, Chacon and Stobbs 1976, Stanley Price 1978). When the green leaf production is high, the leaf/stem ratio is also high and the crude protein concentration of the green leaf is concomitantly high. All these factors imply that the food supply during the short rainy season is best for grazers in the Manyara area, and that the long dry season is worst.

In the long dry season a number of woody species remain in leaf (Table 1.9), and these would, potentially, offer an additional food resource for buffalo.

However, the diet of buffalo in Manyara contains very few browse species, even in the dry season (Table 1.4), except in very dry years. Yet most vegetation types are dominated by shrubs and trees; this implies that there seems potentially to be a superabundance of food. I collected leaf samples of a number of woody species over a year and determined the crude protein content and the acid detergent fibre fraction (Van Soest 1982). It appeared that most browse species have high crude protein levels, even in the dry season (Table 1.10).

And nearly all species that I sampled always had crude protein levels in excess of the maintenance requirements of buffalo (Chapter 2). Domestic cattle of the Iraqw and the Masai utilize many browse species during the dry season, but buffalo do not (Tables 1.10 and 2.3). Apart from the browse species mentioned in Table 2.3, I observed buffalo to consume very small quantities of some other species too, that is they took one or two bites of these plants and walked off. These species were *Sphaeranthus ukambensis, Sida ovata, Heliotropum subulatum, Justicia glabra* and *Tamarindus indica*; in addition they sometimes took young tips of *Salvadora persica, Cordia sinensis* and *Capparis tomentosa*. Apart from the species marked in Table 2.3 as 'not eaten', I have never observed them to eat *Vangueria madagascariensis, Balanites aegyptiaca* and *Senna didymobotrya*. The Manyara buffalo appear to be much less catholic in their diet than Iraqw cattle, and they reject species that are (to the human observer) aromatic, resinous, bitter or spiny. The leaves of browse species that they consume are characterized by low fibre contents (Table 2.3) and do not have any of the mentioned properties. I will return to this issue of browse consumption in Chapter 2.

Table 1.9 Moisture contents of green leaves of different species of shrubs as a percentage of fresh leaf weight

Date/season	*14 Aug 1983 Dry season*	*9 Nov 1983 Dry season*	*26 Jan 1984 Short rain*	*9 April 1984 Long rain*	*27 July 1984 Dry season*	*24 Aug 1984 Dry season*	*25 Sep 1984 Dry season*	*1 Nov 1984 Dry season*	*20 Nov 1984 Short rain*
Deciduous shrubs									
Acalypha fructicosa	–	–	75.7	43.7*	44.8*	–	–	–	74.8
Barleria eranthremoides	51.0	–	70.1	46.5(*)	34.1	32.7	–	–	74.2
Indigofera arrecta	–	–	67.2	50.7*	–	–	–	–	73.4
Indigofera uniflora	–	–	66.0	53.3	–	–	–	–	–
Ocimum suave	–	–	–	75.2	47.9*	48.9*	–	–	78.6
Tephrosia villosa	–	–	64.1	62.0	54.1(*)	–	–	–	72.1
Riverine evergreen shrubs									
Cardiogyne africana	60.5	60.4	62.1	60.9	61.9	60.1	59.4	55.6	77.7
Cordia ovalis	51.5	60.1	65.9	60.5	52.3	56.1	57.1	67.8	74.9
Dovyalis xanthocarpa	57.1	72.1	72.6	62.4	60.2	61.3	58.5	66.2	72.8
Gardenia jovis-tonantus	58.2	60.7	61.8	55.5	54.1	48.0	51.2	54.1	72.2
Justicia cordata	60.5	67.1	74.3	70.2	65.6	60.9	57.3	60.1	73.8
Thylachium africanum	52.7	61.1	60.2	53.2	54.7	50.7	51.5	62.5	56.8
Vangueria madagascariensis	58.9	71.3	70.8	64.1	65.1	62.3	60.7	68.3	74.1
Non-riverine evergreen shrubs									
Acacia tortilis	58.2	60.1	57.9	64.5	53.4	43.8	55.7	59.3	61.0
Capparis farinosa	57.0	58.4	63.1	55.5	55.0	58.7	58.2	58.4	59.3
Cordia sinensis	57.5	69.1	61.4	53.3	61.2	60.4	59.5	65.6	71.9
Maerua triphylla	61.1	62.2	64.4	63.1	61.4	61.0	60.2	61.2	63.8
Ruellia megachlamys	44.3*	51.0	73.0	46.7*	50.0	43.8*	39.9*	53.6	76.6
Salvadora persica	67.0	76.9	78.3	76.3	62.3	74.3	74.2	67.3	74.2
Veronica cinerascens	61.3	60.8	67.9	56.3	64.3	61.5	57.5	59.4	79.0

Pronounced wilting is indicated by an asterisk, while the first signs are indicated by a bracketed asterisk; –, no green leaves present.

Table 1.10 Crude protein contents of green leaves of different species of shrubs. –, no green leaves present because the plant was dormant

	Post-rainy season May–July	*Dry season Aug–Oct*	*Short rains season Nov–Jan*	*Long rains season Feb–Apr*
Deciduous shrubs				
Acalypha fructicosa	18.2	14.8	23.6	20.9
Barleria eranthremoides	16.4	12.9	22.6	20.8
Indigofera arrecta	–	–	32.0	28.5
Indigofera uniflora	–	–	21.9	20.8
Ocimum suave	16.5	15.5	20.3	19.0
Tephrosia villosa	18.7	–	27.8	23.6
Riverine evergreen shrubs				
Cardiogyne africana	13.1	14.0	16.1	11.1
Cordia ovalis	13.9	13.2	20.1	19.8
Dovyalis xanthocarpa	12.3	13.7	18.0	15.8
Gardenia jovis-tonantus	13.5	13.9	15.0	13.1
Justicia cordata	15.9	16.5	19.3	20.6
Thylachium africanum	12.3	15.6	22.5	16.2
Non-riverine evergreen shrubs				
Acacia tortilis	20.8	17.4	19.7	22.7
Capparis farinosa	12.3	19.0	20.4	17.9
Cordia sinensis	16.0	12.9	16.9	14.1
Maerua triphylla	13.0	15.2	24.3	15.7
Ruellia megachlamys	17.8	16.6	19.1	17.2
Salvadora persica	14.1	12.9	15.5	18.9
Vernonia cinerascens	12.3	13.7	18.0	15.8

2

Food for the buffalo

2.1 ENERGY AND PROTEIN REQUIREMENTS FOR BUFFALO

The most critical limiting quality parameter for herbivores is nitrogen (protein) (Sinclair 1977, White 1978, Mattson 1980, Ydenberg and Prins 1981, Berry and Louw 1982, Ketelaars 1986). Buffalo are herbivores, and, thus, to them nitrogen appears to be the most important food quality parameter too (Sinclair 1977). Ruminants are a special category of herbivores to which buffalo also belong. In ruminants digestibility of the food determines the throughput rate of the food in the gut. Digestibility itself depends on the nitrogen content of the food. Hence, in ruminants food intake is related to the nitrogen concentration of the food (Van Soest 1982). The most critical factor regarding the amount of food for herbivores is the density of the food on offer, that is whether it can be collected in the time that is available (Allden and Whittaker 1970, Stobbs 1973a,b, Short 1986). I will come to the problem of time limitation in Manyara buffalo later but will first discuss the buffalo's energy and nutrient requirements.

To live, grow or reproduce, an animal needs energy and nutrients. These requirements have to be met by the ingested food. In Chapter 1, I showed that for the Manyara buffalo this food consists mainly of grasses. Two of the most critical requirements are energy and protein, and intake can be expressed in the same terms, namely as metabolizable energy (kcal or J) and digestible protein (kg). Intake of nutrients and energy can be expressed as the product of the quantity of ingested food (kg day^{-1}) and its quality (concentration of nutrients, kg kg^{-1} or percentage; or of digestible energy, kcal kg^{-1}). Because of the tremendous importance of the livestock industry, much research has been carried out in describing feeding standards for domestic animals such as cattle, and I have drawn on those results for describing the requirements of buffalo. However, this information is also relevant for other herbivores. I have assembled most of the available information in the Appendix. With these formulas, the food intake of buffalo (in terms of both digestible protein and metabolizable energy) can be deduced to depend on the crude protein concentration of the food. The importance of crude protein in the buffalo's food is emphasized by the fact that the crude protein content in fistula samples explains 54% of live weight change in cattle under African range conditions (in Botswana), while the protein content of clipped samples explains 48% of live weight change; digestibility explains less variance (Pratchet *et al.* 1977). Here lies a bone of contention:

> the influence of the crude protein content and digestibility are in disagreement with workers in temperate climates, who found that energy is the major limiting factor, but are in agreement with many workers in the tropics and subtropics. Energy is not the first limiting factor in dry grasslands [in Africa].
>
> (Pratchet *et al.* 1977)

In the Appendix, I have tried to find a means to link the crude protein content to the energy content of the food plants. It will be clear that grazing buffalo have to satisfy their needs for energy and protein simultaneously. The most economic way of achieving this is by reaching an optimal balance between the requirements for energy and for protein. Thus, I have constructed nomograms to determine the most economic crude protein concentration in the food for classes of adult buffalo (Figure 2.1), leading to the identification of a 'balanced diet'.

Adult non-lactating cows need 24 g digestible protein (DP) per 1000 kcal metabolizable energy (ME). For juvenile animals (that is 30% of adult body weight) the balanced diet has to provide 40 g DP for 1000 kcal ME, and for lactating females this has to be 32 g DP for 1000 kcal ME (Crampton and Harris 1969, p. 211 ff., Lloyd *et al.* 1978, p. 435 ff., Barrett and Larkin 1974; Table 6.1). It can now be shown that a low-quality diet provides an excess of energy in relation to protein, and that for maintenance requirements the crude protein concentration in the food has to be approximately 6.2% but that for a balanced diet the crude protein concentration has to be 9.0%. If the crude protein concentration is higher then protein is in excess. For a lactating buffalo cow, a diet containing 8.5% crude protein covers maintenance requirements while at 11.4% crude protein the diet is balanced.

I have assumed that wild African buffalo are physiologically the same as cattle, just as Sinclair (1977) and Demment and Van Soest (1985) did. Hence, I have assumed that the buffalo's needs and digestion efficiency can be legitimately derived from studies in cattle. Although wild bovids such as bison and water buffalo are reported to have a (slightly) higher digestibility efficiency than domestic cattle and zebu (for example Johnson *et al.* 1967, Naga and el-Shazly 1969, Pant and Roy 1970, Verma *et al.* 1970, Hawley *et al.* 1981 but see Chaturvedi *et al.* 1973), this difference is most likely to be due to these wild bovids being larger than cattle.

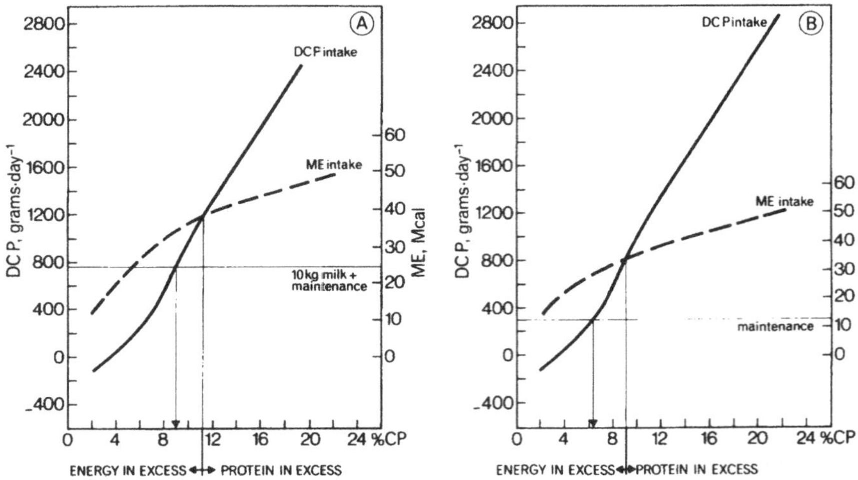

Figure 2.1 Intake of digestible crude protein (DCP; g day^{-1}) and of metabolizable energy (ME; kcal day^{-1}) as a function of the crude protein concentration (CP; %) of roughage feed. (A) The right-hand scale is constructed so that 32 g DCP represents 1000 kcal ME, which gives a balanced diet for a 500 kg adult lactating cow. (B) The right-hand scale is constructed so that 24 g DCP stands for 1000 kcal ME, which gives a balanced diet for a 500 kg cow at maintenance (see Appendix for details). The lines describe the daily intake of DCP (in g) and ME (in kcal) as function of the crude protein concentration of the food, taking into account the reduction in voluntary food intake (kg dry matter day^{-1}) from 3.5% of body weight to 2% when quality decreases (Appendix). Where the two lines cross, the diet is balanced.

Larger bovids have a lower throughput rate so that food will take longer to be digested and the differences in digestibility are not so much species specific as size specific (Bhattacharyya and Mullick 1965, Hoppe 1977, Demment and Van Soest 1985).

With these theoretical insights, the question arises of whether the buffalo in Manyara assemble from available food plants a diet that satisfies their needs for energy and protein in accordance to this 'balanced diet' proposition. I determined total nitrogen concentration and in a number of samples NDF concentration (Van Soest 1982); analyses were done on green leaf, green stem, dead leaf and dead stem of grasses *Cynodon dactylon*, *C. plectostachyus* and *Sporobolus spicatus* and on green parts and dead parts of the sedge *Cyperus laevigatus*. These are the four major food species of the buffalo in Manyara (Chapter 1).

Quality of forage intake was determined by analysis of the faeces samples. Faecal crude protein (FCP%) was calculated as 6.25 × N% in faeces, and dietary crude protein (CP%) was calculated according to Sinclair (1977, p. 233) as CP% = 1.677 FCP% – 6.93. Sinclair calibrated this formula, which was derived from trials with cattle, on results from buffalo in the adjacent Serengeti. The use of faecal nitrogen to calculate dietary nitrogen appears to be a reliable method in a number of free-ranging ruminants (Leslie and Starkey 1985). In Figure 2.2 the crude protein concentration in the buffalo's diet over the year is shown.

The Manyara buffalo exhibit the same pattern over time with regard to the protein concentration in their diet as the buffalo in the Serengeti. The Manyara buffalo's diet, however, is consistently better by about two percentage points than that of Serengeti buffalo (Figure 2.2). A diet as low in quality as reported by Sinclair (1977) from the Serengeti (a minimum content of 2.2% crude protein) was not encountered in Manyara. It is possible, however, that during very dry years the crude protein content of the Manyara buffalo diet could fall to about 2.5% if buffalo were forced to eat dead stalk only (Table 2.1) if the *Cyperus laevigatus* stock were exhausted.

I have not been able to find written evidence that this has ever occurred, but the park rangers report that it has happened, and during the drought of 1994 most stalk was consumed by all herbivores combined. If such a situation arose, available plant nitrogen would become a

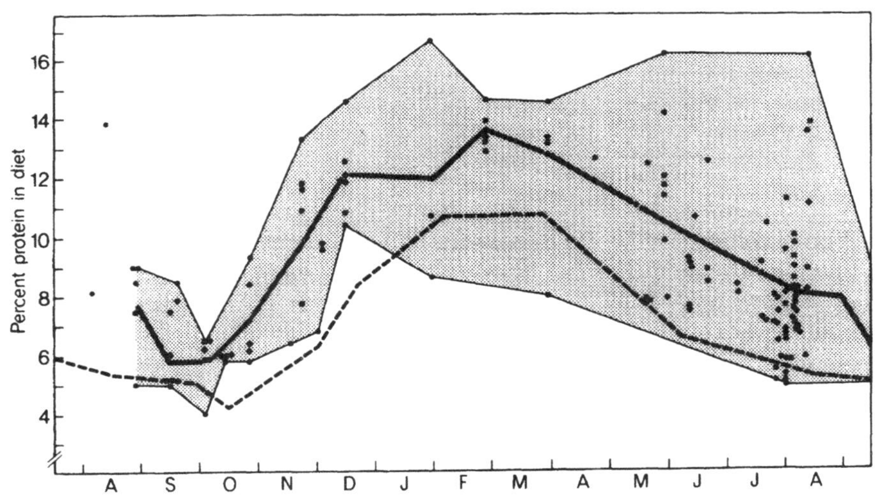

Figure 2.2 The average crude protein concentration in the diet of the Manyara buffalo (solid line) as compared with that of the Serengeti buffalo (broken line; Sinclair 1977, Figure 77). The shaded area describes the range encountered during sampling.

Table 2.1 The crude protein concentration with standard error (sample size in parenthesis) in different seasons for the major monocotyledons as based on 269 samples

Species	*Short rainy season*	*Long rainy season*	*Post-rainy season*	*Dry season*
Sporobolus spicatus				
Green leaves	10.5±0.7 (2)	8.1±2.1 (6)	6.6±1.5 (5)	6.7±1.2 (4)
Green stalk	5.3±0.4 (2)	5.5±1.5 (5)	2.0±– (1)	4.2±– (1)
Dead leaves	6.0±– (1)	3.9±1.1 (5)	3.2±0.8 (2)	3.7±0.2 (3)
Dead stalk	2.4±– (1)	2.8±0.6 (2)	4.8±– (1)	2.5±0.1 (2)
Cynodon dactylon				
Green leaves	15.4±5.4 (12)	15.7±4.0 (10)	10.8±3.6 (4)	11.6±2.5 (23)
Green stalk	11.2±2.7 (11)	9.9±3.1 (9)	5.5±0.7 (3)	5.0±1.4 (22)
Dead leaves	8.8±2.9 (8)	6.1±0.7 (6)	5.8±1.3 (3)	5.8±1.9 (23)
Dead stalk	5.2±1.4 (8)	4.2±1.2 (5)	4.5±0.5 (4)	4.4±1.4 (22)
Cynodon plectostachyus				
Green leaves	21.4±– (1)	20.5±4.6 (6)	9.4±0.5 (3)	9.8±– (1)
Green stalk	12.2±– (1)	10.0±4.4 (6)	7.4±0.8 (2)	2.8±– (1)
Dead leaves	–	6.4±1.5 (5)	6.7±0.1 (2)	4.8±– (2)
Dead stalk	–	4.3±1.4 (4)	5.5±0.7 (2)	2.4±– (1)
Cyperus laevigatus				
Green fraction	7.6±1.4 (2)	6.5±1.8 (6)	6.9±0.5 (3)	5.8±– (1)
Dead fraction	5.4±– (1)	5.9±3.1 (4)	–	–

Note that green leaves always show the highest quality and are most of the time above the maintenance requirements of buffalo (about 7% crude protein).

severe limiting factor for the buffalo, as is typical for the Serengeti. However, I expect that the Manyara buffalo would in such a situation switch to browse, as occurred in September 1973, after which they vacated the park (Warden's Reports 1973).

Further conclusions can be reached on the basis of Figure 2.2. First, the buffalo increase the quality of their diet in the short rains season only slowly, and lag behind the improvement of the vegetation (compare Figure 2.2 and Figure 1.8). Most likely, this reflects the relatively slow adaptation of the rumen to the high quality of the potential food with its concomitant high fermentation rate (Hofmann 1973). A slow adoption of very high-quality food in the diet after a period of poor food may prevent acidosis, which greatly lowers the digestive capabilities of the rumen (Slyter 1976, Van de Veen 1979, p. 138 ff.). Too quick a shift from a poor diet to high-quality food can even cause death in ruminants (Dirksen 1970), and animals in poor physical condition are more susceptible to acidosis than animals in good condition (Slyter 1976). The Manyara buffalo, on average, are in the poorest condition at the beginning of the short rains (Chapter 3) when grass growth becomes vigorous (Chapter 1, Tables 1.5 and 1.6) and potential food improves in quality (Table 2.1). From Figure 2.2 it can also be deduced that the range of protein concentration in the diet of different buffalo individuals at any moment during the short rains is small. This range increases towards the post-rainy season, and decreases again during the late dry season. I think that this reflects, on the one hand, the differences in social standing of buffalo within a herd (Chapter 3) and, on the other hand, the increase in interspecific variability in quality of food on offer during the long rains as compared with that during the short rains (Chapter 1). The pattern of quality of the diet of the buffalo indicates that the short rainy season and the long rainy season are times of affluence (compare Table 2.2 and Figure 2.1 with Figure 3.6).

The crude protein content of the food is then high, which means a high energy content too (about 2250 kcal metabolizable energy per kg dry matter). This results in a voluntary food intake of 3–3.5% of the body weight (Crampton and Harris 1969, Table 14.5). This means that for an average buffalo of 500 kg the food intake during the rainy season is about 15–17.5 kg dry matter per day. During the post-rainy season, the quality of the food declines (Table 2.2), and for some individuals the quality of the food falls below maintenance requirements (compare Figure 2.1). Nevertheless, on average the food quality is still high enough for production of milk or for growth. Indeed, the crude protein content of the food is approximately 10.5%. Such a level results in a metabolizable energy content of about 2150 kcal kg^{-1} food (dry matter) and, thus, the voluntary food intake will be still 3–3.5% of the body weight. So, although food quality declines during the post-rainy season, food intake for an average buffalo of 500 kg is still expected to be 15–17.5 kg dry matter per day.

During the late dry season, the buffalo are not able to select for very high-quality food any more, and their diet falls to 8% crude protein (Table 2.2), but some individuals have a much worse diet (Figure 2.2). This means that, on average, the diet then contains about 2050 kcal metabolizable energy per kg food, and food intake drops to 3% of the body weight (Crampton and Harris 1969, Table 14.5) but in a number of individuals it will have fallen to 2.5%. This implies that for an average buffalo in Manyara food intake during the late dry season decreases to 12.5–15 kg dry matter per day. Food quality declines but, on average, is still above maintenance requirements. During the late dry season, milk production or growth is hardly possible any more. Lactating females have to use body reserves to continue production. Periods of affluence and shortage in the annual cycle of the African buffalo in Manyara show a clear relation to

Table 2.2 The average crude protein concentration expected in the buffalo's diet (second last row) if they were eating indiscriminately all green parts of the major monocotyledons within their home range, as compared with the actually determined crude protein concentration in their diet (bottom row)

	Short rains season (Nov–Jan)	*Long rains season (Feb–Apr)*	*Post-rainy season (May–July)*	*Dry season (Aug–Oct)*
Sporobolus spicatus				
Average CP% in green parts	10.0	7.8	6.4	6.3
Average green standing crop (g m^{-2})	141.9	145.0	91.8	53.8
Approximate total area (ha)	334	334	334	334
Cynodon dactylon				
Average CP% in green parts	13.3	13.2	7.5	7.4
Average green standing crop (g m^{-2})	181.1	151.8	182.5	115.1
Approximate total area (ha)	465	465	465	465
'Woodland species'				
Average CP% in green parts	20.0	15.9	8.5	–
Average green standing crop (g m^{-2})	57.6	86.8	29.8	0.0
Approximate total area (ha)	1374	1374	1374	1374
Cyperus laevigatus				
Average CP% in green parts	7.6	6.5	6.9	5.8
Average green standing crop (g m^{-2})	302.6	325.7	184.5	203.5
Approximate total area (ha)	97	97	97	97
Average CP% in green standing crop	14.2	12.6	7.5	6.8
Average CP% in diet of buffalo	11.4	11.4	10.5	8.0

rainfall, taking into account both the quality and quantity of the food on offer. The main difference between Manyara and the Serengeti is that in the latter area the quality of the diet falls to much lower levels than in Manyara. This is mainly caused by the absence of perennial grasses that remain green during the dry season. The Manyara area is characterized by patches of upwelling groundwater, where the sedge *Cyperus laevigatus* remains green and growing during the dry season, and by patches where the groundwater table is still within rooting depth of the grasses. These areas can be found on the numerous alluvial fans in which the vegetation is dominated by the perennial *Cynodon dactylon*. The ecology of the landscape has a very profound influence on these large herbivores.

At the beginning of this chapter I reviewed evidence that the most critical limiting quality parameter for buffalo is nitrogen (protein). The average crude protein concentration in the diet (Figure 2.2) can be compared with the average crude protein concentration of the potential food within the home range, which I calculated as the weighted average crude protein of the four major species taking into account the green stalk proportions in the swards. When the dietary crude protein concentration and the crude protein 'in the home range' are compared (Table 2.2), it is clear that buffalo are not maximizing the concentration of protein in their diet. On the contrary, during the short rains, and to a lesser extent during the long rains, the buffalo, like red deer (Van de Veen 1979, p. 93), appear to avoid food that is of too high quality. Buffalo nearly always ingest a **balanced** diet (Figure 2.1) and not a diet maximal in protein or energy content. Buffalo in Manyara can achieve a balanced diet by consuming only the four major monocotyledons in their home range, namely *Cynodon dactylon*, *Cyperus laevigatus*, *Cynodon plectostachyus* and *Sporobolus spicatus*, because these vegetation types occur in close proximity to each other, and because each exhibits a different reaction to variations in rainfall. The fine-grained landscape of Manyara (Chapter 1) unlike the Serengeti with its much coarser pattern of sublandscapes lends itself to the provision of a balanced diet for buffalo. Again, research on cattle indicates why such a balanced diet would be advantageous to buffalo. Garrett (1970) reported that beef cattle fed diets containing about 21% crude protein required 20% more feed to maintain energy equilibrium than did those fed diets containing 12% crude protein (cf. Figure 2.1). The cause of this difference is that the energetic cost of metabolizing amino acids ingested in excess of the protein requirements is high (Reid 1974). Perhaps now is the time to pose the question of whether buffalo could have enough time to gather more feed if they chose another diet, because at the beginning of the chapter I mentioned that the most critical factor regarding the amount of food for herbivores is the density of the food on offer, that is whether it can be collected in the time that is available.

2.2 FOOD INTAKE AND SWARD DENSITY

The nutrient intake of buffalo can be impaired because the quantity of food on offer is so low that the buffalo's food intake rate is depressed or because the concentration of the nutrients in the diet (the quality) is too low. At the end of the dry season the nutrient concentration in the food is low, but not so low as to impair food intake. Is there any evidence that the food density in Manyara becomes so low that the food intake is depressed? A number of parameters are known to influence the intake rate of herbivores. These are (a) the biomass standing crop, especially of leaves (Trudell and White 1981, Short 1986), (b) vegetation height (Allden and Whittaker 1970), (c) the ratio between green leaf and green stem (Stobbs 1973b) and (d) the sward bulk density (Chacon and Stobbs 1976).

For the four major food species of the buffalo of Manyara the standing crop of the green leaf biomass per hectare is presented in Figure 2.3 (see also Tables 1.5 and 2.2).

In the tropics, a standing crop of 1000 kg ha^{-1} of green leafy material results in cattle showing maximal grazing times; at 600 kg ha^{-1} grazing time is much lower (Chacon 1976 cited in Whiteman 1980, p. 311). This means that a standing crop of green leaves of 1000 kg ha^{-1} can be considered adequate for cattle or buffalo, but that at a lower standing crop of green leaves the pasture can no longer fulfil the animal's dry matter requirements. The green leaf standing crop of the woodland mixed grass species sward, which consists of *Sporobolus helvolus*, *S. cordofanus*, *Leptocarydion vulpiastrum*, *Urochloa pullulans* and other species, is insufficient to sustain buffalo at any time of the year ('Woodland species' in Figure 2.3). The green leaf biomass of *Cynodon dactylon* is of too low a density for most of the year. During the late dry season, none of the three grass species that are most important for the Manyara buffalo has an adequate green leaf biomass to fulfil the intake rate requirements. In fact, only the sedge *Cyperus laevigatus*, despite its rather low concentration of crude protein (Table 2.1), always has a high enough green biomass to satisfy the intake requirements of buffalo. Buffalo spend much of their grazing time on this sedge, and it constitutes the bulk of their diet during the dry season, as was shown both by faecal analysis (Table 1.4) and by the rumen contents of buffalo that were killed by lions.

The balance between bite size (that is intake rate) and crude protein is, I think, another key to understanding the food species selection of buffalo. Within the range of plant species encountered in the present study, bite size is positively related to sward height, to the green leaf/stem ratio and to the sward bulk density. I calculated a bite size index as an average of these three parameters by considering them as indices also:

log bite size = (log green leaf/stem ratio + log sward height + log sward bulk density)/3

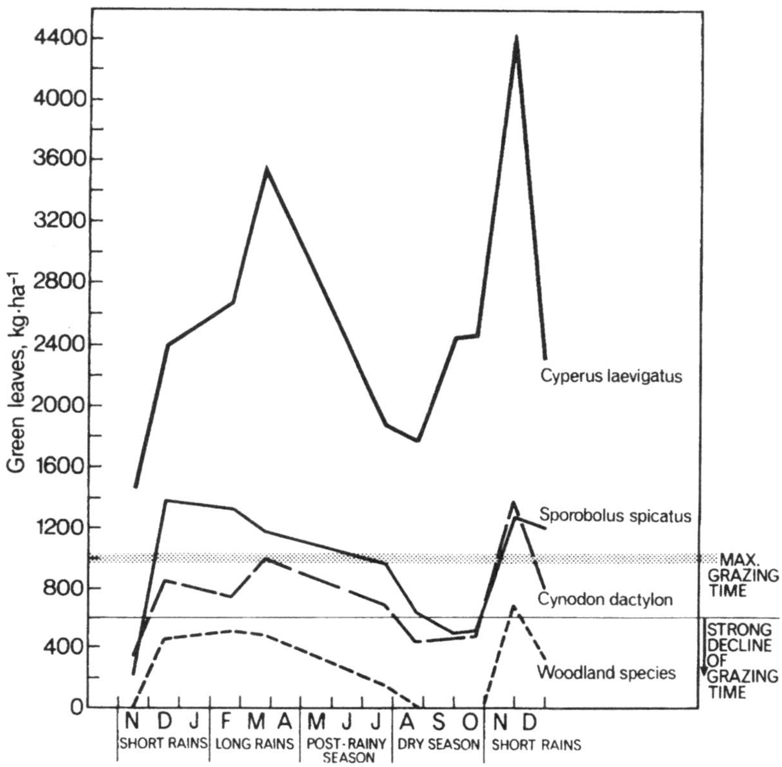

Figure 2.3 Standing crop of live leaves (kg ha^{-1}). If this is lower than 600 kg ha^{-1} (solid line) then the sward is not dense enough to meet intake requirements and there will be a strong decline in the percentage time devoted to grazing. At 1000 kg ha^{-1} (shaded bar) grazing time is maximal but intake requirements can be met. Above this bar, the sward is dense enough for grazing so that grazing time can be lower than maximal.

In Figure 2.4 the relation between bite size index and crude protein concentration is shown, with the preferred food species indicated as based on Table 1.4.

The figure indicates that buffalo prefer species with the highest protein concentration as long as the bite size index is larger than 2.5. When the bite size index of the food species falls below this level in the late dry season, the buffalo start concentrating on *Cyperus laevigatus*, which is low in protein. However, by supplementing the *Cyperus* diet with some of the

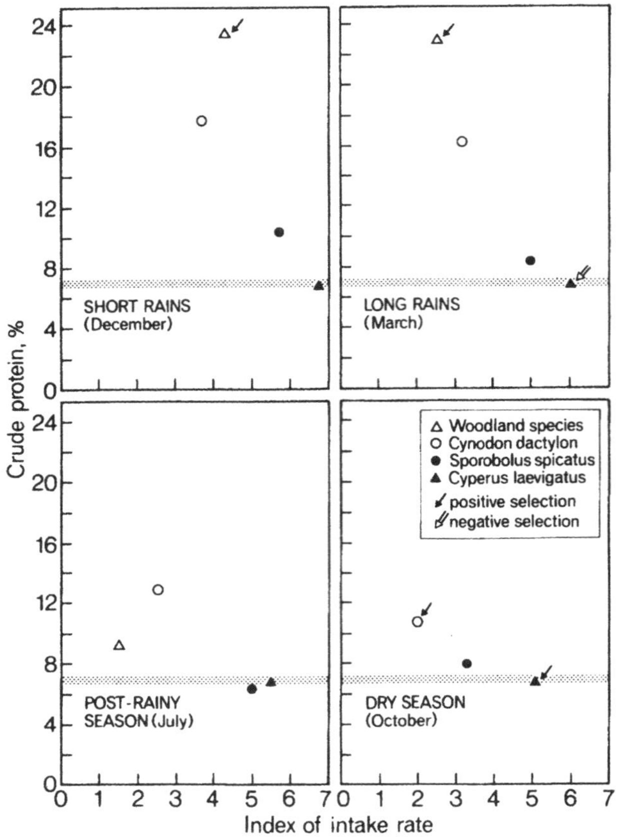

Figure 2.4 Relation between bite size index (see text) and crude protein concentration of the major monocotyledons in Manyara. Species selected by buffalo are marked with arrows. The horizontal bar gives the crude protein concentration at which the buffalo can maintain body weight. The figure shows that buffalo concentrate their efforts on those species that are highest in protein, except in the late dry season, when the food species highest in protein no longer meet the intake requirements: buffalo then switch to the species with the largest bite size index (that is *Cyperus laevigatus*).

other species, the buffalo still achieve a diet with a higher crude protein concentration than *Cyperus* alone can offer (Figures 2.4 and 2.2, Table 1.4). It is clear that only a mixed diet of different food species can ensure that the Manyara buffalo obtain a reasonable total nutrient intake. As was shown in Chapter 1, the Manyara buffalo do indeed have a mixed diet. But are they time limited when it comes to assembling their diet? I will come to this in subsequent sections.

2.3 BUFFALO FEEDING ACTIVITY

Mammalian herbivores feed for a high proportion of the time (that is day and night). Horses, for example, may devote 70–85% of their time to grazing (Tyler 1972, Pratt *et al.* 1986) and cattle as much as 50–60% (Stobbs 1970); because cattle are ruminants they must also allocate time to cud-chewing, which results in a total feeding time of 70–80%. Feeding time depends on (a) the requirements of energy and nutrients, (b) the amount of digestible food that is available and (c) the rate at which food can be ingested. The last is influenced by the distribution of the food. In the previous sections I have discussed the energy and nutrient requirements of buffalo. I have also presented data on the amount of food that is available over the different seasons. In the dry season the density of most Manyara swards is too low to satisfy the intake rate of buffalo. Only *Cyperus laevigatus* swards have a sufficient standing crop all year round. Hence, with the onset of the dry season the time of scarcity begins, because the quality of the *Cyperus* sward is quite low and swards of other grass species show a low standing crop. One of the reasons for the low sward density might be consumption by other herbivore populations (mainly African elephant, Burchell's zebra and wildebeest). In Chapter 6 their possible competitive role is discussed. I now want to address the issue of the amount of time buffalo have available for collecting food and processing it. Flexibility of time budget in periods of food scarcity may help herbivores such as buffalo to maintain intake requirements.

We made numerous, almost daily, random road counts of all herbivores, including buffalo (see Beekman and Prins 1989 for methods). Data are available for the most important periods in the annual cycle. Most observations stem from the daylight period, and care was taken to spread observations equally over the day. Daytime is defined as 06.00 hours to 19.00 hours (East African standard time) and night as 19.00 hours to 06.00 hours. Night observations were few, mainly because buffalo spend the night in habitats with poor visibility for the observers, but I consider the data adequate for quantitative conclusions. In total about 8000 road counts were done, which resulted in more than 10 000 group observations. In total, 480 000 buffalo individuals were scanned (that is group sizes multiplied by number of group observations).

The pattern of grazing activity of buffalo during the daylight period is presented in Figure 2.5A.

The general pattern is a grazing period around midday between 10.00 and 14.00 hours. In the morning and in the afternoon there is a non-grazing period, during which most of the animals ruminate (Figure 2.5B).

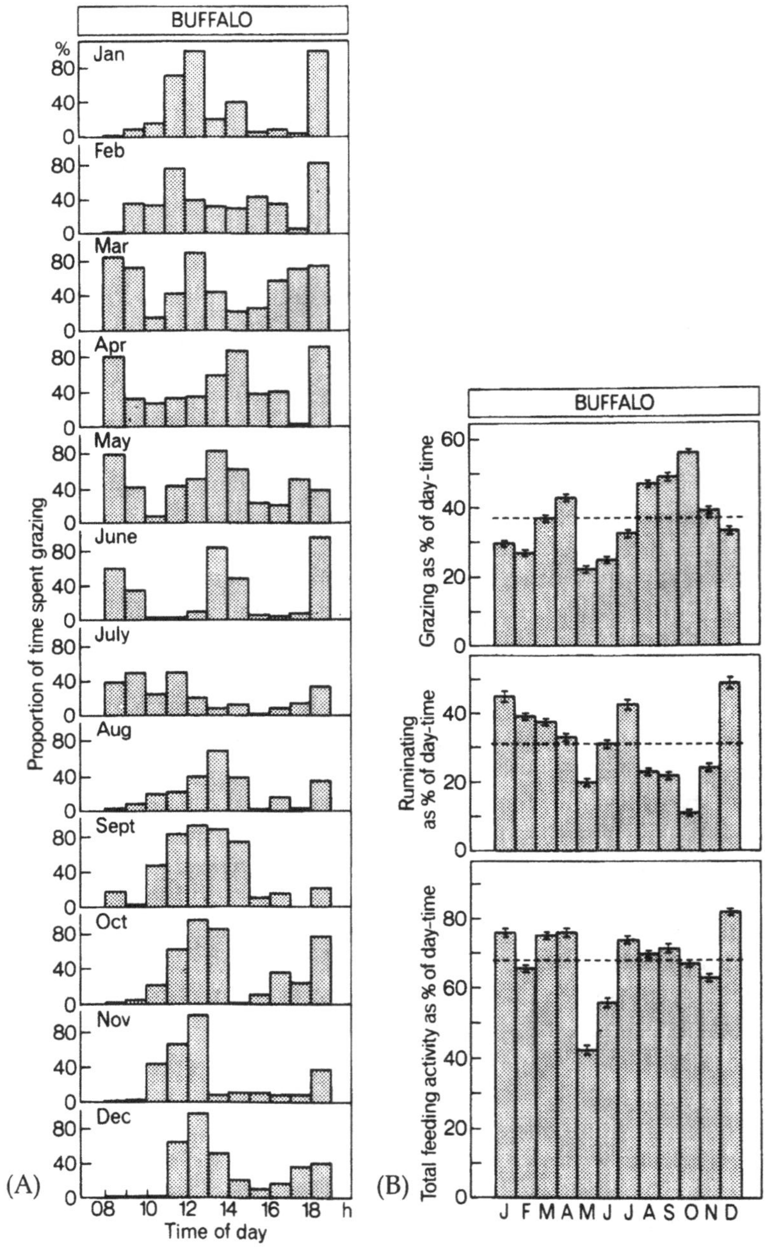

Figure 2.5 Proportion of time spent grazing by buffalo over the day in different months of the year (A), and the average proportion of time devoted to grazing, rumination and feeding (that is the sum of grazing and ruminating) over the year (B). August to October, dry season; November to January, short rains season; February to April, long rains season; May to July, post-rains season. Vertical bars represent 99.9% confidence limits.

This pattern corresponds to that as described by Sinclair (1977) for buffalo in the Serengeti. The midday grazing period varies from 1.5 hours in July to 4.5 hours in April. Similarly, the resting period in the morning varies from 1 hour in October to 3 hours in January and that in the afternoon from about 2 hours in August to 4 hours in February. During the dry season additional grazing bouts were observed in early morning and late afternoon, but these become less distinct owing to the individual buffalo becoming less synchronized in their behaviour. The time spent grazing during the day shows two peaks in the course of the year (Figure 2.5B), which can be explained by the prevailing food conditions and the nutrient and energy requirements. The first peak falls in the long rains season when food is abundant, with both annual and perennial grasses on offer, and of a fairly good quality. This is also the period in which buffalo accumulate body reserves (Chapters 3 and 4). In the post-rains season, the grazing percentage drops to a minimum but after that, through the long dry season, buffalo increase this percentage again. The peak in October, which is the year's maximum, reflects the bad food conditions in the dry season, and it appears that buffalo have to graze longer and longer to meet their requirements. With the onset of the first rains in early November, the perennial grasses start to grow and food becomes abundant and of good quality (Chapter 1). The grazing percentage drops again (Figure 2.5).

Extensive observations from the wet season revealed some 70–80% grazing during the night, while in the dry season at least 50% but mostly over 80% of the individual buffalo were grazing at night. We think that night grazing in Manyara is at least as common (but probably more common) as grazing during the day, and that there are no important proportional differences between day and night grazing during the course of a year. Mloszewski (1983) found negligible differences between day and night grazing in the buffalo he studied, but those in the Serengeti mainly graze at night in the wet season and show more daytime grazing in the dry season (Sinclair 1977). In Uganda, buffalo grazed 3–4 hours during the day and 5–6 hours at night (Grimsdell 1969). All data combined give an average yearly proportion of the total time budget devoted to grazing of 37%, with a monthly range of 22–56%. In the Serengeti the average is 41% with a range of 38–44% (Sinclair 1977). The variation in Manyara may thus be wider than in the Serengeti, but Sinclair unfortunately does not report data throughout the year, and especially not from the periods when the Manyara buffalo show the highest and lowest grazing time proportions (October and May, respectively). Mloszewski (1983) reports a range of grazing times between 36 and 43%, which is close to the values mentioned by Sinclair (1977).

Intake of herbage by a grazing animal is proportional to both the time spent grazing and the rate of grazing; the two components of rate of grazing are rate of biting and bite size. If grazing rate is reduced, this has to be compensated for by increasing grazing time (Stobbs 1973a,b, Duncan 1992). Bite rate and bite size depend on the structure of the sward and the morphology of the animal's mouth. In Manyara grazing time is increasingly lost during the dry season as the buffalo reject patches of vegetation (indicative of food selection). During this period buffalo change from plucking with their tongue to cutting with their incisors, which results in observably smaller bite sizes. Hence, it is likely that they increase their bite rate to compensate for the reduction in bite size. This, together with the time lost in finding acceptable patches, indicates that intake rate is decreased in the dry season, which in turn forces the buffalo to increase their grazing time during the day.

One would expect rumination time to increase when food quality decreases and, concomitantly, total fibre intake to increase under the condition that total feed intake remains equal (or increases) (Welch and Smith 1969, 1970). For buffalo, there are enough data on rumination to study the effect of season on rumination, and hence, indirectly, on total fibre intake. Buffalo ruminate on average 31% of the daytime, but there is a large monthly variation with a range of 11–49% (Figure 2.5B). Rumination is maximal during the short rains. It then decreases in May, but in June and July the percentage increases again, followed by a sharp decline in the long dry season to a minimum value in October. Serengeti buffalo show a very different pattern, with an increase from 31% in the long rains season to 37% in the early dry season. Sinclair (1977) concluded that with the deterioration of the food quality (that is a progressively higher cell wall content in the diet) Serengeti buffalo have to spend more time ruminating. This explanation does not appear likely since rumination time is proportional not to the relative but to the **absolute** amount of fibre intake (Welch and Smith 1969, 1970), and an increase in the fibre content of the food causes intake reduction (Van Soest 1982). In Manyara, buffalo also have to deal with decreasing sward quality throughout the dry season, but the Manyara buffalo do decrease their rumination time.

2.4 BUFFALO AS MIXED FEEDERS?

Hindgut fermenters probably have a selective advantage when food quality is a limiting factor, whereas ruminants perform better when the quantity of food is in limited supply (Janis 1976). As was shown in Beekman and Prins (1989), the compensation for deteriorating food quality in zebra is, as expected, mainly through a higher intake rate, which speeds up the throughput rate in the gut, and so zebra obtain more energy and nutrients per unit of time. This option is generally considered impossible for grass-eating ruminants such as buffalo because they are constrained by the minimum protein concentration of 7–8% in the diet to make fermentation in the rumen possible (section 2.1), and because of the retention of large and coarse food particles in the rumen by the ostium ruminoreticulare (Hofmann 1973, p. 22). Options for ruminants when food quality declines include selecting for less fibrous food, so that they have to become more selective in the dry season in order to keep fermentation going. In that case, ruminants have to increase grazing time; wildebeest appear to do this (Chapter 6).

Another option is switching to browse, as is done by elephant in the dry season in Manyara (Beekman and Prins 1989). However, like the Serengeti buffalo, Manyara buffalo continue eating grass in the dry season as their diet contains only 5% at that time (Table 1.4; Lamprey 1964a, b). As browse is relatively high in protein (Table 1.10), while grass is not during the dry season, one would expect Manyara buffalo to supplement their diet to a larger extent with browse, as they do in some other East African areas (Jarman 1971, Mloszewski 1983). Browse is always available to buffalo (Chapter 1; Vesey-FitzGerald 1969), so there must be another reason why they do not adopt a diet-switching strategy during the dry season as do the cattle of Iraqw farmers and Masai pastoralists in the vicinity (Homewood and Rogers 1991, p. 155 ff.; cf. Rees 1974). The reason that Manyara buffalo do not switch to browse may be found in either the assemblage of gut microbes or the morphology of their rumen, and especially in the plasticity of this morphology under influence of the diet (Hofmann 1973, p. 38 ff.), but buffalo also may avoid browse because the levels of toxins, tannins or other polyphenolic compounds are too high in many browse species (Owen-Smith *et al.* 1993); the low production of saliva by buffalo (about 100 l day^{-1}: R.R. Hofmann, personal communication) could mean that they do not produce enough tannin-binding proteins in their saliva to bind browse tannins. This would then preclude inclusion of too much browse in their diet. However, consumption by local cattle of leaves of shrubs and trees in the area around Manyara during the dry season pleads against accepting this argument too easily, just as does the fact that buffalo populations elsewhere in Africa can include much browse in their diet (MacLatchy 1932, Mloszewski 1983). Whatever the explanation of the avoidance of browse by the Manyara buffalo, the general rule appears to be that they avoid browse species that are resinous, spiny or aromatic or taste bitter (Table 2.3).

The browse species that they include in their dry season diet have a low content of acid detergent fibre. Van Soest (1982) showed a very strong negative correlation between the digestibility of the food of cattle and its acid detergent fibre content. Thus, when buffalo are constrained by low grass quality during the dry season, they can upgrade the food mixture in the rumen by including some browse high in crude protein. At the same time, buffalo should avoid increasing the fibre content of this

Table 2.3 Preference of buffalo and of cattle of the Iraqw for green leaves of shrubs during the dry season in relation to the crude protein (CP) concentration and acid detergent fibre (ADF) concentration, both based on the means of five sampling dates

	Buffalo	*Iraqw cattle*	*Dry season*		*Remarks*
			%CP	*%ADF*	
Justicia cordata	+++	++	15.8	17.2	None
Gardenia jovis-tonantus	++	+++	13.3	27.6	None
Maerua triphylla	++	+++	17.4	13.7	Slightly sclerophyllic
Ocimum suave	++	–	16.0	n.d.	Strongly aromatic
Abutilon mauritianum	+	+++	19.7	n.d.	None
Acacia tortilis	+	+++	18.2	25.6	Thorny
Capparis farinosa	+	++	17.2	20.7	Thorny, slightly sclerophyllic
Cordia sinensis	+	++	14.3	39.6	Slightly sclerophyllic
Salvadora persica	+	–	13.9	13.6	Slightly sclerophyllic, salty and bitter
Tephrosia villosa	+	+++	18.7	n.d.	None
Acalypha fructicosa	–	+++	16.5	n.d.	Strongly aromatic, slightly sticky (resinous?)
Cordia ovalis	–	++	15.1	35.8	Tough with silica crystals on epidermis
Dovyalis xanthocarpus	–	+++	13.6	20.8	Few thorns, bitter
Ruellia megachlamys	–	+++	17.0	21.0	Aromatic, sticky (resinous?)
Vernonia cinerascens	–	++	13.3	21.2	Aromatic
Barleria erantyhremoides	–	–	14.0	n.d.	Thorny
Cardiogyne africane	–	–	13.2	11.9	Thorny, sclerophyllic, latex
Thylachium africanum	–	–	17.0	22.9	Sclerophyllic, latex

Buffalo preference is based on observations but that of cattle on interviews. n.d., Not determined; –, not eaten.

mixture as this would further impair digestibility. Inclusion of only browse species low in acid detergent fibre appears to be a good strategy.

A third way of dealing with low food quality, namely increasing the throughput rate, as in hindgut fermenters, is generally thought to be impossible for ruminants. However, Van Soest (1982) voiced some doubts about this, and a number of observations collected on the buffalo of Manyara appear to confirm his doubts. We collected data on the number of chewing movements during rumination and on time spent per regurgitated bolus. Increasing the rumination time per amount of food means that food particles become smaller, which in turn means that microbes have more 'sites of attack', resulting in a higher fermentation rate. In buffalo, the chewing time per bolus increases in the early dry season, but decreases again after August (Figure 2.6A), when food quality further decreases to below maintenance level (8% crude protein) and to below the level which is needed for microbial growth in the rumen (also approximately 8% crude protein; Appendix).

Both the increase and the decrease are significant ($F_{1,1} = \infty$, $P < 0.025$, and $F_{1,1} = 712.8$, $P < 0.025$, respectively). However, the chewing rate (number of chewing movements per second) does not increase during

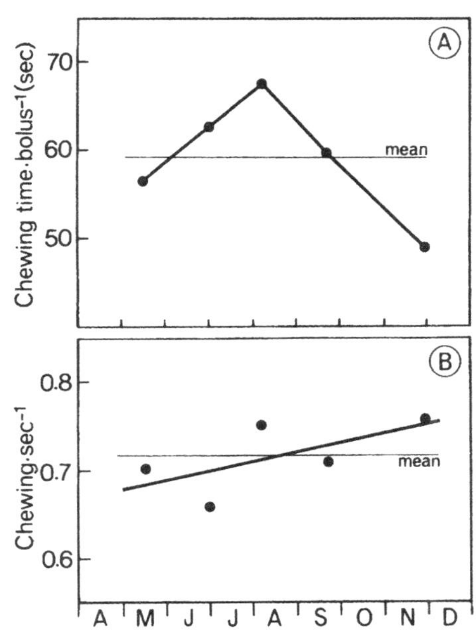

Figure 2.6 Chewing time per bolus during rumination (A) and chewing rate during rumination (B) in relation to the time of year.

the dry season (Figure 2.6B; $F_{1,3} = 3.0$), so apparently chewing efficiency does not change. Hence, the total number of chewing movements per regurgitated bolus decreases. This means either that the size of a bolus decreases or that a bolus is less well chewed. A decrease in bolus size means that more boluses have to be produced from the same rumen content; the implications of this are unclear. The hypothesis that the bolus is less well chewed at the end of the dry season implies that food particles remain larger after chewing and thus that coarser material has to pass through the orifice, as has also been suggested by Van Soest (1982). This would benefit a ruminant on low-quality diets, when overall fermentation rate is low, because cell content can still be fermented quickly, that is the rumen has to be emptied as quickly as possible. In this sense, buffalo resemble 'concentrate selectors' in the problems they face, and, hence, react more as 'intermediate feeders' than as typical roughage grazers (Hofmann 1973, p. 43 ff.).

2.5 ARE BUFFALO TIME-LIMITED OR RESOURCE-LIMITED?

Total feeding activity is defined as grazing plus rumination. In the Manyara buffalo this averages 68.3% over the year (Figure 2.5B). The relatively low percentages in May and June may be explained by the fact that most matings take place then. As there appears to be competition between the bulls (Chapter 4), this may cause social unrest in the herd at this time of the year, which could reduce the feeding time. The high value in December could be explained by the buffalo working hard to regain condition when food circumstances are optimal. Thus, there clearly is flexibility in the buffalo's time budgeting. However, because they must spend a considerable proportion of time feeding during some parts of the year – some 70–80% – the question arises whether they can compensate for poor feeding conditions with low standing crop and/or low quality of forage by adjusting their time budget sufficiently. If they cannot, then they are considered to be time limited. Stobbs (1970) considered this to be the case in cattle that spent some 55% of the time grazing: as a result of an unfavourable vegetation structure their bite size was too small and they could not increase the number of bites, and hence their intake did not meet their requirements.

Buffalo spend much time grazing and ruminating. This implies that they cannot easily increase both grazing and ruminating simultaneously, so these activities can be thought of as competing for time. The same has been suggested by Van Soest (1982) for other ruminants. The upper limit of grazing time proportion in cattle is estimated to be 55–60% (Stobbs 1970), while the maximum for buffalo is 56% (Figure 2.5B). The maximal rumination time proportion for cattle is estimated to be 40% (Welch and

Smith 1969, 1970, Van Soest 1982) and the maximum in buffalo is 49%. Hence, under different food conditions buffalo can choose different ways of partitioning their time budget, namely devoting more time to grazing or more to ruminating. This choice can be expressed in the ratio of rumination time to grazing time (R/G ratio; Figure 2.7).

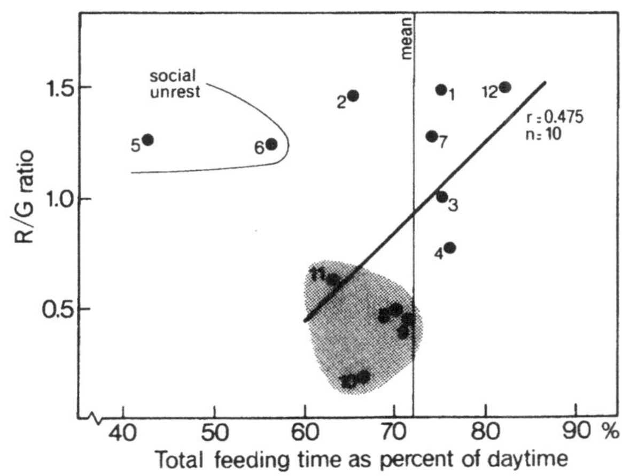

Figure 2.7 The ratio between proportion of time spent on rumination and grazing (R/G ratio) in relation to the total feeding time. Numbers refer to months: 1 = January, 2 = February, etc. During May and June there is much social unrest in the herd because of mating activities (Figure 5.5).

With the onset of the short rains, food becomes abundant and of very good quality, and buffalo spent relatively little time on grazing and much time on rumination (R/G ratio > 1). In this period they regain condition. During the long rains, the R/G ratio slowly decreases because more time is spent grazing and less ruminating (R/G ratio > 1). This can be explained by the fact that during the long rains food is abundant but some parts of the vegetation are becoming old and low in quality and other parts of the vegetation are still high in quality (because, for example, it was recently grazed). Hence, buffalo have to increase their selectiveness during the long rains, and consequently increase grazing time. During the post-rains season, all grasses mature and food is still abundant, but of decreasing quality, so that the buffalo have to ruminate more. Consequently they have to reduce grazing time and the R/G ratio increases to above unity. Thus, buffalo seem to have two different ways of partitioning their time budget, namely an option in which they stress bulk feeding (proportionally little grazing and much rumination) and an option in which they emphasize selection (proportionally much grazing and little rumination). The 'bulk feeding option' is chosen when food is abundant and of good quality **or** when food is abundant and of relatively poor quality. The 'selection option' is applied when food is abundant but of mixed quality **or** when food is scarce and of poor quality. During the long dry season food becomes scarce in Manyara as there is no production to compensate for removal by all herbivores and simultaneously food quality decreases and the crude protein content of certain vegetation stands drops to even below maintenance level (Chapter 1). At this time buffalo are forced to become very selective. Grazing time increases and rumination time decreases sharply. The observation that rumination time decreases implies that total fibre intake in buffalo becomes low at this time, and this suggests that buffalo are constrained by rumen fill regardless of the inferred throughput rate (section 2.4) at the end of the dry season. Accepting lower quality food that takes less grazing time but requires more rumination time would not help since the rate of fermentation would become too low. Consequently the R/G ratio falls far below unity. This confirms the inference made in section 2.1 that the actual energy intake is low at this time of the year, and, as will be shown in Chapters 3 and 4, buffalo lose condition during the end of the dry season. However, the time devoted to feeding is not maximal at all during this period of scarcity (Figure 2.7). Furthermore, their idling time does not decrease during the dry season (Table 2.4).

My interpretation of this plethora of information on the behaviour of the Manyara buffalo is that they are not time limited but appear to be constrained by their inability to gather enough high-quality food, which is no longer present in the late dry season (Chapter 6). This analysis based on the buffalo's behaviour confirms Sinclair's (1977) conclusion, which was based on bone marrow analyses and diet composition of buffalo in the Serengeti, that the buffalo of Manyara are resource limited.

Table 2.4 Daytime percentages of moving and idling (resting) for the four major herbivores

	Buffalo		*Wildebeest*		*Zebra*		*Elephant*	
Month	*Moving*	*Idling*	*Moving*	*Idling*	*Moving*	*Idling*	*Moving*	*Idling*
Jan	5.9	18.7	–	–	–	–	–	–
Feb	5.6	28.8	4.0	19.3	3.3	35.2	3.6	21.2
Mar	4.9	20.0	3.2	31.2	1.6	25.0	8.0	13.8
Apr	13.5	10.5	5.6	23.3	7.3	35.2	7.5	16.4
May	13.6	44.0	9.9	31.1	7.8	28.2	–	–
Jun	11.8	32.4	9.1	20.9	10.9	26.5	–	–
Jul	1.5	24.5	0.9	27.6	3.6	18.3	7.6	12.9
Aug	8.6	21.5	5.0	18.6	2.4	11.8	17.5	6.7
Sep	2.3	26.3	0.4	16.5	4.6	12.7	9.1	7.8
Oct	6.6	26.7	1.6	9.9	4.3	11.9	19.7	7.3
Nov	4.7	32.0	3.1	34.2	5.6	28.3	8.2	17.8
Dec	2.2	16.0	0.9	25.7	4.4	42.2	12.0	15.2
$\bar{\chi}_{late\ dry}$	5.8	24.8	2.0	15.0	3.7	12.1	13.5	7.3
$\bar{\chi}_{rest}$	7.1	25.2	5.1	26.7	5.8	29.9	7.9	16.2
$\bar{\chi}_{post\ rainy}$	13.0	29.0	8.2	25.1	8.7	30.0	–	–
$\bar{\chi}_{rest}$	4.7	23.8	2.4	22.9	3.7	20.6	–	–

Moving is walking while not feeding, and idling is a rest activity meaning all but moving, grazing, browsing and ruminating. 'Late dry' refers to August to October and 'post-rains' to April to June.

3

Social organization of buffalo cows

3.1 DISTINGUISHING SEX AND AGE CATEGORIES AND ASSESSING CONDITION

When a buffalo is 3 years old its sex can be recognized from the shape of its head and horns. In younger buffalo sex is more difficult to determine from a distance since it depends on seeing testicles or penis in males. I define cows as 'adult' if they are 6 years or older, and as 'subadult' if they are 3–5 years old. I define 'cows with dependent offspring' as adult cows with calves or young juveniles at foot and 'cows without dependent offspring' as adult cows to which neither calves nor juveniles are attached. I have adopted Pienaar's (1969a) definition of calves as young buffalo that do not reach the inguinal fold at its highest point on the flank of the cow. Juveniles by my definition are older than calves but younger than subadults. This definition of juvenile is different from Sinclair's (1977, p. 163), and in my terminology all categories are mutually exclusive. To determine the age of dead buffalo I use a set of photographs of teeth of buffalo of different ages prepared by Grimsdell (1969), which was also used by Sinclair (1977). Figure 3.1 was prepared from Sinclair's data (Tables 13, 19, 20) and shows that males and females have the same weight development up to 6 years of age. I assume that the Manyara buffalo show the same development.

I have devised the following scale to measure the condition of buffalo based on discussions with farmers and cattle salesmen at a large cattle market ('Zuidlaardermarkt') in The Netherlands and inspection of scores of cattle at this market. Physical condition is assessed at a market by prospective buyers, and my rationale was that the prices cattle fetch are closely related to physical condition (and breeding state). This was confirmed in the discussions. Hence I judged the condition of buffalo in the same way. Later Herd and Sprott (1986) published their body condition score (BCS) for cattle in Texas, and I give their scores in parenthesis:

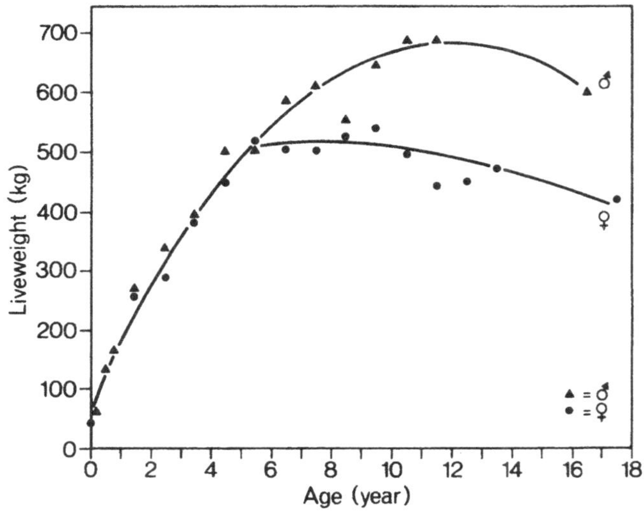

Figure 3.1 Relation between body weight and age of buffalo, based on data from Sinclair (1977). Buffalo are considered adult at the age of 7 years.

Very poor	Ridge on shoulder blades, spinal ridge and ribs visible (BCS = 1)
Poor	Flank, pelvis and tail base concave, ribs visible (BCS = 3)
Moderate	Pelvis less concave, ribs just visible (BCS = 5)
Good	Ribs not visible, flank not concave, tail base 'filled' (BCS = 7)
Excellent	Body contours convex, neck shows fat ridges when bent (BCS = 8–9)

In the calculations I assign index values to the condition scores, namely 1.0 for very poor to 5.0 for excellent. I then treat these index values as ordinary numerical values, which appears to be legitimate as there is a linear relation between body condition score and percentage fat in cattle: a BCS of 1 corresponds to 0% fat in the empty body (that is live weight less the contents of the digestive tract), a BCS of 2 corresponds to 4% fat, a BCS of 3 corresponds to 8% fat, and so on (Herd and Sprott 1986). Body condition or changes in body condition, rather than live weights or changes in weight, are thought to be a more reliable guide for evaluating the nutritional status of cattle, and body condition scoring provides a reliable measure of an animal's nutrition reserves (Herd and Sprott 1986, but see Monro and Skinner 1979). A similar scale is used to assess the condition of horses (Duncan 1992).

During the observations I mentally divided the moving herd into herd sectors. I defined the 'front' as the first 10% of buffalo in a herd, the 'subfront' the next 10%, the 'centre front' as the third 10%, the 'centre' as the middle 40%, the 'centre rear' as the next 10%, the 'subrear' as the second last 10%, and the 'rear' as the last 10% of buffalo. It was difficult to subcategorize the central animals since some of the herds were too large. This explains why I made the central group relatively large compared with the other ones. Four large herds and two small ones occupied Manyara; most of the observations were done on the Central Herd (section 3.3). During 1981–85 approximately 1000 days were spent in the study area and, after a short visit in 1990, about 250 days in 1991. I identified individual buffalo by scars, broken horns, eyes that were blinded, lost tails and especially natural ear markings. By using a telescope buffalo could successfully be identified from a distance up to 150 m and sometimes even further. The identifications resulted in 189 ID cards of cows (Figure 3.2). The idiosyncrasies of buffalo appeared to be stable for at least 2 and possibly even 3 years.

3.2 FITNESS AND SOCIAL INEQUALITY

When buffalo in a herd were grazing or moving from one feeding ground to another, the herd profile revealed a repeatable structure of age–sex classes, as individuals belonging to these classes were not distributed randomly over the herd sectors (Table 3.1).

The distribution of the age–sex classes deviated highly significantly from an even distribution ($\chi^2 = 127.5$, d.f. = 30, $P < 0.001$). This unevenness was mainly caused by adult bulls concentrating in the (sub)front and (sub)rear of the herd, and by adult cows without dependent offspring concentrating in the subfront and centre front and in the centre rear and subrear. The same analysis has been carried out on data from two other total counts, with the same result. Classification of nearest neighbours within the moving herd did not reveal any significant association between age–sex categories (except for mothers and their calves), and one has to speak in terms of 'concentrations' of these particular classes.

Between 1981 and 1985 I was able to establish in 143 cases a cow's location in the herd during identification and subsequent reidentification. Table 3.2 shows that there was a very significant constancy of herd position.

A cow first seen in the front of a herd was almost certain to be seen in the front at the next sighting, and one at the rear was usually at the rear next time. I think that buffalo cows show a similar spatial pattern as domestic cattle under traditional management practice. Reinhardt (1983a) observed that zebu have a rather fixed location in traditional herds, and Galaty (1989) reported that the moving cattle herd of the Masai have a definite structure, with certain individuals and cohorts

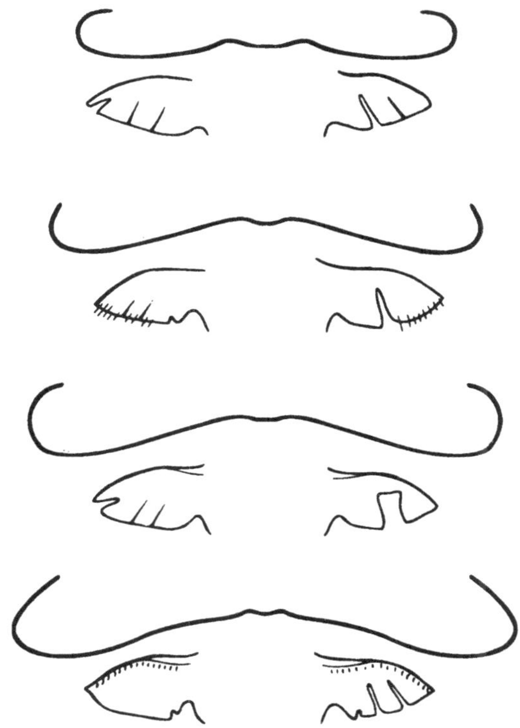

Figure 3.2 Identification sketches of four adult buffalo cows with their ID numbers. N, Northern Herd; C, Central Herd; BE, notches in both ears.

walking in the front, the sides, and at the back of the herd. Buffalo cows also show this fidelity of position, and, thus, a buffalo herd is a highly organized social unit. This is in contrast to wildebeest, which live in aggregations whose constitution changes continuously in terms of numbers and individual members (Watson 1967). Not all buffalo cows kept the same position at all times. In Table 3.2 there are 114 cases of cows having the same position between two sightings and 29 cases of cows changing position. All cases in which a change in position was observed were related to changes in 'breeding status', that is birth of calf, death of calf or weaning, while of the 114 cows that maintained their position only 24 showed changes in breeding status and 90 did not. From this I concluded that an adult buffalo cow keeps her rather fixed position in her herd under normal conditions but can change her position as a result of changes in reproductive status.

Table 3.1 Location of buffalo age–sex classes in the Central Herd

	Front	*Subfront*	*Centre front*	*Centre*	*Centre rear*	*Subrear*	*Rear*	*Total*
Percentage of buffalo at that location	10	10	10	40	10	10	10	
Adult bulls	1.47*	1.99***	0.66	0.44***	0.74	1.54**	1.84***	136
Adult cows without dependent offspring	1.00	1.48***	1.40***	0.61***	1.48***	1.37***	0.85	351
Adult cows with calves	0.32**	0.96	1.38	1.12	1.38	0.85	0.64	94
Adult cows with juveniles	1.41	0.87	0.87	1.03	0.98	0.76	0.98	92
Subadult cows	1.00	1.00	2.50*	0.38*	2.00	1.50	0.50	20
Subadult bulls	1.84*	1.02	1.22	0.61*	0.61	1.22	1.63	49

The figures are index values (that is number of individuals of a particular class divided by the expected number if the individuals of that class were distributed equally over the whole herd) and the asterisks indicate significant deviation of the equal distribution (* $P < 0.05$, ** $P < 0.01$, *** $P < 0.001$) as calculated by Habermann's method of adjusted residuals after chi-squared testing (Everitt 1977).

Table 3.2 Location of adult buffalo cow in the herd at time of identification

	Front	*Subfront and centre front*	*Centre*	*Centre rear and subrear*	*Rear*	*Total*
Percentage of buffalo at that location	10	20	40	20	10	100
Location of adult buffalo cow at subsequent identification						
Front	51^{+***}	5^{-}NS	0^{-***}	0^{-**}	2^{-***}	58
Subfront and centre front	3^{-}NS	6^{+***}	1^{-}NS	0^{-}NS	0^{-*}	10
Centre	2^{-***}	3^{-}NS	28^{+***}	1^{-}NS	3^{-}NS	37
Centre rear and subrear	1^{-}NS	0^{-}NS	0^{-}NS	4^{+***}	2^{-}NS	7
Rear	3^{-***}	0^{-*}	0^{-***}	3^{-}NS	25^{+***}	31
Total	60	14	29	8	32	143

The constancy of individual cows between resightings is always highly significant and independent of place within the herd. For example, an individual cow seen in the front has the highest chance of being seen in the front again. The figures are absolute numbers, and deviations of a random distribution are calculated with Habermann's method of adjusted residuals after χ^2 analysis (taking into account the size of the different segments of the herd). $^{+}$ Indicates more than expected and $^{-}$ less than expected; NS, not significant at the 5% level; * $P < 0.05$, ** $P < 0.01$ and *** $P < 0.001$.

Figure 3.3 serves as an example of the relation between food intake (expressed as bites per step) and location of an adult cow in the herd when on 14 April 1984 the Central Herd was grazing on a *Cyperus laevigatus* sward.

The figure clearly shows that animals in the **subfront** obtained the highest number of bites per step. Buffalo to the fore of these subfront animals were often displaced by cows coming from behind, but buffalo cows in the subfront were never displaced by animals walking behind them. The lowest number of bites per step was observed in the rear of the herd, as the buffalo there had to switch at short intervals from one patch of food to the next. Patch depletion through grazing is clearly demonstrated in Figure 3.4.

From this it can be inferred that herd-living buffalo, because they have

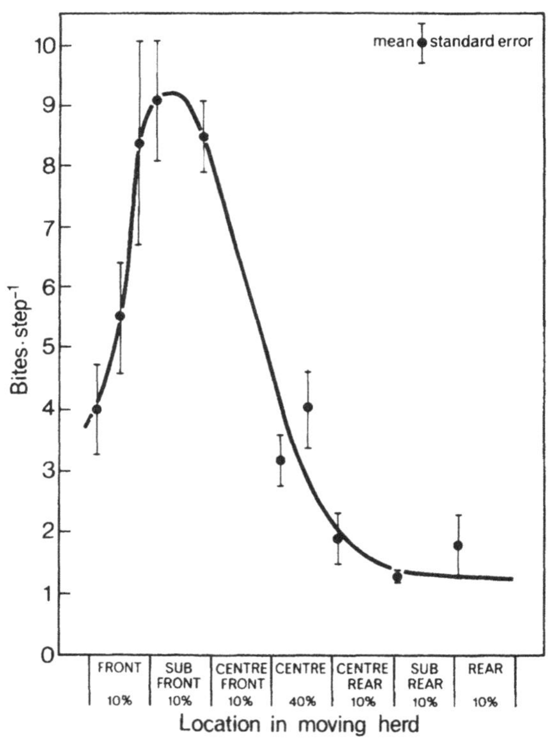

Figure 3.3 Number of bites per step (as measure of intake rate) of adult buffalo cows grazing *Cyperus laevigatus*, in relation to their position in the grazing herd. The percentages under the abscissa give the number of individuals as proportion of all herd members (excluding adult bulls) in a given herd section.

a more or less fixed position in the herd are always in a good or in a poor place with regard to food intake. This should be reflected in the relation between buffalo condition and position within the herd. On a number of dates we scored enough buffalo with regard to their body condition to enable an analysis of the relation between location in the herd and condition (Table 3.3).

From a condition point of view, the best place in a herd was indeed in the subfront of the moving herd, the second best place was in the front, the next best between the subfront and the centre (called the centre front), and from there the condition of the adult cows steadily declined towards the rear of the herd (Figure 3.5).

These observations were repeated in 1991, and again this pattern of differential condition of individuals over the herd was maintained. It gave me a thrill to discover that Burchell described the same for springbuck almost 150 years ago:

> Their numbers are absolutely incalculable. The **foremost of the mighty host are fat, and the rear ones meagre and lean**, from the utter consumption of all herbage and water that may be met with by the former before the latter can reach it... [emphasis added].
>
> (Williams 1859)

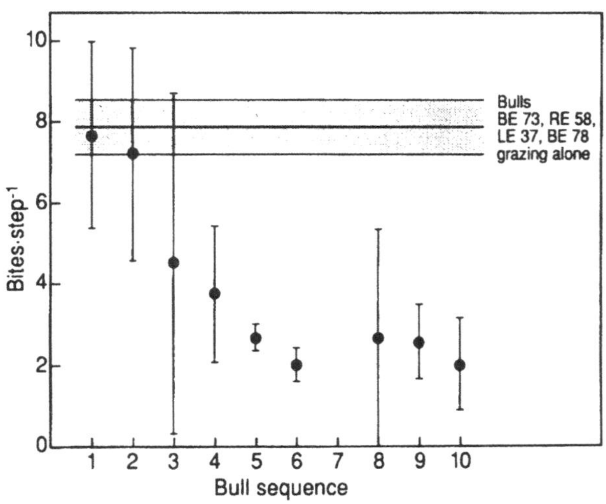

Figure 3.4 Patch depletion is shown by buffalo bulls grazing *Cyperus laevigatus*: the first two bulls entering the patch have the same intake rate as individual bulls grazing the same patch during a period of 4 days before this particular visit (horizontal bar; mean and 95% confidence limits) but bulls entering the patch behind the first two had a lower intake rate (vertical lines; 95% confidence limits). In all cases $n = 50$.

Table 3.3 The average condition of buffalo cows without dependent offspring and of cows with calves or juveniles in relation to location in the Central Herd on different dates through the year

Date/category	*Front*	*Subfront*	*Centre front*	*Centre*	*Centre rear*	*Subrear*	*Rear*	*Total*
27 June 1983								
Cows without dependent offspring	3.555 (n=18)	3.643 (n=21)	3.636 (n=22)	3.265 (n=37)	3.300 (n=20)	3.218 (n=22)	3.122 (n=18)	3.344 (n=158)
Cows with calves or juveniles	3.200 (n=15)	3.344 (n=18)	3.500 (n=18)	3.250 (n=16)	3.200 (n=10)	3.233 (n=12)	3.100 (n=10)	3.283 (n=99)
24 July 1983								
Cows without dependent offspring	3.281 (n=32)	3.364 (n=33)	3.048 (n=21)	3.071 (n=14)	2.938 (n=16)	2.815 (n=27)	2.801 (n=35)	3.062 (n=178)
Cows with calves or juveniles	3.067 (n=30)	3.115 (n=26)	3.000 (n=21)	3.053 (n=19)	2.889 (n=18)	2.708 (n=24)	2.529 (n=17)	2.929 (n=155)
6 December 1983								
Cows without dependent offspring	2.603 (n=39)	2.636 (n=43)	2.523 (n=22)	2.453 (n=32)	2.300 (n=20)	2.278 (n=9)	2.222 (n=18)	2.448 (n=173)
Cows with calves or juveniles	1.929 (n=14)	2.300 (n=10)	2.250 (n=6)	2.143 (n=7)	1.800 (n=5)	1.917 (n=6)	1.800 (n=5)	2.075 (n=53)
31 August 1984								
Cows without dependent offspring	3.077 (n=13)	3.181 (n=11)	3.100 (n=10)	2.892 (n=37)	3.000 (n=8)	2.900 (n=10)	2.889 (n=9)	2.906 (n=106)
Cows with calves or juveniles	2.500 (n=2)	2.800 (n=5)	3.00 (n=3)	2.750 (n=16)	2.818 (n=11)	2.800 (n=10)	2.714 (n=7)	2.778 (n=54)

1 December 1984								
Cows without dependent offspring	2.875 ($n=16$)	3.000 ($n=11$)	2.565 ($n=23$)	2.403 ($n=26$)	2.375 ($n=16$)	2.167 ($n=21$)	2.160 ($n=25$)	2.449 ($n=138$)
Cows with calves or juveniles	2.700 ($n=10$)	2.735 ($n=17$)	2.885 ($n=13$)	2.125 ($n=4$)	2.267 ($n=15$)	2.125 ($n=6$)	2.220 ($n=10$)	2.527 ($n=75$)
Estimate of average condition of								
Cows without dependent offspring	3.086	3.174	2.973	2.818	2.779	2.668	2.634	2.842
Cows with calves or juveniles	2.675	2.871	2.933	2.661	2.577	2.547	2.463	2.718

The average condition is calculated as a weighted mean. Note that the positions between the front and the centre are best, and the rear is worst from a condition point of view.

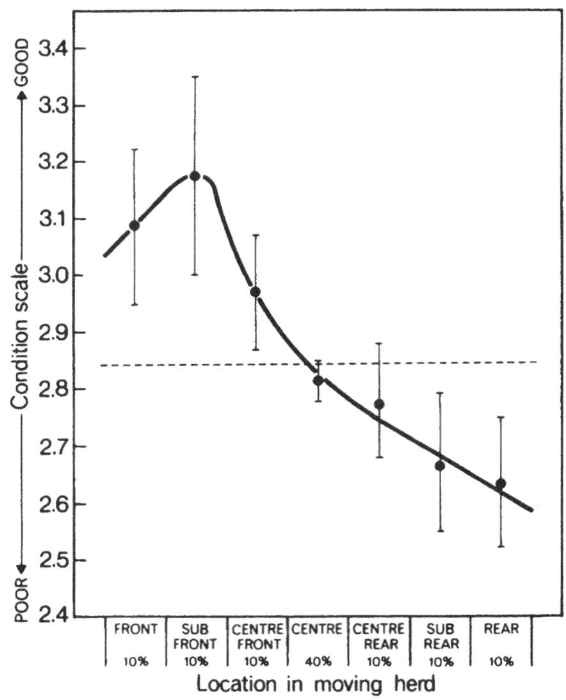

Figure 3.5 Average physical condition of adult cows without dependent offspring (with 95% confidence limits) in relation to their position in the moving herd. The horizontal broken line gives the average condition. See text for definitions of the condition scale.

Condition loss and gain depended on position within the herd too. For cows without dependent offspring I compared the relation between condition and position in the herd at the end of June (the data set closest to the maximum condition) and beginning of December (closest to the minimum condition) (Figure 3.6).

I deliberately took the category of cows without dependent offspring because lactation can have a masking effect on condition gains and losses. The results show that all cows lost condition regardless of location, but that cows in the best location, that is the subfront, lost condition less rapidly than those in the worst location, that is the rear (Figure 3.7B; n = 130, $t_{128} = 11.41$, $P < 0.001$, $R^2 = 0.51$).

These results are confirmed by the relation between the position of known individual cows and their condition changes. Loss of condition in

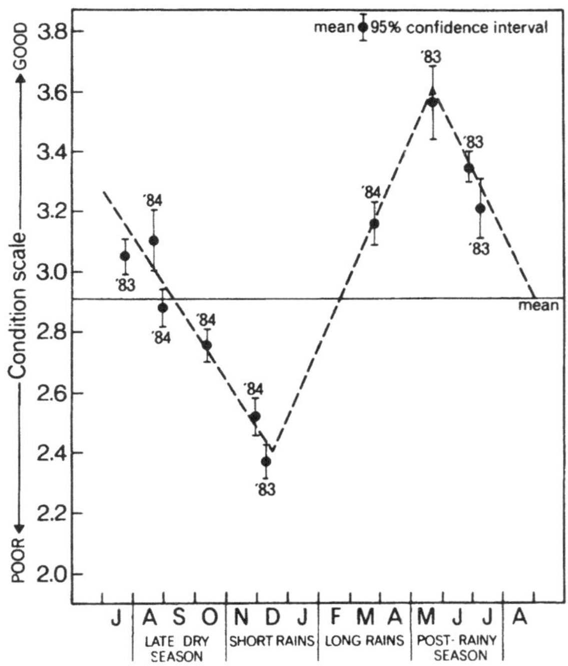

Figure 3.6 Average physical condition of adult cows in relation to time of year. Cows reach their poorest condition when the short rains have already started and food quality has increased.

1991 was a near exact repeat of that in 1983 and 1984 (Figure 3.8B). It is, *inter alia*, remarkable how the 'condition profiles' over the herd at the end of the post-rains season of 1983 was mirrored in that of 1991, and the condition profile at the end of the dry season of 1984 in that of 1991 too (Figure 3.8A).

Condition assessment in 1983/1984 was done by myself, but the other assessments were carried out independently from me by, respectively, Jan Beekman and Henk van der Jeugd. This indicates the repeatability of the condition assessment. The results of these different periods clearly indicates that for individual buffalo cows occupying the best position in the herd mitigates the negative effect of the dry season on condition.

Grimsdell (1969) concluded that the duration of the calving interval in buffalo is largely dependent on nutrition, and that buffalo cows in a good condition produce more offspring than those in a poor condition.

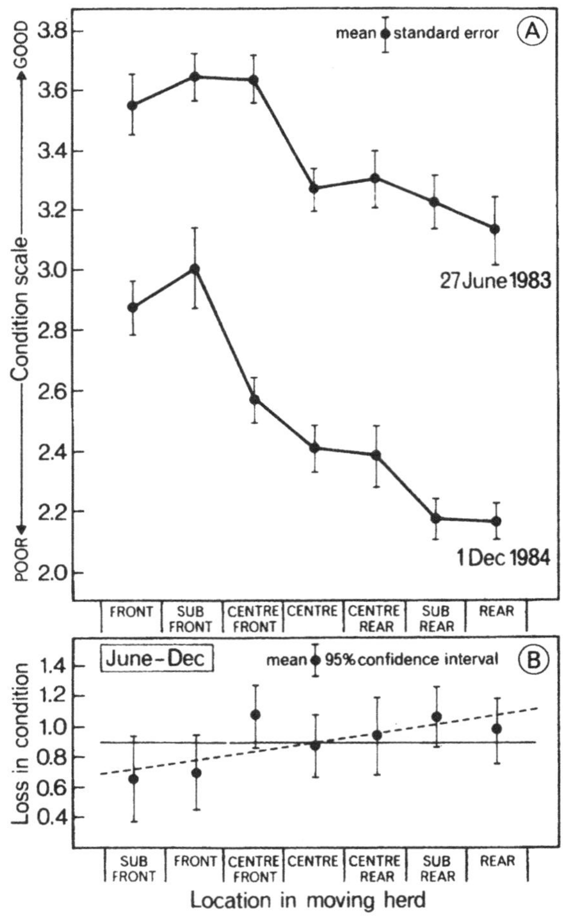

Figure 3.7 Average physical condition of adult cows without dependent offspring in relation to their location in the moving herd (A) at the time when they are on average in their best condition (27 June) and poorest condition (1 December). The location immediately behind the front is the best place. Condition loss (B) is least at the subfront and most severe in the rear of the herd.

The first part of his statement, regarding the calving interval, might be true, but the results from Manyara are much more strongly indicative of social processes playing a role in the determination of calving interval (see below). To prove the second part of his statement, data on individual life histories have to be collected, and hence it will take several years to determine empirically whether the 'fattest are the fittest' in large mammals under field conditions (Reiter *et al.* 1978).

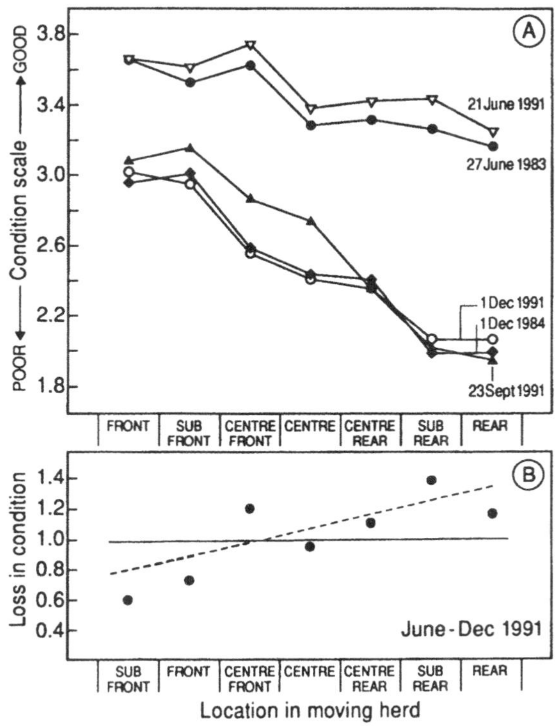

Figure 3.8 The average physical condition of adult cows without dependent offspring of the Central Herd in relation to their position in the moving herd (A) compared for 1983 and 1984 with 1991 shows that the pattern of differential condition is markedly similar for different years. (B) Shows that in 1991 condition loss was as in 1984 (Figure 3.7).

I place much emphasis on body condition as it is thought that condition, that is nutrient reserves stored in the body, plays an important role in deciding between failure or success in reproduction. To assess these effects of body condition in African buffalo, I rely to a large extent on observation on related bovids, namely zebu and cattle, as much research has been carried out on these domesticated species. It has been found that in cattle calving interval (which determines lifetime reproductive success) depends primarily on environmental factors and has a very low heritability (Dhillon *et al.* 1970). Date of weaning is important (Dhillon *et al.* 1970, Holness and Hopley 1978), which depends on the plane of nutrition of the mother (Barrett and Larkin 1974). Dhillon *et al.* (1970) found that the service period (which determines the calving interval) is longer in the dry season, when the food quality is poor, and that heavier cows have a shorter period. The differences in weight between successful and unsuccessful cows within one herd can be rather small, as was demonstrated by DeVaccaro *et al.* (1977), who found cows that stayed barren were only 10% lighter than those which became pregnant. So in tropical cattle weight and body condition to a large extent determine the calving interval, which is confirmed by the results reviewed by Herd and Sprott (1986) for beef cattle under more temperate conditions. Under range conditions, the fattest cows are the individuals with the highest reproductive success. I have used this as a working hypothesis because it is probably true and no data militate against it. Circumstantial evidence can be found in the observation that buffalo cows with dependent offspring, that is cows with suckling calves or juveniles, had a slightly worse condition than cows without dependent offspring (paired data with regard to location and date: 35 pairs of means, of which in 32 cases the mean condition of cows without dependent offspring was better than the condition of those with dependent offspring, and in only three cases it was the opposite: $\chi^2 = 24.03$, d.f. = 1, $P < 0.001$; see also Table 3.3).

This result agrees with the finding of Grimsdell (1969) on carcass analyses of barren cows versus lactating buffalo cows in Uganda and indicates that there is a cost associated to reproduction. Hence, females that have a better condition (because of a better position) can better afford 'to pay these costs' than females in a worse condition. In that case herd buffalo in the subfront position, because they are generally in a better condition than buffalo further to the rear, and because they lose condition more slowly during the dry season, can be expected to have a higher fitness. This has important consequences for the interpretation of the buffalo herd and the social inequality within the herd.

If good condition is a necessity for high reproductive success, one would expect cows without dependent offspring in the (sub)front of the herd to be more attractive to breeding adult bulls than cows without dependent offspring in the rear. In Figure 3.9 this is made clear by delineating the relation between condition and pregnancy rate for cows with a condition above and below condition score 3.

Cows above this score (which approximately coincides with the herd average) are expected to have a pregnancy rate of 85–95%, while those below condition score 3 are expected to have a pregnancy rate of less than 60%. I based these pregnancy rates in relation to condition on data from over 1000 head of cattle reported by Herd and Sprott (1986). Observations on buffalo bulls (Chapter 4) showed that adult bulls in a good condition, which are expected to have a high place in the bull hierarchy, are concentrated in the (sub)front of the herd, while adult bulls after losing their prime condition and their high rank, are found espe-

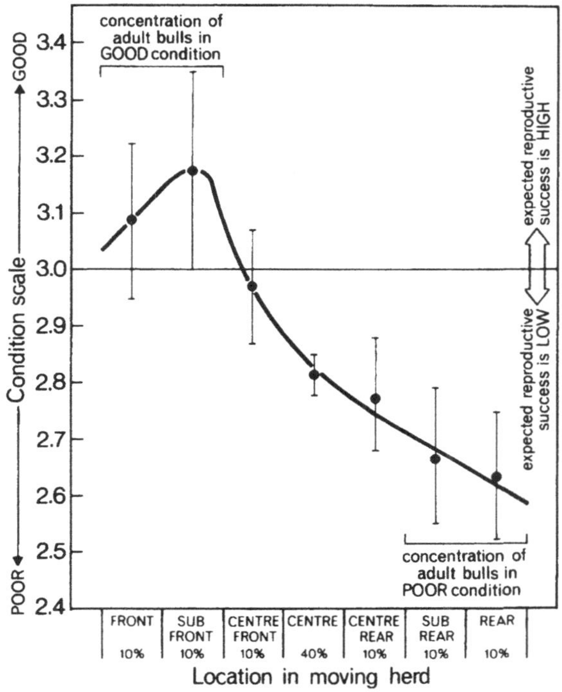

Figure 3.9 Average physical condition of adult cows without dependent offspring in relation to their location in the moving herd and the inferred effects on reproduction (based on Herd and Sprott 1986) and attractiveness to adult bulls (Table 3.1 and Chapter 4). Bulls in a good condition are in the best position to compete and concentrate themselves where cows with a high reproductive success can be expected. Because of competition, bulls lose condition and move to the rear of the herd before they leave the herd to recoup condition.

cially in the rear of the herd. The analysis with regard to adult bull behaviour fits with the inference from the present results on the relation between position and condition of buffalo cows. Thus buffalo in the rear of the herd are in a less favourable position than those nearer the front, and social inequality is an important feature of a large buffalo herd (compare Teunissen *et al.* 1985 and also Banks *et al.* 1979, Kikkawa 1980a, b, Baker *et al.* 1981; for a review of the effect of social structure on individual fecundity and mortality see Dunbar 1985). The cause of the differences in condition within the herd appears to be correlated with, on the one hand, competition through food depletion by buffalo grazing in front of other buffalo and by displacements and, on the other hand, by the aggravation of this effect of food competition on body condition through intestinal nematode infestation (Chapter 5). It appears, then, that an average rate of reproduction, as used by Grimsdell (1969) or Sinclair (1977), is not detailed enough to understand the relation between density and reproduction since one has also to take into account the size and structure of the social unit in which buffalo live (compare Dunbar 1985 on baboons). I will come to this in Chapter 5.

3.3 THE BUFFALO HERD AS A DISCRETE UNIT

The mixed herd of buffalo consists of females and males. A mixed herd in Manyara in the early 1980s typically consisted of adult cows (66.2%), subadult cows (6.4%), subadult bulls (9.6%), juveniles (11.6%) and calves (6.2%) (average of 12 counts over the year of the Central Herd in Manyara; $n = 8900$). In October and November 1991 this same herd had more or less the same age–sex class constitution, namely adult cows (69.0%), subadult cows (7.3%), subadult bulls (4.4%), juveniles (7.3%) and calves (11.9%) (average of two counts; $n = 820$). The percentage of adult males is variable (Chapter 4) and these are excluded in these calculations. From 1981 to 1985, 119 identified adult cows were resighted 296 times. Not a single resighting involved herds other than the herd of first identification or a fragment of that herd (barring seven cases which involved sightings outside any herd). The cows which we identified outside their herd were either in the process of giving birth (five cases) or isolated from their herd by a tending bull (two cases). Several times we observed that two herds mingled for a few hours. Subsequently the two herds separated again, all known individuals going to their own herd, as was also observed by Grimsdell (1969) in Uganda. On the basis of these observations I concluded that the herd is composed of a fixed set of cows and subadult bulls.

In Manyara a number of herds occurred, some of them very small but others very large. In 1985 these herds were the Northern Herd (280 individuals), Msasa Herd (12 individuals), Central Herd (980), Escarpment Herd (180), Southern Herd (450) and Yambi Herd (50 individuals). The number of individuals that these herds contained had only slightly changed by 1991, but in 1994 the herds were seriously depleted, mainly because of poaching. I have been able to trace the history of the present buffalo herds in Manyara over the last 35 years. The basis for this work was laid during the fieldwork of 1981 to 1985, and during subsequent fieldwork in Manyara in 1990, 1991 and 1994. Another important source of information was the Warden's Reports, which contained valuable information from 1959 until 1974. I also used field notes of D.F. Vesey-FitzGerald (end 1960s) and Dr I. Douglas-Hamilton (early 1970s), some reports and letters of Dr A.R.E. Sinclair (from the same period), and had discussions with Mzee Mhoja Burengo (ranger in Manyara since 1963). From this historical information it is clear that a herd is a permanent entity which exists much longer than one generation (about 7 years) or than the average lifespan of buffalo cows (about 12 years – Chapter 5) as some of the present-day herds have existed at least since 1958. Their numbers and home ranges are reconstructed in Figure 3.10. The history of the different herds is as follows.

The **Yambi Herd** was a large herd originally of several hundred head with a home range between Yambi River and Garufa River at the southern end of Manyara. At the end of the 1950s, the largest part of their range became cultivated by European settlers. In 1959 a game fence was erected between the park and the estates, but this did not completely prevent the buffalo from visiting the part of their original grazing grounds that lay outside the reserve. The Warden observed in October 1959 that 'the fence is having an effect on the movement of elephant [but] this is not the case with buffalo who just get on their knees and slide under the lowest strand of the cable'. The area available to them was at its smallest in the 1960s and because of the estates they became cut off from access routes to the Marang Forest on the Mbulu Plateau and they were heavily hunted. This situation became even worse at the beginning of the 1970s when the remaining part of their home range was to a large extent flooded by the lake during a series of wet years (Chapter 1) and they were reduced to only 30 or 40 animals. The herd got some relief during the 1970s as a farm north of the Yambi River was deserted and was incorporated into the park. By 1985 this herd had slightly increased to 50 head. In 1990 and 1991 this herd was very shy, presumably because of poaching, and I had difficulty in assessing its numbers. By 1994 it had grown to about 230 individuals and occupied the former Martha Estate, which was incorporated into the park in 1991. It is noteworthy that this herd did not readjust its northern home range boundary to the north at times when it was under such heavy pressure. I never observed this herd together with another one.

The **Southern Herd** is a split-off of the Central Herd. At the end of the 1950s, the Southern Herd did not exist as a herd, but at that time it was increasingly often remarked in the Warden's Reports that a number of buffalo of the Central Herd spent their time as a group in the southern part of the home range of the Central Herd and stayed there for weeks on end. By 1965 the Southern Herd had permanently split off and numbered 250 head. By 1985 it had nearly doubled in size, which can be accounted for by reproduction only. In 1991 the herd had further grown to about 600 individuals which, again, can be accounted for by reproduction only. This herd mingles freely with the Central Herd when both herds are grazing in the same area. By 1994 it was seriously depleted.

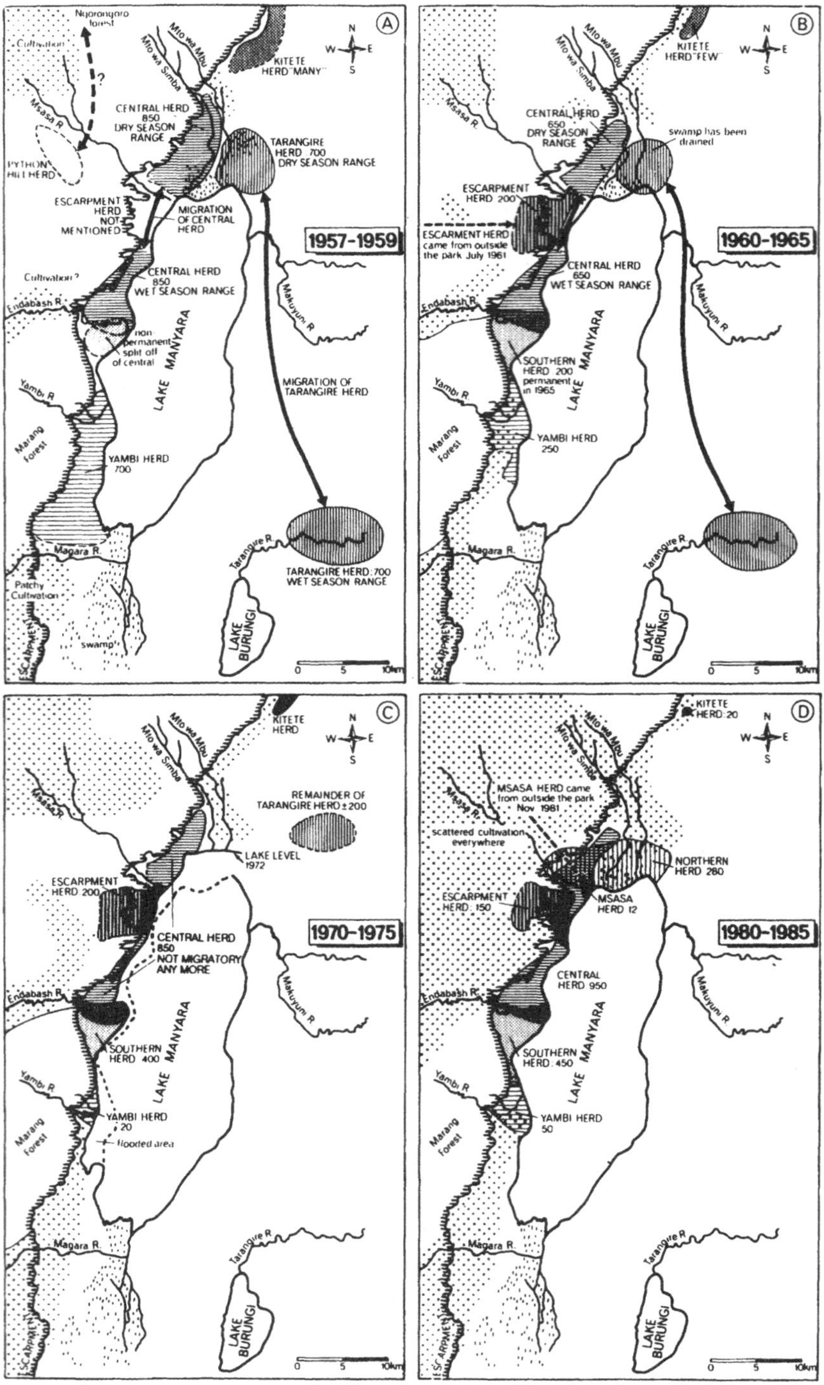

Figure 3.10 Home ranges of the different buffalo herds in Manyara over time. Spread of cultivation is based on aerial photographs (Loth and Prins 1986). In 1994 the Northern Herd, the Kitete Herd, and most likely the Escarpment Herd too, were extinct as a result of poaching and hunting.

The **Central Herd** consisted of approximately 900 head of buffalo in the 1950s but lost some 150 members because of rinderpest in 1959 (Chapter 5) and became even further diminished in numbers by splitting off of the Southern Herd in the 1960s. Since the 1960s, it has grown to slightly over its original size, and in 1985 it numbered 980 head. It remained at the same number of individuals to 1991, but by 1994 it was reduced to about 300 individuals. The Central Herd lost some of its home range at the extreme north of the park owing to cultivation around the village of Mto-wa-Mbu spreading into the forest at the foot of the escarpment in 1958 to 1960 (Chapter 1). In the 1950s and 1960s, this herd had a migratory habit, spending the wet season in the Endabash area and the dry season in the forest glades in the north. This habit was lost around 1971 when the lake flooded a large part of the northern home range. Since that time the herd has moved up and down through the home range approximately every 2 weeks.

The **Escarpment Herd** is new to the area. It was first mentioned in the Warden's Report for July 1961: 'during the month a large breeding herd of over 200 animals was noted in the area. This herd is **not** resident in the park... '. It is possible that the Escarpment Herd came from the Marang Forest, above the escarpment at the southern end of the park. However, it might have come from anywhere between the Ngorongoro Forest, the Marang Forest and Lake Eyasi (50 km to the west). I think it likely that this herd was displaced because of bush clearing and settlement on the plateau (Chapter 1). This herd settled in an area on the plateau near Ndala River and in the park down the escarpment, and the same range was occupied between 1981 and 1985, and still in 1991. In the period 1981 to 1985 the herd came down into the park every 2–3 weeks, but later slightly less frequently. The herd did not grow in size in the period between 1961 and 1985. Lack of herd growth might have been caused by hunting and poaching since this herd spent most of its time outside the park on the Mbulu Plateau, or, perhaps, it lost members and increased to the same number again. Between 1985 and 1991 it increased to about 280 buffalo. The herd never mingled with the Central Herd with which it shares a part of the home range. In 1994 it appeared to be extinct.

The **Msasa Herd** is also new to the area. I first noted these animals in the groundwater forest in the northern part of the park in March 1983. At that time it numbered only five adult cows, two subadult cows and one subadult bull and two juveniles. These animals were very black, as is typical for buffalo from above the escarpment. It is not known where these buffalo came from, but I think it likely that they are a remnant of the Python Hill Herd (see below). When the Msasa Herd came into the park, it wandered around for 2 weeks, and then settled on the riverine delta of Msasa River. The herd adopted a fixed home range and did not mingle with the two herds with which its home range overlapped. Since its arrival in Manyara this herd has had a high annual growth rate (Chapter 5); in 1991 it consisted of 50 individuals.

The **Northern Herd** was a remnant, perhaps the only one, of the Tarangire Herd. This was a large herd of 600–750 head of buffalo in the 1950s which migrated between the swamps near Mto-wa-Mbu, where it spent the dry season, and the Tarangire River near Lake Burunge, where it lived in the wet season. These buffalo were not hunted by Masai or Mbugwe in the 1950s, apparently because Masai in the rift valley hardly hunted at the time and the Mbugwe preferred to hunt zebra. In the 1960s the swamp near Mto-wa-Mbu was drained for cultivation and the migration ceased. I have not been able to discover whether this herd then split or was reduced by hunting, but some of the remaining 200 buffalo settled in the northern area of Manyara and occupied a new home range in the beginning of the 1970s (contrasting with the Yambi Herd). Since approximately 1965 it increased from 200 to 280 individuals in 1985. In the 1980s this herd rarely mingled with the Central Herd. In 1991 the Northern Herd was reduced in numbers to maximally 160 buffalo. We saw it only rarely in the park and observed this herd most times at the eastern side of the Simba River, that is outside the park, and also in the *Typha* swamp at the northern extremity of the lake. It was extinct in 1994.

There were a number of other herds around Manyara in the 1950s and 1960s. Most of these have disappeared. The **Kitete Herd** lived around the Kitete swamp in the rift valley north of Manyara. It was reduced first by rinderpest in 1959 (Chapter 5), and then further by spreading of cultivation and by hunting. In 1984, an aerial census revealed that its numbers had shrunk to only 20 individuals. I could no longer find any evidence of this herd in 1991. The **Python Hill Herd** I named after Giekurum (the Iraqw name for an extinct small volcano on the Mbulu Plateau) at the centre of its home range. This herd consisted of several hundred animals in the 1950s and 1960s, and it moved between Giekurum and the Ngorongoro Crater rim forest. Owing to the spreading of cultivation (Chapter 1) and its associated killing for crop protection, this herd nearly became extinct in 1985, and in 1991 very few, if any, individuals survived. In 1994 it was extinct.

That the buffalo of the Escarpment Herd, Msasa Herd and Northern Herd took refuge in the protected area is not unprecedented. In April 1960, after prolonged heavy rains on the Mbulu Plateau, there was a massive influx of buffalo allegedly from this plateau. The Warden (A.M. Morgan-Davies) observed that

> Buffalo numbers have increased by nearly double. The Reserve is covered with them... Never in my life have I ever seen so many buffalo in such a small area – it really is most impressive and

> reminds me of the days, or rather the photos one sees, of the once large herds of American bison that used to roam the Kansas prairies. Head Game Scout Kalala, who has been stationed in the area for the past five years... has never before seen so many buffalo in Manyara. On a foot patrol to Maji Moto [some 40 km from HQ] no less than 4000 were counted – there must have been at least another 1000 not been seen for the thick cover. Patrols in this area have been most stimulating as a result. The increase in buffalo figures is due, I think, to an influx from the Marang Forest Reserve immediately above Maji Moto.

Numbers had decreased to normal again during May and June 1960.

The above shows that buffalo herds are entities with a long history. The only similar example I know is female roan antelopes (Joubert 1974). The history of the Manyara buffalo herds also suggests the possibility that the social structure can be even more permanent than the home range. In a geographically defined area herds can come into 'existence' after immigration from elsewhere, but herds also arise as a result of fissiparity ('budding off') after a period of fusion–fission. In Manyara immigration was mainly caused by displacement through the spread of cultivation. This proceeds quickly, as can be deduced from the increase in the human population of Mto-wa-Mbu, for example (a village which was founded between 1917 and 1920), which doubled between 1950 and 1969, and again between 1970 and 1979 to 8000 (Arens 1979). In 1991 it counted about 23 000 people [Chama Cha Mapinduzi (Tanzanian Socialist Party) village secretary, personal communication].

3.4 FUSION–FISSION SOCIETY

In small herds (or small herd fragments), the distance between the grazing buffalo was significantly greater than in big herds (big groups), both in the dry season ($F_{7,132} = 5.26$, $P < 0.001$) and in the wet season (Figure 3.11, $F_{7,183} = 5.34$, $P < 0.001$).

I assume that a large distance between individuals reflects better grazing conditions for an individual than a small distance. As buffalo grazed closer together in the dry season than in the wet season competition between individuals was more severe during the dry season than during the wet season. Given the relation between group size, nearest neighbour distance and inferred competition, it appears that there will be a stronger tendency for large herds (groups) to split than for small herds (groups). The effect of group size on nearest neighbour distance, however, is just as strong during the dry season as during the wet season. Hence, on the basis of these data, there is no reason to expect seasonality in splitting frequency, which contrasts with Sinclair (1977, p. 141 ff.), who

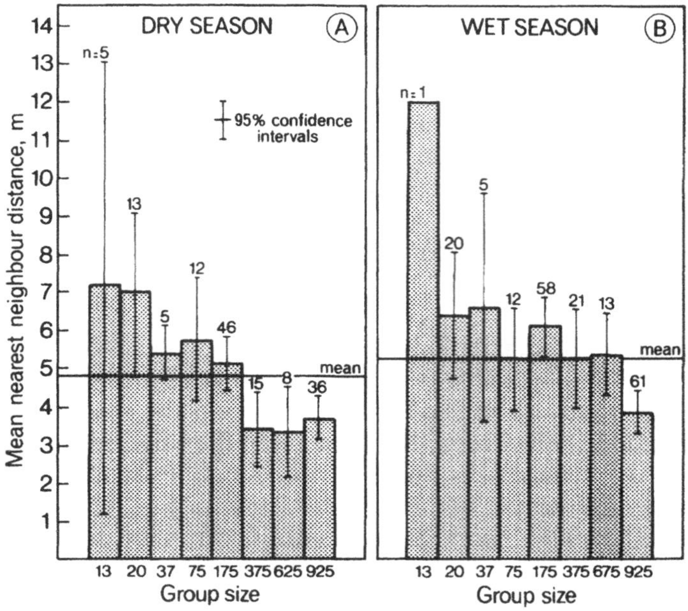

Figure 3.11 Mean distance between nearest grazing neighbours in relation to herd size. Bachelor bull groups are excluded. Both in the wet season (B: here defined as November to June) and in the dry season (A: July to October) the inter-individual distance decreases with herd size (F-test, both $P < 0.001$).

observed more small herds or 'fragments' in the dry season than in the wet season in the Serengeti. But Sinclair's inference that a small average herd size is caused by herd splitting is supported by the linear relation between the monthly proportion of days that the Central Herd was split and the mean size of fragments of that herd. The average number of buffalo in a herd fragment of the Central Herd equalled 933 minus 4.74 times the percentage of days that the herd was split ($n = 18$, $t_{16} = 14.11$, $P < 0.001$, $R^2 = 0.75$). In 1991, when it was drier than in the previous research period, the Manyara buffalo behaved more like the Serengeti buffalo. The number of days that the herd was split in the dry season of 1984 and 1985 was 50.0%, and in 1991 it was 55.0%. However, in the post-rains season of 1984 and 1985 the herd was in a fission state on only 35.4% of the days, but during the much drier post-rains season of 1991 this had increased to 64.4%. Although there was no relation between the proportion of days that the Manyara herds were split and month or season in the period 1981 to 1985 (Table 3.4), there was a very significant interaction between splitting frequency and herd size:

Table 3.4 The proportion of days that the Central Herd (980 individuals) and the Northern Herd (280 individuals) were split into fragments (fission state) or were grazing as one herd (fusion state) over the seasons

	Number (percentage) of days the herd is			
	In fission state		*In fusion state*	
Central Herd				
Short rainy season	17	(28.3%)	43	(71.7%)
Long rainy season	32	(52.5%)	29	(47.5%)
Post-rainy season	28	(35.4%)	51	(64.6%)
Dry season	27	(50.0%)	27	(50.0%)
Northern Herd				
Short rainy season	13	(31.0%)	29	(69.0%)
Long rainy season	5	(17.2%)	24	(82.8%)
Post-rainy season	12	(33.3%)	24	(66.7%)
Dry season	8	(19.5%)	33	(80.5%)

Short rainy season = November to January, long rainy season = February to April, post-rainy season = May to July, dry season = August to October.

Msasa Herd	12 individuals: never split (n = 68)
Yambi Herd	50 individuals: never split (n = 22)
Escarpment Herd	180 individuals: was split twice (n = 109)
Northern Herd	280 individuals: was split 26% of the days (n = 148)
Southern Herd	50 individuals: was split four times (n = 13)
Central Herd	980 individuals: was split 41% of the days (n = 254)

Thus, there appears to be a relationship between the size of a herd and its propensity to split: large herds have a fusion–fission society but small herds do not, which is underscored by the fact that splitting of the Northern Herd occurred significantly less often than by the Central Herd ($\chi^2 = 9.663$, d.f. = 1, $P < 0.001$).

The frequency distribution of the sizes of the herd fragments identified as parts of the Central Herd shows that splitting is complicated. This herd was observed 150 times as one single herd and broken up 104 times: 44 times in two parts, six times in three parts, once in four parts; and in at least 53 instances one fragment was not located by the observer. It is likely that the notorious breakaway animals consisted of a nucleus of approximately 100 individuals (which were involved in 100% of the split-offs; see

Figure 3.12), and another group of approximately 100 buffalo which nearly always joined the first group (in 92% of the split-offs).

Given the fact that the intake rate of the rear animals in the herd was lower than that of the front buffalo, I have arbitrarily assigned the 'role of notorious splitters' to the animals in the rear.

The data on known individuals and their position in the intact Central Herd in relation to the identification of these animals in fragments of this herd (Table 3.5), support the notion that the rear animals tended to break away from the main herd.

This might have been a way for animals in the rear of the herd to overcome social inequality. Breaking away from the main herd can perhaps, ultimately, lead to formation of a new herd through fissiparity. Normally, it leads to the formation of a temporary fragment grazing on its own. In the tiny Msasa Herd, social inequality was not significant, and every herd member was equally often in the front, but even in such a small herd there was a tendency for inequality (Figure 3.13; compare Reinhardt 1983a).

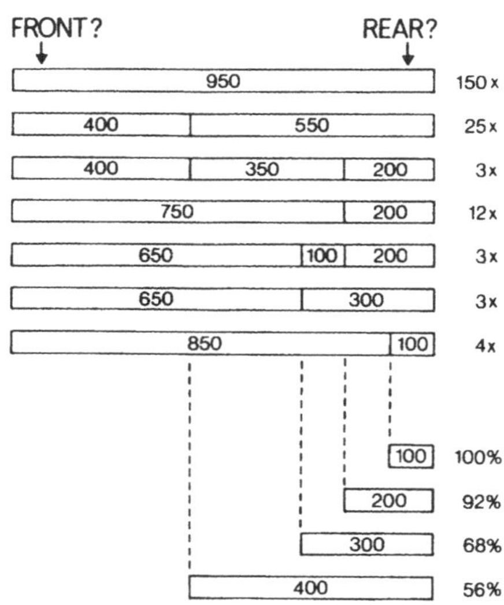

Figure 3.12 The Central Herd shows a fusion–fission pattern. All observations in which the whole herd or all fragments of this herd were located in Manyara are plotted in this figure with the leading criterion that as few division lines as possible should emerge. Individuals in the rear of the herd, where conditions are worst, split off most frequently from the main herd.

Table 3.5 The location of known individual adult cows in the Central Herd in the main herd (that is in fusion state) or in a fragment (that is in fission state) support the splitting pattern of individuals as presented in Figure 3.12

Individual	*Location in main herd*	*Location in fragment*	*Size of fragment*
CR 14	Rear	Centre	200
CO 45	Rear	Centre	100
CB 14	Centre rear	Unknown	100
CR 52	Centre rear	Front	100
CR 52	Centre rear	Centre	100
CL 42	Centre rear	Rear	100
CB 18	Centre rear	Unknown	100
CO 26	Centre	Centre front	350
CO 18	Centre	Centre rear	550
CR 36	Centre	Unknown	850
CR 7	Centre	Front	350
CR 3	Centre front	Unknown	100
CO 36	Front	Front	400
CO 22	Front	Front	850
CR 51	Front	Unknown	100

The only two observations which militate against that figure is the presence of individuals CR 3 and CR 51 in the fragments. CR 3 cannot be explained, except if we accept that she was misidentified (but that is possible for all individuals). CR 51 had a newborn calf at foot which she had delivered outside the herd, and it appears that she has joined the fragment later, most likely as a protection against predators.

Yet, even in 1991, when this herd had grown to about 50 animals, I still could not detect social inequality through assessing the condition of the individuals in this herd; the average condition scores of divisions within the Msasa Herd were not significantly different in May, August or October. Large herds offer less favourable feeding opportunities than small ones, as the inter-individual spacing is smaller in big herds than in small herds (Figure 3.11; compare Watts 1985). In Manyara, large herds split more frequently than small herds, and the animals in the rear have the highest tendency to split off. In at least one case, one of these fragments became increasingly isolated from the original herd and formed a new herd, when the Southern Herd ultimately split off from the Central Herd at the beginning of the 1960s. This phenomenon has also been described by Sinclair (1977, p. 141) for a large herd of 1100 individuals in the Serengeti, which formed two nearly equally sized herds.

Fusion–fission societies are considered typical of primates. Struhsaker and Leland (1979) defined such a society as follows:

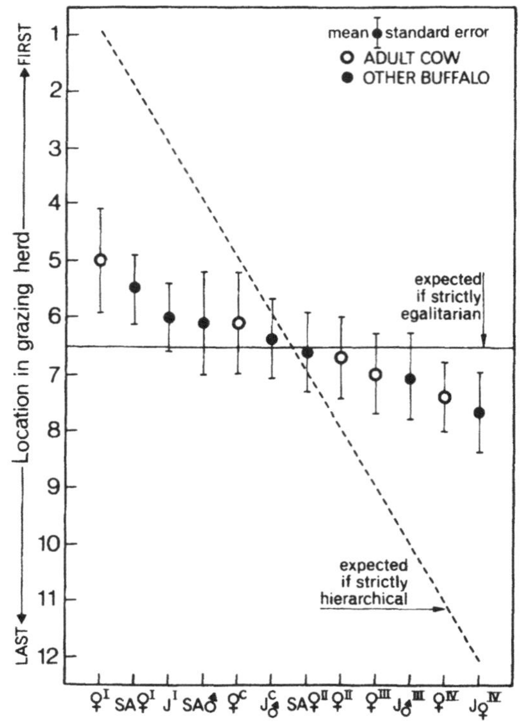

Figure 3.13 Location of individual herd members in the very small Msasa Herd shows that herd members already have a tendency to maintain the same position but position fidelity is still low (based on 130 observations). SA, subadult; J, juvenile; superscripts: I, II, III and IV, individuals from the same category; c, calf.

> Members of a fusion–fission society are (1) of the same social network, and (2) all members of the network will disperse (a) as units or individuals, and/or in variable combinations of associations of units or individuals, and (b) generally for the major part of the day but often much longer.

My results clearly show that in large herds the buffalo exhibit this fusion–fission society pattern. Evidence from the literature reveals that this may be quite normal for buffalo and that competition between individuals in the society may be more important than the effect of season. Grimsdell's (1969) four study herds in Uganda were apparently too small to show this pattern. Sinclair (1977, p. 119) suggested that small groups in the Serengeti were fragments of large herds. He observed herd splitting mainly in the dry season (which is the season with poor quality food). We observed the same pattern for the Manyara buffalo in 1991, which was a drought year (with only 364 mm of rain as compared with the average annual rainfall of 678 mm). However, there was no clear pattern in herd splitting with season in 1981 to 1985 in Manyara (during this period the rainfall was slightly higher than the long-term average). In Kruger National Park herd splitting took place during the summer/wet season (Anonymous 1960), which was considered to be the favourable season for buffalo (Pienaar *et al.* 1966). In Gabon, herd splitting was reported for the dry season, which was considered the favourable season there (Maclatchy 1932). Herd fusion has also been reported for Zaire (Dalimier 1955). A clear relation between season, or its inferred favourableness, does not emerge from these reports, but a fusion–fission pattern is reported from the whole geographical range of the African buffalo. Apart from buffalo, a number of bovids appears to have a fusion–fission organization, such as bison (McHugh 1958, Fuller 1960), water buffalo (Tulloch 1969) and gaur (Schaller 1967). It can be expected that the fusion–fission pattern in bovids is in some way related to the dispersion of food items in the home range of the herd, but especially to the inequality as a result of social 'stratification' in herds beyond a certain size. The repeated fusion–fission pattern can be understood as a means of staying in contact with the information from the successful members of the herd. I will come to this again in Chapter 7, where I will show evidence for this information sharing.

4

Social organization of buffalo bulls

4.1 BULL MOVEMENTS ON A MACRO-SCALE

Both cows and bulls stop growing at between 6 and 7 years, but the development of the typical male secondary characteristics such as large horns, starts at the age of 7 years (Figure 4.1).

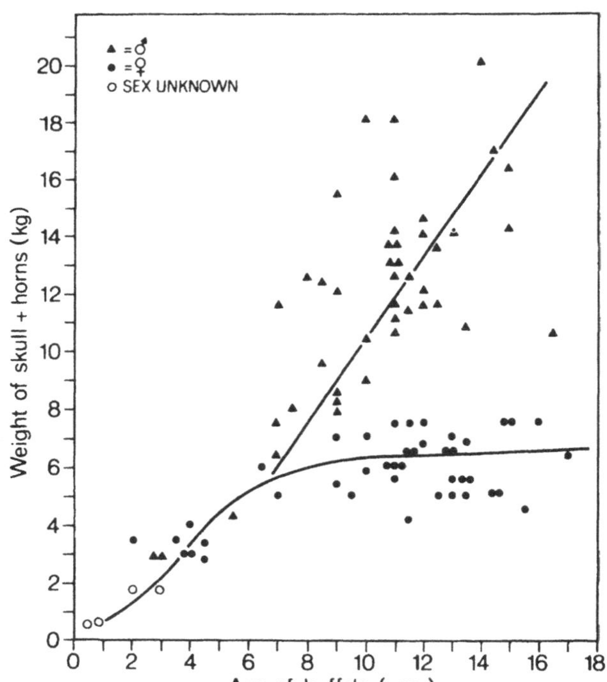

Figure 4.1 Relationship between age and weight of skull plus horns of Manyara buffalo. Clear sexual dimorphism starts at the age of 7, when buffalo are adult.

In the field, however, in contrast to what many professional hunters claim, it is impossible to estimate the age within the category of adult buffalo bulls from body characteristics because individual variation with regard to horn parameters is too great. For this reason I classified bulls simply as either adult (7 years or older) or subadult. In the field, one can easily distinguish adult bulls from subadult ones. In subadult bulls the testicles are small, the gap between the horn bases is wide and the skin between the horns is covered with hair. Adult bulls have much larger testicles and generally the horn bosses nearly touch each other (Figure 4.2).

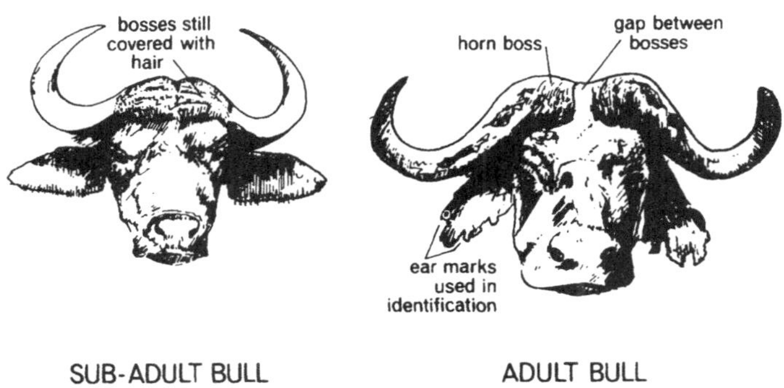

Figure 4.2 Sketches of a typical subadult bull (between 4 and 7 years of age) and an adult bull (older than 7 years). Note the ear notches, which were used for identification.

Matings are carried out by adult bulls only (Grimsdell 1969, Sinclair 1977) and even cow tending ('guarding' or 'consorting' in other species) is never carried out by subadult bulls. The behaviour of buffalo bulls towards oestrus buffalo cows is similar to that of cattle (Reinhardt 1983b) or American bison (McHugh 1958, Lott 1981). Presumably nearly all matings occur at night in buffalo because I have only twice, in 6 years in Manyara, seen matings by adult bulls.

From observations on known individual cows it is clear that the mixed herds are structural entities, as cows were never observed in herds other than their own (Chapter 3). The same holds for subadult bulls, as they were only rarely seen outside a mixed herd (of approximately 1000 sightings of 'bachelor bulls' only three were classified as subadult bulls). Thus, it appears that bulls stay in the mixed herd in which they were born until they are adult, but from this moment onwards they also can be encountered in 'bachelor groups'. Identification of bulls resulted in 240 ID cards of bulls (Figure 4.3).

Individuals remained identifiable for at least 2 and possibly even 3 years as was the case with cows (Chapter 3).

I started my research on the social organization of adult buffalo bulls assuming that there were a finite number of bulls in Manyara and that after a given time I would have seen them all if these bulls were sedentary within the boundaries of the study area. My expectation was, thus, that after a certain time 'new' bulls would no longer be added to the ID set, apart from those which would become identifiable, for example by

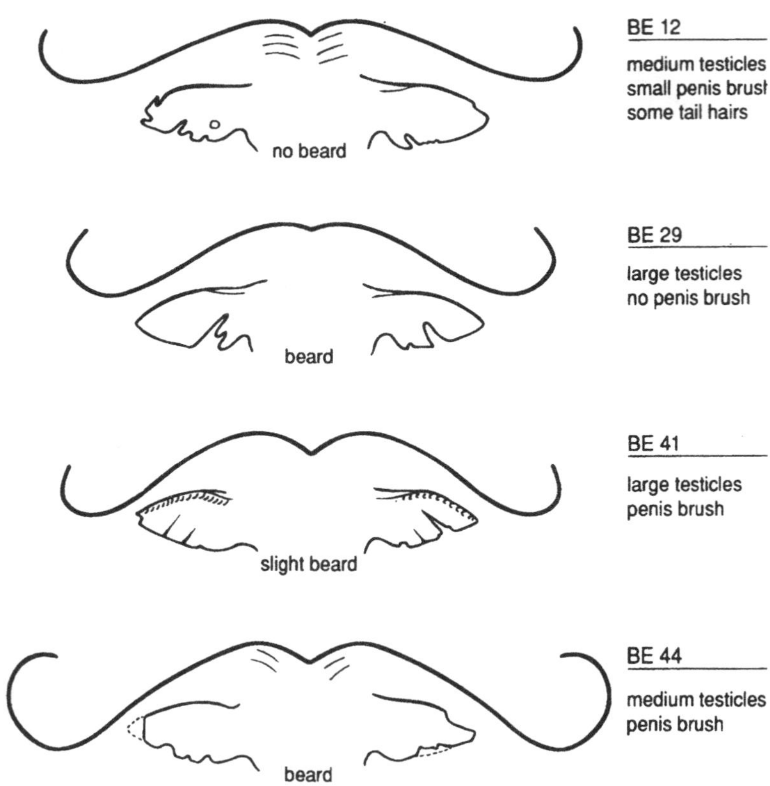

Figure 4.3 Identification sketches of four adult buffalo bulls with their ID numbers. BE stands for notches in both ears. Sizes of testicles, beard and penis brush were also used in the identifications, just as the form of the horns.

acquiring scars. An analysis of the cumulative number of identifications over time of cows and bulls (Figure 4.4) reveals that after 3 months I had described the cows but that the addition of new bulls continued.

The mortality of adult bulls was 15.9% per year (Chapter 5). I assumed that the adult bull population would grow at the same rate as the total population, that is at a rate of 2% per year (Drent and Prins 1987). Hence, I would expect the rate of individual bull identifications to be equal to the sum of the annual mortality rate and the population growth rate. However, the cumulative increase in identified bulls during phase II in Figure 4.4 was 53.5% per year, which is much higher than the expected 17.9%. In contrast, the cumulative increase in identified cows in phase II of the research period (8.5% per year), corresponded to an inferred recruitment of subadult cows into the adult cow population of approximately 5.4% per year (the adult cow mortality; Chapter 5) plus an estimated population growth of 2% per year. Hence, the observations on known individuals confirm that cows are sedentary, but they indicate that bulls in the early 1980s showed a constant flux through Manyara.

Another series of observations also suggests that bull movements took place on a large scale. The total number of adult bulls in the area north of Endabash River at any moment was approximately 150 individuals

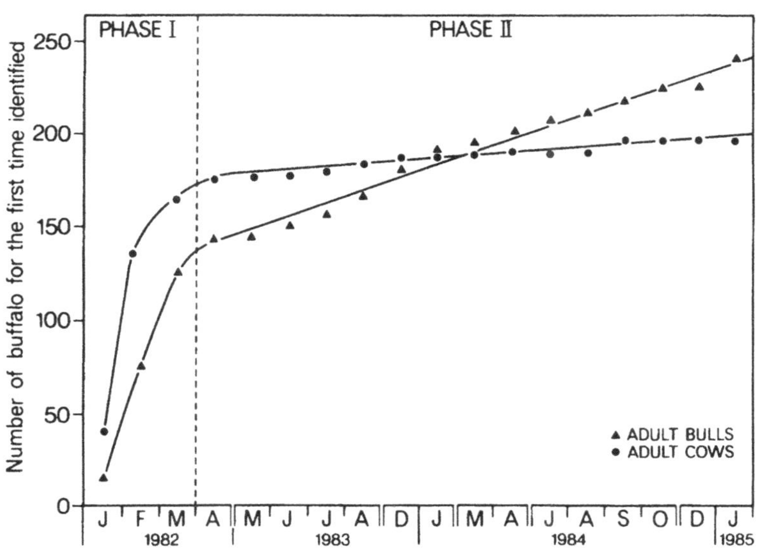

Figure 4.4 The cumulative number of initial identifications of individual adult buffalo. Phase I is the starting period of the research, arbitrarily distinguished from phase II, the subsequent period.

(mean of 92 counts), while the ratio of identifiable bulls to non-identifiable ones (that is those without characters to be used in visual identification) equalled 77:126 (mean of 54 group scans). Hence, the average number of identifiable bulls in the study area at any moment was approximately 55 [that is 77/(77+126) × 150]. The total number of bulls identified during the study, however, was 240.

A third line of reasoning, again, pointed in the direction of bull movements. Since 240 bulls were identified, and as the ratio between identifiable and non-identifiable bulls is known, the total number of adult bulls in the study area during the study period must have been at least 240 × 203/77 = 633. The discrepancy between this number and the number of adult bulls in the study area at any time, which was 150, cannot be explained by recruitment of subadult bulls from local herds into the adult bull population, as this was estimated to be approximately 25 per year, nor does it appear to be explained by acquisition of new identification marks by adult bulls.

Fourthly, the distribution of (re)sighting frequencies of identified bulls (Table 4.1) deviates highly significantly from a binomial distribution ($\chi^2 = 94.1$, d.f. $= 4$, $P < 0.005$).

Table 4.1 Frequency distribution of sightings and resightings of individual bulls in the Manyara area and its surroundings from 1982 to 1985

Number of times identified	*Number of individual bulls*
1	136
2	59
3	16
4	9
5	7
6	8
7	1
8	2
9	1
10	1
11 and more	0
453 identifications	240 individual bulls

This means that most bulls were identified only once. This indicates that they moved out of the study area before they were resighted.

The conclusion then is that adult bulls were moving into Manyara and disappeared again partially because of death (Chapter 5) and partially because of emigration. Positive proof of emigration was obtained when two bulls first sighted in Manyara were resighted in the Ngorongoro Crater (40 km NW of Manyara), while another one was resighted in Tarangire National Park (50 km SE of Manyara; Figure 9.5). Local information also pointed to extensive buffalo bull movements at that time: bulls were encountered in areas where mixed herds had not been observed in recent years. Aerial reconnaissance between Manyara and the Ngorongoro area and Tarangire National Park in 1983 revealed, especially in dense bush, some small bull groups. By 1991 this situation had changed. Both agricultural activities and hunting outside the park became much more intense between 1985 and 1991. I will come to the possible change of the social organization of the buffalo bulls at the end of section 4.3.

4.2 BULL MOVEMENTS BETWEEN SOCIAL ENVIRONMENTS

Buffalo bulls were most often identified while outside a mixed herd, that is as 'bachelors' (Table 4.2), but also in mixed herds.

Table 4.2 Total number of identifications of individual bulls in the Manyara area and its surroundings from 1982 to 1985

	Number of identifications
Bachelors (in bachelor group)	178
Bachelors (in mixed group)	74
Bachelors (combined)	252
Herd bulls (in Central Herd)	92
Herd bulls (in Escarpment Herd)	49
Herd bulls (in Northern Herd)	29
Herd bulls (in Msasa Herd)	1
Herd bulls (outside Manyara)	2
Herd bulls (combined)	173
Bulls (all combined)	425

In 28 cases the social context was not noted, which explains the difference with Table 4.1.

Individual bulls shifted between these environments: one day a bull could be found in a specific mixed herd, the next day the same individual could be bachelor, and vice versa (Table 4.3).

Table 4.3 Resightings of known bulls in the Manyara area and its surroundings

	Next sighting as					
	Bachelors in bachelor groups	*Bachelors in mixed groups*	*Herd bulls in Central Herd*	*Herd bulls in Escarpment Herd*	*Herd bulls in Northern Herd*	*Herd bulls in other herds*
Bachelors (in bachelor group)	46	6	16	10	6	1
Bachelors (in mixed groups)	11	19	0	3	4	1
Herd bulls (in Central Herd)	11	2	18	4	4	1
Herd bulls (in Escarpment Herd)	5	4	1	7	2	0
Herd bulls (in Northern Herd)	6	2	2	3	1	1
Herd bulls (in other herds)	0	0	0	0	0	0

The social environment at a sighting and at a subsequent sighting could be established in 197 sightings of 139 different bulls. 'Other herds' means a combination of the very small Msasa Herd and herds outside the Manyara area. A small mixed group is a group of bachelors with some females temporarily attached to it. The distribution of all resightings is not random (based on the distribution of all sightings in Table 4.2) ($\chi^2 = 82.01$, d.f. = 28, $P < 0.005$). There is a clear tendency for the bulls to be resighted in the same social environment in which they were seen previously (91 resightings in the same and 106 resightings in a different social environment, $\chi^2 = 27.18$, d.f. = 1, $P < 0.005$).

Changes in social environments were not random within the time frame of resightings. However, it was rare to obtain more than two resightings of an individual bull within the same social environment (three times: n = 4; four times and more, n = 4). The distribution of resightings within the same social environment does not deviate significantly from the distribution of all resightings (χ^2 = 9.422, d.f. = 5, not significant at the 5% level), indicating that bulls had a low fidelity to the groups in which they were seen. The time interval between resightings was short compared with the total study time (Table 4.4).

Resightings of bachelors were shortest, while it took longest to resight bachelors which had been mixed herd bulls first. The chance of resighting a bull (43.8%) did not depend on the social environment in which it was first identified (first identified as bachelor: never seen again, n = 65; or subsequently reidentified, n = 80; first identified as herd bull, never seen again, n = 47; reidentified later, n = 64). This implies that the difference in resighting intervals of Table 4.4 (particularly those between bachelor and herd as initial state) cannot be ascribed to chance. Very few adult bulls were ever resighted after 200 days, implying that there was only a small chance that an adult bull moved back into Manyara after an extended period of stay outside Manyara. This suggests that the number of adult bulls met during adult lifetime was very large under the condition that bulls did not move in small subsets of associate bulls which were moving together.

If adult bulls associated on a completely random basis during the period that they were outside a mixed herd, the distribution of group sizes should follow a Fisher's logarithmic series distribution (Pollard 1977), which it does more or less (Figure 4.5).

Yet, single bulls were encountered less often than expected in 1983 and 1984, while bull groups of three to five were seen more often than expected. In 1991 the same picture emerged, and bulls still occurred more frequently in groups of three to five (Figure 4.6).

Thus bulls preferentially associated in small groups when they were outside a mixed herd, and preferred the company of two to four other bulls. It is unlikely that bulls associated because of preference for the same spot at the same time since the average patch size is of the order of 7 ha while I estimated that an average bull group is contained within 1 ha. Hence, an adult bull apparently prefers the close presence of only a few others as an anti-predator strategy (Chapter 5). Anti-predator strategy is best illustrated by a real-life example. A bull (B35) was attacked by three lionesses on the *Sporobolus spicatus* grassland and was severely wounded because two lionesses were perched on his back and biting his shoulder. He remained standing, however, and threw them off. He then stood motionless for a few minutes with his horns presented in their direction while they stood facing him at a distance of about 5 m, after which he

Table 4.4 Resighting interval of bulls in the Manyara study area

First seen as	*Resighted as*	*Mean number of days elapsed since previous sighting (±SD) for observations shorter than 200 days*	*Mean number of days elapsed since previous sighting (±SD) for observations longer than 200 days*
Bachelors	Bachelors	11.7 ± 17.5 (*n*=40)	406.5 ± 36.9 (*n*=4)
Bachelors	Herd bulls	15.0 ± 16.8 (*n*=30)	465.0 ± 134.6 (*n*=6)
Herd bulls	Herd bulls	31.2 ± 51.2 (*n*=34)	468.1 ± 200.6 (*n*=8)
Herd bulls	Bachelors	51.8 ± 59.2 (*n*=14)	487.1 ±186.0 (*n*=8)
		$F_{3,113}=5.288$ $P<0.005$	$F_{3,22}=0.209$ NS Mean = 463.8 days

A distinction has been made between intervals shorter than 200 days and intervals longer than 200 days. This distinction is arbitrary, but it was felt that after a very long interval there would be an increasing chance of misidentification and, secondly, bulls which have not been seen for such a long time have almost certainly been outside the study area in the interim.

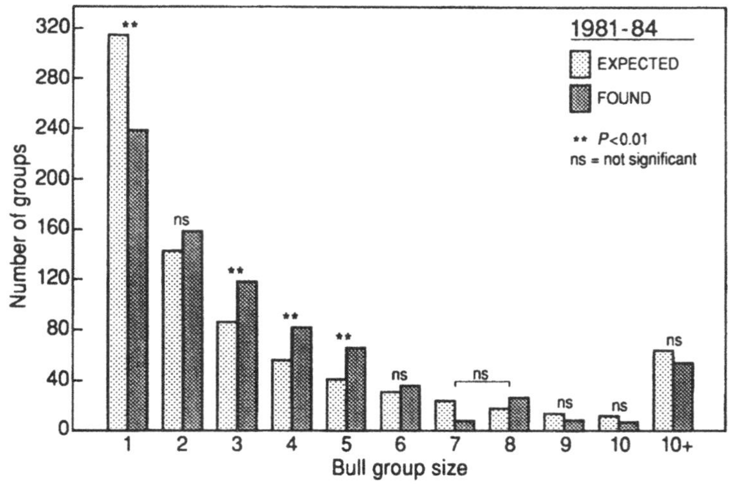

Figure 4.5 In the period 1981 to 1984 bachelor bulls in Manyara were observed in groups of three to five individuals more frequently than expected, and less often as singletons.

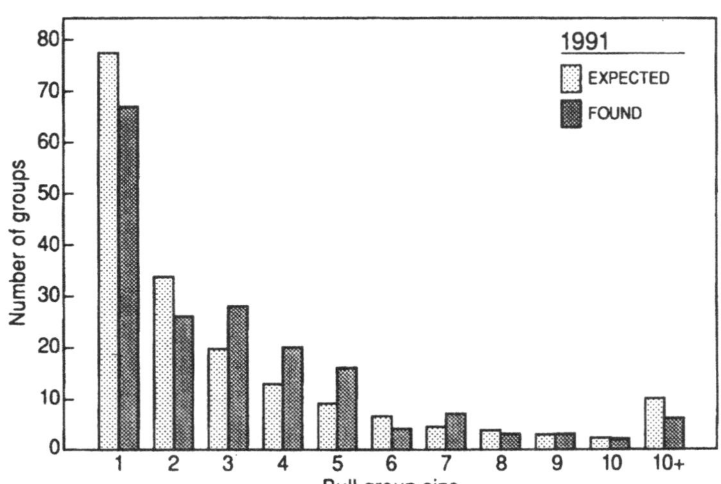

Figure 4.6 In 1991 group size distribution of bachelor bulls in Manyara was nearly the same as in the period 1981 to 1984 (Figure 4.5).

turned and walked away a few metres. He then turned to the lionesses again, and faced them for about 10 minutes. Apparently, the lionesses did not dare to attack again. Then, bleeding severely, he walked for 4 km until he met a small group of bachelor bulls, whereupon he collapsed, and lay on his side for 2 hours. This could be interpreted to mean that he felt safe. He then resumed grazing and had fully recovered after 2 weeks.

Morgan-Davies (Warden's Report 1961) describes another instance when a bachelor bull obtained safety from other bulls.

> On the 6th [of July 1961] an adult lion was seen attacking a full grown buffalo bull in the Endabash area. The lion sprang upon the flank of the moving buffalo bringing the latter to a grinding halt. Another male buffalo near by rushed into the fray and, with a hefty biff, hurled the lion to the ground. The vanquished lion then sloped off...

However, such assistance is not universal: I have twice seen lions attack a buffalo without any other buffalo within about 40 m paying the slightest attention.

The average number of bulls in a bachelor bull group was 3.92 in 1983 and 1984 and 3.33 in 1991. These average bull group sizes are very close to Hamilton's (1971) theoretical 'primary group' of 3.22. However, both Hamilton (1971) and Vine (1971) state that if several groups arise as a result of the individuals banding together in order to minimize their detection and/or predation, it is advantageous to the individuals that these groups amalgamate because the largest group gives the greatest safety. However, associating buffalo bachelor bulls rarely if ever went beyond the stage of the primary group, regardless of the increased risk in certain vegetation types or at certain times of the day (Chapter 7). This indicates either that there was a disadvantage to being in larger groups or that the amount of time needed for the amalgamation was longer than the average duration of group membership of the individual bulls. The latter explanation seems most likely, as bulls rarely associated for more than 1 day, and most observations indicate that bull–bull associations lasted a few hours only (cf. Grubb 1974 on Soay sheep, *pace* Mloszewski 1983, p. 132 ff.).

An association analysis on known individuals over time indicates that individual adult bulls did not associate with specific other adult bulls. With 240 (= n) known bulls, 28 680 (= $[n^2 - n]/2$) pairs of bulls (dyads) can be arranged, of which 5%, that is 1434, could be expected to show a significant association purely on the basis of chance in a pairwise test of association (Mueller-Dombois and Ellenberg 1974). However, only 23 bull dyads showed significant association on the 5% level, indicating that these bulls did not form stable associations with other adult bulls.

The very short associations between adult bulls means that they are often alone or in small groups. For this behaviour the bulls pay the price of increased risk of predation. I have noticed that bulls are much better than cows at defending themselves against lions, and they often attack lions. Sometimes they would spot lions from as far as 1 km and would walk deliberately in their direction. They ran after lions and chased them into the trees. If the lions failed to find cover, the buffalo sometimes even killed them. However, perhaps partly because of this, the predation risk of adult bachelor bulls is some ten times higher than that of cows (Chapter 5). This increased risk appears to be an unavoidable cost of an increase in condition during the bachelor stage (see below).

Average bull group size was in the early 1980s independent of time of day (day versus night observations, *t*-test on 798 observations, not significant at the 5% level) and of the vegetation type in which the bulls occurred (Table 4.5).

Also, the mean nearest neighbour distance between bulls was independent of time of day, group size, season and vegetation type (*t*-tests, *F*-tests and Scheffé's multiple comparisons, all not significant at the 5% level; $n = 200$). These results are contrary to expectation if predation were an important behaviour modifying factor. Because lions show their peak hunting activity just after sunset and before sunrise, and hunt more during the night than during the day, one would expect larger group sizes at night than during the day, but this is not borne out by the data. By the same token, one would expect larger groups in riskier vegetation types (Chapter 7) and, again, the data do not show this (Table 4.5). However, bachelor bulls often withdrew during the night in adjacent mixed herds, which is in accordance with the inference from Chapter 5 that in these mixed herds adult bulls were safer than if they were in a bachelor group.

Bulls, then, formed short-term associations with an ever-changing array of other individual bulls, and these lasted minimally an hour and maximally a few days. In 1991 we saw bachelor bulls much less frequently than in the early 1980s. We were in the field for 161 days during May to November in 1991 and encountered bachelor bulls on 43.5% of these days. During the same period in 1983 and 1984 we encountered bachelor bulls on 80.8% of the 167 days. Hence, the frequency of encountering bachelor bulls had declined ($\chi^2 = 48.817$, d.f. = 1, $P < 0.0001$). Even during the dry season of 1991 bachelor bulls became increasingly rare. In May to July we saw bachelors on about 60% of the days, but in August to November this figure had declined to about 30%. In 1983 and 1984 these figures were, respectively, 75% and 90%. At the same time, the average bachelor bull group was slightly smaller in 1991 than in 1983 to 1984. This implies that the total number of bachelor bulls was about three times lower in 1991 (about 50) than in 1983 to 1984 (about 150). As one could

Table 4.5 Number of adult bulls in bachelor groups in different vegetation types during the wet season (November to May) and dry season (June to October) in the study area

Vegetation (landscape ecological unit)	*Safety rating*	*Bachelor bull group sizes (±95% confidence limits)*		*Difference between seasons at 5% level*
		Dry season	*Wet season*	
Bare mud (A0)	S	3.0 (0.6), n=76	4.5 (1.0), n=131	$t_{205}=2.058$, $P<0.05$
Cyperus laevigatus (A1)	S	5.3 (1.2), n=104	7.0 (1.4), n=93	$t_{195}=1.774$, NS
Typha swamp	S	3.4 (2.2), n=11	3.1 (2.7), n=10	$t_{19}=0.165$, NS
Sporobolus spicatus (A2)	D	2.7 (0.5), n=56	3.3 (0.8), n=75	$t_{129}=1.203$, NS
Cynodon dactylon (G1, A3, A4, A5, R1)	D	2.5 (0.5), n=61	2.7 (0.4), n=122	$t_{181}=0.459$, NS
Woodland and bush (all other types)	N	2.9 (1.1), n=19	2.5 (0.5), n=40	$t_{57}=0.864$, NS
All vegetation types combined		3.6 (0.5), n=327	4.1 (0.5), n=471	$t_{769}=2.279$, $P<0.05$

The differences in group size in the different vegetation types are generally not dependent on season (t-tests) and the differences between the vegetations are not significantly different at the 5% level (Scheffé multiple comparison after F-tests on original data). Bachelor bulls tend to occur in slightly larger groups in the wet season than in the dry season (t-test). The safety rating of the different vegetation types is taken from Chapter 7: S, safe, D, dangerous, N, neutral, risk according to expectation.

expect from the interpretation that bulls randomly associate for a short period of time only, the inferred lowered density of bachelor bulls resulted in relatively more lone bachelor bulls in 1991 than in the 1980s (compare Figures 4.5 and 4.6). The frequency distribution of bull group sizes, however, was still not random in 1991 ($\chi^2 = 19.603$, d.f. $= 9$, $P < 0.05$) and bachelor bulls still preferred the company of two to four other bulls.

I think bachelor bulls were rarer for two reasons. First, the meteorological year 1990/1991 was much drier, with only 364 mm of rain compared with 953 mm in 1982/1983 and 626 mm in 1983/1984. Could the relatively low number of bachelor bulls have been compensated by a higher number in the herds in 1991? The average adult bull frequency in the herds ranged between 15 and 18% in the dry seasons of 1983 and 1984; in the 1991 dry season this figure was 19–20%. Thus, the total number of herd bulls was about 280 and the number of bachelor bulls was about 150 in the early 1980s (totally 430). In 1991 the total number of herd bulls was about 340 and the number of bachelors about 50 (totally 390). This suggests that the absolute number of adult bulls declined even though the total number of buffalo increased slightly. Hence, bachelor bull numbers did not decline because they sought 'refuge' in the herds during the drought. Secondly, according to professional hunters the hunting pressure outside the national park had increased greatly between 1985 and 1991. Adult bulls moved between Manyara, Tarangire and Ngorongoro Crater, and had to pass through hunting concessions in the game-controlled areas, where poaching is also rampant. On their way between the protected areas their chance of being killed was higher in the early 1990s than in the early 1980s. This would explain the rarity of bachelor bulls in Manyara in 1991. In 1994, when hunting and poaching pressure had even further increased, buffalo bulls became very rare and started banding together in Manyara. Average bachelor group size had increased dramatically to 5.6 ($n = 47$).

4.3 DOMINANCE INTERACTIONS BETWEEN ADULT BULLS

Adult bulls could engage in sparring, in dominance interactions and in outright fights. Sparring is seen in calves, and rarely in adult cows. When two sparring bulls meet they do not charge each other, and there appears to be very little pushing. As Sinclair (1977, p. 110) observed, a male will initiate sparring by approaching another slowly, putting his horns down and waiting for the other to do likewise. While sparring they twist their horns from side to side for several minutes, and in general the context is one of play rather than aggression (Reinhardt and Reinhardt 1982). In contrast to Sinclair (1977, p. 110), I have observed sparring relatively frequently between adult bulls, as has Mloszewski (1983, p. 123). Juvenile bulls can be engaged in sparring for up to an hour in one single session, but sparring sessions in adult bulls are much shorter (seven sparring bouts per session, with an average bout duration of approximately ten seconds). Before and after sparring bulls can graze side by side, and during sparring sessions they can rub the face or the body of the sparring partner. I never observed sparring to escalate into outright aggression, and only once it did end in a short chase. My interpretation of this behaviour is that sparring should only be viewed as a friendly contest.

The behaviour associated with dominance interactions has been described by Sinclair (1977, p. 108 ff.). The different behaviours are so rare that all the presently described observations on bull–bull dominance interactions (except outright fighting; see below) have been lumped together since as during the study period of more than 200 000 'buffalo–hours', dominance interactions were seen only 22 times. Typically, overt dominance interactions occurred between bulls that were of the same size and of the same condition, more so than would be expected on a random basis ($P < 0.01$) or on the basis of the availability of bulls in the group in which a particular dominance encounter took place (Table 4.6).

Table 4.6 The relation between the condition of the dominant bulls and of the opponents with which these bulls were engaged in dominance interactions, and with the condition of the potential opponents (that is the condition at that time of the other bulls in the bachelor groups with which the bull could choose to interact)

	Condition of the opponent relative to the winner's condition						
	−1.5	*−1*	*−0.5*	*0*	*+0.5*	*+1*	*+1.5*
Actual dominance interactions between adult bulls (n=21)	0	0	0	19	1	1	0
Potential dominance interactions of the same bulls, i.e. possible choice of opponents in relation to condition range available in the groupings (n=168)	0	4	20	102	21	18	3

The bulls enter into dominance interactions especially with bulls equal in condition (χ^2 = 7.799, d.f. = 1, $P < 0.01$ after dividing the sample into bulls with equal condition and those with unequal condition because of the small sample size). Dominance is not related to absolute condition.

In these interactions, one bull (B 35) is omitted as it was dominated by two other bulls (B 41 and L 42, both in a worse physical condition) immediately after B 35 was severely wounded during a failed lion attack.

I assessed the physical condition (Chapter 3) of many bulls during the observations: being dominant in a dyadic interaction was dependent not on condition itself but on the condition difference between the contestants, since the distribution of conditions of dominant bulls does not deviate significantly from that of all bulls in the groups of bachelors in which the interactions took place ($\chi^2 = 2.154$, d.f. = 2, not significant at the 5% level).

Sometimes the dominance interaction does not lead to the establishment of an actual dominance and outright fighting between adult bulls can ensue. These fights are extremely rare. To my knowledge only one fight has been recorded in Manyara during the last 35 years, namely on 20 May 1966. After fighting for several hours, the winner gored his opponent from the side into his heart, killing him. The winner, however, could not dislodge his horn from the rib cage of his, now dead, opponent, and was also found dead the next day (M. Burengo, personal communication). A similar case took place in the Ngorongoro Crater in May 1983 (T. Iwaku, personal communication): after a short fight the winner hurled his opponent over his shoulder, causing the death of his opponent. The winner again died a short while later, presumably from spinal damage. A third case was reported from Amboseli, Kenya (R. Noë, personal communication) and again both contestants died. In these three cases, the observers were of the opinion that the fighters were of equal size and condition. An entry in the Warden's Report for 24 September 1966 states 'One buffalo was found dead at Ndala River, the carcass indicated that it had been injured by another buffalo when fighting'; the actual fight, however, was not witnessed. Simultaneous death of contestants during fighting has also been reported in musk ox (Wilkinson and Shank 1976).

The observations on bull–bull interactions suggest a mechanism of strength assessment by individual bulls of other bulls. As demonstrated, meeting between bulls is random and any other adult bull in the population can become a competitor for females in a mixed herd. The strength of a strange bull has preferably to be assessed before outright fighting, as fighting appears to be very dangerous for any individual involved. In the dominance interactions between bulls, lateral display (as described by Sinclair 1977, p. 109; Geist 1971) plays a prominent role.

One feature of adult bulls is conspicuous when viewed from alongside, namely the 'beard' (comparable to the dewlap in *Bos* spp. but much smaller). The size of the beard is clearly positively associated with testicle size (medium and large testicles versus no beard, small beard or large beard: $\chi^2 = 12.064$, d.f. = 2, $n = 110$, $P < 0.005$), which cannot be seen from the side in a lateral display. In a submissive bull the showiness of the beard is diminished by the bull's posture, but the stance adopted during the lateral display emphasizes the beard. Hence, beard size can be a clue to the tendency to attack a strange bull if there is a relationship between aggressiveness, testosterone levels and testicle size in African buffalo.

In several species an increase in testis weight with age occurs in juvenile and subadult males, but this increase is less clear in adult males (reindeer, Leader-Williams 1979, fallow deer, Chaplin and White 1972). A good correlation between testes weight and plasma testosterone levels has been found in reindeer (Leader-Williams 1979). Testosterone has a direct positive effect on social aggressive behaviour and dominance but not on sexual behaviour in red deer (Lincoln *et al.* 1972). High testosterone levels are positively related to dominance and aggression, especially during initial encounters (Perret 1985, 1986, Sachser and Pröve 1986). Hence, it is unlikely that adult age differences can be determined from testes size (or beard size), but probably the tendency to attack can be estimated from these observable features.

The strength of a strange bull can be assessed from his size and from his condition. Size I have found very difficult to estimate, but relative condition is easy and, because most of the preponderance of dominance interactions are between bulls of the same condition, it appears that bulls are also able to assess one another's condition. This makes it important to have a look at the condition of the Manyara bulls.

4.4 RE-ENTRANT CONSECUTIVE POLYGYNY

Movements of adult bulls into and out of mixed herds were clearly related to changes in physical condition. Bulls entered a mixed herd when their condition was good. During their stay in the herd they lost condition until they dropped out and became bachelor. During the bachelor stage they regained condition, which indicates better 'living conditions' for bulls with bachelor status than for bulls in a mixed herd (Table 4.7). Apart from the data in Table 4.7, 40 bachelor bulls and 13 herd bulls stayed in the same condition.

The total comparison between bachelor and herd states of bulls then yielded a χ^2 value of 5.762 (d.f. = 2, P = 0.06), indicating that proportionally more bachelors than herd bulls gained condition and proportionally fewer bachelors lost condition. In the same vein, a Mann–Whitney U-test on all data on condition changes gave U_{max} = 2136.3 and T = 11.739 (n_1 = 68, n_2 = 24), which is highly significant. All these analyses point to condition gain during the bachelor phase and condition loss for adult herd bulls, resulting in a strategy of switching between the two social environments (Figure 4.7).

Table 4.7 Relation between condition and 'living conditions'

	Bulls gaining condition (increase/10 days)		*Bulls losing condition (decrease/10 days)*	
	Bulls stay bachelor	*Bulls stay herd bull*	*Bulls stay bachelor*	*Bulls stay herd bull*
	5.00	5.00	1.67	3.33
	5.00	0.09	0.56	3.33
	5.00	0.03	0.28	1.67
	0.83		0.28	1.11
	0.33		0.26	0.33
	0.24		0.19	0.19
	0.24		0.16	0.15
	0.23		0.12	0.14
	0.19		0.11	0.03
	n_1=9	n_2=3	n_1=9	n_2=9
Mann–Whitney *U*-test	U_{max}=44.25		U_{max}=59.00	
	P<0.025		P<0.05	

Bulls staying bachelor between two observations gain condition faster than bulls remaining in a herd between two observations (Mann–Whitney test, $P < 0.025$), while within the group of bachelors other bulls lose condition more slowly than bulls in the group of herd bulls ($P < 0.05$). Both data sets point at a more favourable nutritional balance for bachelor bulls than for those in herds.

The traditional view, proposed by some authors, was that old bulls are driven out of the mixed herds by young and vigorous bulls and afterwards associate in bachelor groups. This was questioned by Maclatchy (1932). He asked 'How can this be the case if they are so much larger than the young ones?'. Maclatchy then proposed (tongue in cheek, I think) that it was perhaps hypochondria that caused the large males to become solitary. Many years later Grimsdell (1969) proposed that old bulls lost libido and left the mixed herds. This idea was adopted by Sinclair (1977), but he added that bulls could leave mixed herds temporarily during periods of reduced food availability. Neither in the Manyara area nor elsewhere in the Masai ecosystem is the bachelor phase a permanent one, contrary to the suggestion based on observations elsewhere in Africa (Grimsdell 1969, Sinclair 1977, Mloszewski 1983, p. 124).

Being in a bachelor group has two conspicuous disadvantages for an adult bull, namely no recourse to adult cows – and hence no direct contribution to reproductive success – and an enhanced risk of being killed

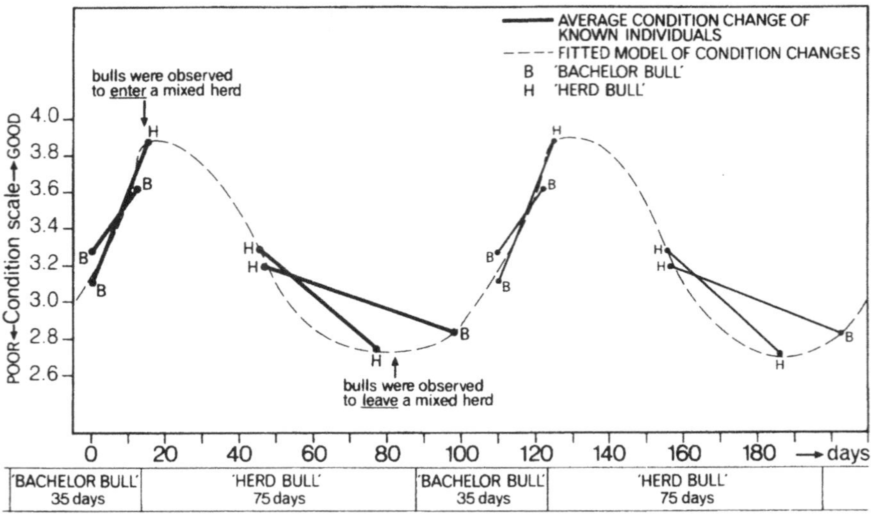

Figure 4.7 Bulls increase their physical condition during the period that they are bachelor and lose condition in the mixed herds. They enter these herds at peak condition and leave the herds after a few weeks when their condition is at its worst.

by lions (Chapter 5) with concomitant termination of reproductive success. From the observations in Manyara a different picture of the social organization of buffalo bulls emerges. In this system adult buffalo bulls steadily moved between the two social environments (the herd and bachelor phases) depending on a relationship between body condition changes, dominance relationships between adult bulls and breeding opportunities. The same situation is reported to occur in male Coke's hartebeest (Stanley Price 1974), and perhaps in male African elephant (L. Lindsay and P. Lee, personal communication). This type of social organization can best be described as a re-entrant consecutive polygyny.

These findings beg the question: 'Why did bulls lose condition in mixed herds?' Clearly, the answer is not that they often fought with other bulls, as red deer do (Clutton-Brock *et al.* 1979), because fights and overt dominance interactions between bulls were rare. Nor was there overt competition for food, as there was between adult cows and between adult cows and subadult bulls (Chapter 3). Bulls adopted the bachelor state simply by remaining on a grazing ground which was being deserted by the mixed herd of which he was a member without any observable behavioural interaction. There was also no striking difference in the time devoted to grazing during the day by adult cows and adult bulls in a mixed herd (20.9 and 21.1%, respectively), as was reported for Soay sheep (Morton-Boyd and Jewell 1974). However, compared with other categories of buffalo, adult bulls spend on average some 20% less time feeding during the night because of their increased vigilance (Table 7.3), and this increased level of vigilance is thought to be related to watching for predators, competitors and mates (Chapter 5).

The pattern of condition changes in adult buffalo bulls is similar to that of male ungulates in temperate areas except that in this case the males regain condition during the summer and lose condition during the autumn rut (Geist 1971, Morton-Boyd and Jewell 1974, Clutton-Brock *et al.* 1982) or, if there is no clear rut, they lose condition in winter (Duncan 1992). In contrast, in the Manyara buffalo bulls there is a cycle in condition changes of only 3 months (Figure 4.7). The major difference between the temperate region and Manyara appears to be that at higher latitudes females of most ruminant species come into oestrus for only during a limited period of the year (with the exception of *Bos* and *Bison* species when kept under good feeding conditions; Sinclair 1977), while in Manyara buffalo cows, as a group, can come into oestrus all year round and there is no clear calving season (Chapter 6). This means that an adult buffalo bull, if in a good condition to mate or to compete with other bulls, can always expect some cows to come in oestrus.

If all cows in a mixed herd are pregnant or are lactating, very few if any cows would be expected to come into oestrus soon, and the bull is better to leave the herd in search of another. If relatively few cows are pregnant or lactating, the bull should stay and compete with other bulls for access to the cows (compare Manzolillo 1986). However, by staying, the bull loses condition (Figure 4.7) which means that he loses his competitive advantage over other bulls, that is he becomes less attractive to cows (cows clearly refuse to be mated by some bulls; see Le Boeuf 1974, Hausfater 1975, Lott 1979, Clutton-Brock *et al.* 1982). Although the exact mechanism is only yet surmised (for example, in domestic bulls there is a relation between undernutrition and the titre of a number of sex-related hormones; Gauthier and Coulard 1986), it appears likely that an adult bull's chances of successful matings diminish during his stay in a mixed herd (compare Anderson and Fedak 1985), and so there comes a time that it is better for a bull to leave the mixed herd in order to regain condition, during which time he has to accept an increased risk of predation (Chapter 5).

Adult bulls in the Masai ecosystem start their wandering life at the age of about 7 years and die at an average age of 11.5 years (Chapter 5). During this period of 4.5 years they wander around in a 'bull pool', which in the Manyara area alone was estimated to be at least 633 from 1981 to 1985.

Hence, during their adult life they meet a large number of other adult bulls. Because of this most bulls are actually strangers to each other, contrary to the situation with African elephant bulls (Leuthold 1977, p. 200). These other bulls are potential competitors when cows in oestrus are around.

It appears that bulls are continuously assessing the fighting ability of other bulls. As explained above, bulls are likely to use a number of clues to assess the fighting ability of another bull, namely the size of the beard (indicating the tendency to attack) and the condition (which can be an indication of the ability to persist in fighting). In domestic cattle other predictors of dominance are variably related to horn size, weight, body size, age and especially length of stay in the herd (Schein and Fohrman 1955, Beilharz and Mylrea 1963, Wagnon *et al.* 1966, Brantas 1968, Clutton-Brock *et al.* 1976). In buffalo, age is closely related to weight (Figure 3.1) except in adults. It appears unlikely that a buffalo bull will know the age of his opponent (they may never have met before) and there is a very weak correlation between horn dimensions and age (Figure 4.1). Hence, it is impossible (at least for the human observer) to estimate the age of an adult bull on the basis of size or horn morphology. This would imply that the ranking of adult bulls is determined not by age (contrary to the opinion of Leuthold 1977, p. 191) but by strength and perhaps aggression, as is the case with in American bison (Lott 1979). However, since outright fighting is very dangerous the best strategy open to an individual adult bull would be to improve his condition as much as possible, as dominance contests, with their inherent possibility of lethal escalation, are carried out between animals of the same condition (Table 4.6). Condition then appears to be a reliable measure of 'resource holding power', and an individual should withdraw without damage when its resource holding power, and thus condition, is exceeded by its opponent (Parker 1974). An advantage of using condition as a measure of resource holding power is, that unlike size, it reflects the momentary fighting ability.

The social organization of buffalo bulls in Manyara, is then quite different from that described by Sinclair (1977) for the nearby Serengeti. It is quite possible that both types of social organization occur because of the very different ecological circumstances and conditions in these two areas (Chapter 9). Here I want to re-emphasize that, to gain further insight into the selective forces that operate, comparative research **between** populations of the same species is of great importance in order to discover functional and causal explanations of social organization and its adaptiveness. The observations on the Manyara buffalo bulls also imply that the decisions taken by these bulls appear to be especially geared to maximizing the number of matings they can achieve. This is in strong contrast to the decisions that buffalo cows have to take (Chapters 3 and 8). Their decisions are aimed, I think, at maximizing, or at least satisfying, their nutritive requirements. Yet, individuals of both sexes have to avoid diseases and predation that would terminate their reproductive output. Because of this, I will now first turn to the influence of disease in this buffalo population.

5

Population dynamics, catastrophes and stability

5.1 ARE BUFFALO THE PREFERRED PREY OF LIONS?

Three main external factors are important in the regulation of wild animal populations: food supply, predation and disease. The first two have been considered often, but population regulation by disease has received considerably less attention. Indeed, disease can play a dramatic role, as is demonstrated by the effect of the myxomatosis virus on rabbits. Yet, these catastrophes are rare events and can be studied only occasionally in wild animal populations, although some epidemics have been reported in, for example, white-tailed deer (Marburger and Thomas 1965), howler monkey (yellow fever; Chivers 1969, Mittermeier 1973), African elephant (in Manyara *Mycoplasma* pneumonia caused the death of approximately 23% of the elephant population in 1977; Weyerhaeuser 1982), or jackal and wild dog (Serengeti, probably due to rabies; J.G. Grootenhuis, personal communication). Whether disease has to be regarded as a population regulatory factor in mammals is not clear, but models and some observations strongly suggest that it is possible (Anderson and May 1979, 1985, May and Anderson 1979, Anderson 1981, Leader-Williams 1982). In birds disease is possibly more important (Anderson and May 1979; see also Potts *et al.* 1984). In this chapter I will discuss the impact of diseases on mortality, but first, I will discuss the impact of predation.

The buffalo of Manyara live in an area with the highest known density of lions (Mills *et al.* 1978). Buffalo are formidable animals that are well able to protect themselves against large cats, much better than, for example, bison (Guthrie 1990). I have several times seen interactions between lions and buffalo which resulted in the buffalo chasing the lions into deep cover or even up the trees. In May 1991 one adult lioness was killed outright by buffalo and one was severely wounded in the belly and died several days later. Also, in July 1984 I saw a buffalo gore an adult male lion in the stomach, but apparently this male survived. It is not only lion cubs that are killed by buffalo (Schaller 1972, p. 189) but adults too. These observations confirm the cases that Schaller (1972, p. 260 ff.) cites. Thus, buffalo seem to be a dangerous prey for lions. Yet, the lions of Manyara have been reported to prefer buffalo (Makacha and Schaller 1969); but do the data really support the idea of **preference**?

I considered dead buffalo to have been killed by lions when I had seen the hunt myself or when somebody reported it to me; or when during the night I heard a death bellow and the next day found a dead buffalo in the direction from which the bellow had come; or when I found scratch marks on the back of the dead buffalo (Kruuk and Turner 1967, Schaller 1972; cf. Guthrie 1990). I considered lion likely to have killed those dead buffalo which I observed eaten by lions but from which I had not heard a death bellow or on which I did not see scratch marks. Spotted hyena can kill buffalo but do so rarely (Schaller 1972, Kruuk 1972), and the number of hyena in Manyara is low. Twice I have seen buffalo attacked by hyenas; in both cases the buffalo was stuck in the mud and a hyena jumped on top of his back and started tearing at the flesh. A few hours later the buffalo was dead.

Over the years quite some information has been collected on the food of the Manyara lions. Rangers recorded all lion kills from 1966 to 1969, and C.M.K. Malima of the University of Dar-es-Salaam did the same between 1985 and 1988. At the end of the 1960s the reported prey species were, in decreasing order of importance, buffalo, Burchell's zebra, calves of giraffe, baboon, impala and warthog. At the end of the 1980s the lion's preferred prey had changed slightly; in decreasing order of importance the prey species were now buffalo, zebra, wildebeest, warthog, ostrich and giraffe. Because the numbers of the different prey species in Manyara were by and large known for the different periods (Table 5.10) I could calculate the relative risk of predation by lions for these species of prey. Relative to buffalo, zebra, wildebeest and giraffe, the remains after lion consumption of the small prey (impala, baboon, ostrich and warthog) were hard to find (see also Kruuk and Turner 1967). Hence I made a distinction between these two groups in the calculations. Table 5.1 shows that the risk of the smaller prey animals being killed by lions was not significantly different for the different species, but that within the group of larger prey animals buffalo ran a lower risk of predation, particularly compared with zebra.

The vulnerability of zebra to lion predation has often been documented. Iain Douglas-Hamilton's (unpublished) observations in Manyara during his study on elephant (Table 5.1) underscore the vulnerability of zebra. It should be noted that the data sets in this table partially overlap, and that they have cases in common with those published by Schaller

Table 5.1 Relative risk for different species of prey in Manyara and the proportions of these species in the diet of lions

	Large prey					*Small prey*				
	Buffalo	*Zebra*	*Giraffe*	*Wildebeest*	*Total*	*Impala*	*Warthog*	*Ostrich*	*Baboon*	*Total*
Period 1966–69										
Estimated population size	1524	87.5	60	–	1671.5	451	20.5	–	400	871.5
Reported number of kills by lions	45	20	5	–	70	4	1	–	5	10
Preference index (95% confidence limits)	0.71* ±0.13	5.46* ±2.24	1.99 ±3.26	–	$\chi^2_2=80.85$ $P<0.001$	0.77 ±0.60	4.25 ±13.17	–	1.08 ±0.68	$\chi^2_1=2.79$ NS
Period June 1967 to July 1968										
Estimated population size	1548	100	–	–	1648	451	–	–	400	851
Reported number of kills by lions	28	9	–	–	37	5	–	–	5	10
Preference index (95% confidence limits)	0.80* ±0.17	4.00* ±2.66	–	–	$\chi^2_1=18.55$ $P<0.005$	0.93 ±0.58	–	–	1.05 ±0.65	$\chi^2_1=0.02$ NS
Period 1985–88										
Estimated population size	2097	255	66	675	3094	–	110	18	–	128
Reported number of kills by lions	44	30	1	27	102	–	10	3	–	13
Preference index (95% confidence limits)	0.64* ±0.14	3.57* ±1.13	0.46 ±6.33	1.21* ±0.20	$\chi^2_3=66.25$ $P<0.001$	–	0.90 ±0.216	1.64 ±1.88	–	$\chi^2_1=0.87$ NS

Individual weight (kg)	420	164	100	120	32	40	40	22
Number of kills (1966–69)	45	20	5	0	4	1	0	5
Proportion (%) of killed prey weight	82.3	14.3	2.2	0.0	0.6	0.2	0.0	0.5
Number of kills (1985–88)	44	30	1	27	0	10	3	0
Proportion (%) of killed prey weight	67.8	18.0	0.4	11.9	0	1.5	0.4	0

The data for the period 1966–69 (28 months) are based on Warden's Reports, the data for June 1967 to July 1968 are unpublished data of I. Douglas-Hamilton and those for 1985–88 are unpublished data of C.M.K. Malima. A distinction has been made between 'large' (> 50 kg) and 'small' prey because the remains of the former are more likely to be found. All giraffe kills were reported to be calves, and an estimated weight for this prey has been used, just as for ostrich. Other estimated weights per individual are from Schaller (1972, Table 74), apart from baboon, for which the weight was calculated according to the method of Schaller from Haltenorth and Diller (1979). An asterisk indicates that the observed preference index with its 95% confidence interval deviates from unity.

(1972, Table 37). Malima's (unpublished) data for 1985 to 1988 are similar – zebra were preferentially killed and buffalo not (Table 5.1). However, although fewer buffalo were killed by lions than expected by chance considering their number, buffalo formed the largest part of the diet of the Manyara lions in the period 1966–69 and again from 1985 to 1988.

In Manyara buffalo took the brunt of the predation by lions, more so than in other areas in East and South Africa. In Manyara 43.6% of all lion kills were buffalo (56.3% in 1966–69 and 34.5% in 1985–88: Table 5.1). Were Manyara buffalo living under risky conditions when compared with buffalo of other areas? In Kruger National Park, South Africa, buffalo as a proportion of all lion kills was 1.0% in the 1940s and 9.2% in the 1950s (Pienaar 1969b). Other reported figures are in Kafue National Park, Zambia, 30.5% (Mitchell *et al.* 1965), Nairobi National Park, Kenya, and Serengeti National Park, Tanzania, 3.3% in the 1950s (Wright 1960); Serengeti National Park 8% (Kruuk and Turner 1967), Serengeti woodlands 2.4% in the 1960s and 20.0% in the 1970s (Hanby and Bygott 1979). Finally, in Nairobi National Park buffalo have again been reported to constitute a negligible proportion of lion kills (Rudnai 1974). Manyara was different from these other areas in that it probably had the highest density of mammalian biomass in Africa (Chapter 1), made up by some 37% buffalo and some 35% elephant (rarely available to lions). It has an extremely high density of buffalo of approximately 20 per km^2 (section 5.9). Simultaneously, the lion density in the study area was also the highest recorded in Africa (Mills *et al.* 1978). However, when expressed as density per herbivore, the density of lions in Manyara was not exceptionally high (Delany and Happold 1979).

Contrary to Mloszewski's (1983, p. 147) conclusion, lions appear to prey on buffalo in accordance with the latter's relative abundance in an area. The 'preference index' (calculated as the proportion of a species in the kills divided by the proportion of the same species in the population) was 0.94 in Manyara in the 1960s and 0.91 in the 1980s (from Table 5.1), while in 14 sections of the Kruger National Park it was on average 1.06 without showing a relation with buffalo relative abundance (Pienaar 1969b). All three indices did not deviate significantly from 1.0. Hence, the suggestion that lions prefer buffalo in Manyara (Makacha and Schaller 1969) does not seem to be borne out by the data. I conclude that buffalo in Manyara do not live under exceptional conditions, and individually they experience approximately the same risk as those in other populations.

The Manyara lions, however, were in a sense atypical of lions or other large felids. Generally, large cats take prey smaller than themselves (Packer 1986, Sunquist and Sunquist 1989). The data summarized in Table 5.1 show that the modal prey item was buffalo, and adult at that. A modal adult buffalo is about 420 kg, while the body weight of adult lions ranges between 120 and 200 kg. Buffalo were apparently not an easy prey to tackle by lions, as indicated by the reports on lions killed by buffalo. Malima (unpublished) recorded in Manyara the number of lions feeding on killed prey between 1985 and 1988. If the number of lions feeding can be accepted as an indication of the number of lions involved in the kill, then the observation that there was a positive relation between pride size and proportion of buffalo in the kills (Table 5.2) indicates that the coordinated attack of several lions increased the chance of lions killing a buffalo.

The assumption, however, that there was a clear relation between feeding group size and number of lions involved in the hunt needs extra research. Apparently, individual lions can abstain from hunting when they are in a group and leave the hunt to other members of the pride (Schaller 1972, p. 242), yet the observations in Table 5.2 corroborate Schaller's findings (1972, Table 59) that hunting success is related to the size of the hunting group. This appears to be a general pattern in felids (Packer 1986) and all carnivores (Gittleman 1989).

5.2 DIFFERENT CAUSES OF DEATH

Of all dead buffalo that I encountered between 1981 and 1985, I could ascertain the cause of death in 82 cases (45.8%) out of a total of 179. Of these, 57 were certainly lion kills and 10 very likely; I have combined these groups simply as lion kills. Of the remaining 97 cases five were not used in the analyses because of the advanced state of disintegration: in 92 cases I could not establish the cause of death. This group of 92 mainly contained the remains of buffalo which had died more than 3 months before I discovered their remains. From 1959 to 1969, cause of death was established in a large number of a large number (328) by the Manyara rangers (Table 5.3). During that period, most buffalo (250 = 78.0%) died from rinderpest, in an outbreak that occurred during a few weeks in 1959 (section 5.6). If these deaths due to the rinderpest epidemic are excluded from the data set, then there was no significant difference between the incidence of lion predation in the set of known causes of death in the periods 1959–69 and 1981–85 ($\chi^2 = 0.547$, d.f. = 1, NS). I concluded that, as long as there was no rinderpest epidemic, lion predation accounted for nearly all known deaths, and was thus the chief proximate cause of death: 88.4% of the mortality appears to have been due to these large cats. The role of other predators was negligible (Table 5.3). However, during a serious outbreak of disease, disease was a greater cause of death than predation. Indeed, the single rinderpest outbreak of 1959 was responsible for twice as many known deaths as the known number of lion kills in some 25 years. The effects of getting stuck in the mud, hyena

Table 5.2 Proportion of kills in relation to lion pride size, based on unpublished data of C.M.K. Malima, collected between 1985 and 1988

	Lion pride size								
	10		*7*		*4*		*1*		*Per cent of all kills*
	Per cent	*(n)*	*Per cent*	*(n)*	*Per cent*	*(n)*	*Per cent*	*(n)*	
Buffalo	43.8	(21)	30.6	(11)	25.0	(4)	14.3	(1)	34.6
Zebra	16.7	(8)	33.3	(12)	43.8	(7)	28.6	(2)	27.1
Wildebeest	18.3	(9)	30.6	(11)	31.3	(5)	28.6	(2)	25.2
Warthog	14.6	(7)	2.8	(1)	0.0	(0)	28.6	(2)	9.3
Ostrich	6.3	(3)	0.0	(0)	0.0	(0)	0.0	(0)	2.8
Giraffe	0.0	(0)	2.8	(1)	0.0	(0)	0.0	(0)	0.9

Larger prides include more buffalo in their diet.

Table 5.3 Known causes of death of buffalo in Manyara from 1959 to 1969 (from Warden's Reports 1959–73) and from 1981 to 1985

Period	*1959–69*			*1981–85*			*1959–85*			
Known cause of death	*n*	*a (%)*	*b (%)*	*n*	*a (%)*	*c (%)*	*n*	*a (%)*	*b (%)*	*c (%)*
Rinderpest	250	77.6	–	0	0.0	0.0	250	61.9	–	–
Other diseases	6	1.9	8.3	2	2.4	2.7	8	2.0	5.2	5.5
Lion	62	19.3	86.1	67	81.7	90.5	129	32.0	83.8	88.4
Poachers	0	0.0	0.0	8	9.8	–	8	2.0	5.2	–
Stuck in mud	0	0.0	0.0	3	3.7	4.1	3	0.7	1.9	2.1
Fighting between bulls	3	0.9	4.2	0	0.0	0.0	3	0.7	1.9	2.1
Calf abandoned, taken by leopard	0	0.0	0.0	1	1.2	1.4	1	0.2	0.6	0.7
Hyena	0	0.0	0.0	1	1.2	1.4	1	0.2	0.6	0.7
Fell down a cliff	1	0.3	1.4	0	0.0	0.0	1	0.2	0.6	0.7
Total known causes	322			82			404			
Total, rinderpest excluded	72			82			154			

Observed and inferred deaths of calves are certainly underrecorded. a, all known causes included; b, rinderpest excluded; c, poaching excluded; d, poaching and rinderpest excluded.

predation, fighting and diseases other than rinderpest appeared to be negligible; and poaching accounted for a small fraction of deaths (Table 5.3). There were signs in 1991 that poaching had increased. Buffalo at the southern end of Manyara were very shy, and ran away when they saw a vehicle. Yet the number of tourists had increased from about 30 000 per year to over 60 000. By 1994 the total number of buffalo had decreased considerably, most likely due to poaching (section 5.13).

Quite a few studies have assessed the relative importance of predation as cause of death of wild ungulates. The overall impression one gets from these data (Table 5.4) is that under most natural conditions predation is very important during periods when major epidemics are absent or at times that mortality is not caused by a reduction in the carrying capacity of the system, as during drought.

The findings from studies in the Serengeti ecosystem form an exception to this rule. The explanation might be sought in the low number of sedentary predators in relation to the number of herbivores in that area (Table 11.20 in Delany and Happold 1979), most of which are migratory (Schaller 1972, Bertram 1973, Hilborn and Sinclair 1979, Houston 1979, Sinclair 1985). However, for buffalo, which is more or less sedentary in the Serengeti, risk of predation appears to be low as well: of a total annual mortality of approximately 10% (Sinclair 1977), only one-quarter was due to lion predation (Sinclair and Norton-Griffiths 1982). The annual predation pressure in the Serengeti was, thus, about 2.5%, which is about half that in Manyara (section 5.4).

5.3 RISK, SEX AND AGE

Between 1981 and 1985 26 adult bulls and 35 adult cows were recorded as killed, while the adult sex ratio in the population was one bull to 4.5 cows (n = 2327 calculated from 29 censuses over the period). Hence, the risk for the two sexes was different, with bulls running a significantly higher risk than cows (χ^2 after Yates' continuity correction = 22.88, d.f. = 1, $P < 0.001$). Assuming an identical sex ratio during Schaller's (1972) study in Manyara, his findings confirm those for 1981 to 1985, as he reported 44 adult bulls killed by lions and only six adult cows (χ^2 = 159.18, d.f. = 1, $P < 0.001$). Also, Malima's (unpublished) data of lion predation between 1985 and 1988 reveal the same pattern again. He reported 21 killed bulls and 14 killed cows. Still assuming the same sex ratio, this again was a significant bias towards males (χ^2 = 20.903, d.f. = 1, $P <$ 0.001). These three studies of predation in Manyara all confirm the general pattern that males ran a higher risk than females. This difference in risk for the two sexes has been found in other species too: in wildebeest (Schaller 1972), hartebeest, Burchell's zebra and eland (Rudnai 1974),

Table 5.4 Importance of predation as a cause of death of wild ungulates

Prey species	*Predator species*	*Percentage of mortality accounted for by predation*	*Remarks and study area*	*Reference*
African buffalo	Lion mainly	89	Period without rinderpest, Manyara	This study
African buffalo	Lion mainly	19	Period with rinderpest, Manyara	This study
African buffalo	Lion mainly	23–28	Serengeti	Sinclair and Norton-Griffiths (1982)
Wildebeest	Lion mainly	93	Period without drought, Kalahari	Mills (1984)
Wildebeest	Lion mainly	38	Period with drought, Kalahari	Mills (1984)
Wildebeest	Lion mainly	96	Timbavati, SA	Hirst (1969)
Wildebeest	Lion mainly	13–21	Serengeti	Sinclair and Norton-Griffiths (1982)
Wildebeest	Lion mainly	30	Serengeti	Sinclair (1985)
Burchell's zebra	Lion mainly	90	Timbavati, SA	Hirst (1969)
Burchell's zebra	Lion mainly	59–74	Serengeti	Sinclair and Norton-Griffiths (1982)
Serengeti ungulates	Diverse	30	Serengeti	Houston (1979)
Moose	Wolf	89	Isle Royale	Peterson (1977)
Springbok	Leopard mainly	87	Kalahari	Mills (1984)
Gemsbok	Diverse	84	Kalahari	Mills (1984)
Waterbuck	Lion mainly	82	Timbavati, SA	Hirst (1969)
Kudu	Lion mainly	63	Timbavati, SA	Hirst (1969)
Impala	Leopard mainly	55	Timbavati, SA	Hirst (1969)
Giraffe	Lion mainly	34	Timbavati, SA	Hirst (1969)
Warthog	Lion mainly	29	Timbavati, SA	Hirst (1969)

In natural ecosystems mortality of large herbivores due to predation can be high in the absence of epidemics or droughts. In the table rounded figures are given. The low values for the Serengeti are discussed in the text.

gemsbok, springbok and again hartebeest (Mills 1984). The same was found by Sinclair (1977) in a sample of 75 buffalo kills from the Serengeti and by Mloszewski (1983, Table 8.3) in southern Africa. Because of the killing of animals with a low reproductive value, Slobodkin (1974) was led to call predators 'prudent'. The higher predation rate on males fits into the idea of 'prudent predator', as from the predators' point of view many prey males are expendable for a sustained yield. However, it is not necessarily true that males have a lower reproductive value than females.

I did not find a significant difference between the proportions of subadult bulls and subadult cows killed by lions (Table 5.5). Apparently they were exposed to the same degree of risk, so I classified them together as subadult buffalo (more than 3 years and younger than 7 years). Also, subadult buffalo and adult cows were equally prone to lion predation. As outlined in Chapter 3, adult cows and subadults of both sexes together formed the permanent core of the herds. The herd-living buffalo ran a significantly lower risk than the adult bulls (Table 5.5), that live in the herds for limited periods of time only, alternating herd life with the bachelor state (Chapter 4). From my calculations it appears that buffalo in the herd experienced only approximately one-third the risk of adult bulls.

Bulls did not appear to be more vulnerable to lions *per se*: twice I observed lion attack fail as a result of vigorous defence by bulls although cows were never seen to be successful. I have also seen attacks on bulls that have failed because of interference by another bull (Chapter 4). Also the horns of buffalo bulls are much more formidable than those of buffalo cows (Figure 4.1). The explanation for the high risk of bulls must then be sought in the difference in living conditions of adult males and the herd-living adult females, subadult bulls and subadult cows. Bachelor adult bulls were often without the protection of conspecifics (Figures 4.5 and 4.6; cf. Hamilton 1971, Bertram 1978), and I thus conclude that living in herds affords buffalo protection against predation by lions. If this is true, then for adult bulls there must have been a trade-off between the benefits of staying in the herd or leaving it. I think this trade-off was loss of condition in the mixed herd, and condition gain, which is necessary for successful competition with other bulls, during the bachelor phase (Chapter 4). A higher risk of predation was then balanced by this necessary condition gain outside the herd.

Because the predation pressure on cows was less than that on bulls, one would expect cows to grow older than bulls, or, if this is not the case, to have higher mortality in the juvenile period. I recorded the ages of dead buffalo from their tooth characteristics (Chapter 3). This method appears to be quite reliable, but see Caughley (1980) for a critical discussion of ageing

Table 5.5 Number of buffalo of different age–sex categories killed by lions in Manyara from 1981 to 1985

	Subadult bulls	*Subadult cows*	*Subadults (bulls+cows)*	*Adult cows*	*Subadults (bulls+cows)*	*Adult bulls*	*Herd buffalo*	*Adult bulls*	*Total*
Number killed by lions	4	2	6	35	6	26	41	26	67
Number in the population	100	67	167	689	167	153	856	153	1009
	$\chi^2_1=0.006$, $P > 0.95$		$\chi^2_1=0.349$, NS		$\chi^2_1=13.030$, $P < 0.001$		$\chi^2_1=27.272$, $P < 0.001$		

It is concluded that the predation risk for subadult bulls does not differ from that for subadult cows, and that the risk for subadult buffalo does not differ from that of adult buffalo cows. These categories form stable mixed herds. Adult bulls switch between herds and are often encountered as bachelors and run a higher risk of predation.

by tooth wear. The average age at death was 11.1 years in females (SD = 3.5 years, n = 78), while it was 10.6 years (SD = 2.6 year, n = 60) in males. This difference is not significantly different (t = 1.059, d.f. = 136, $P > 0.1$). Females were killed by lions at higher average age (12.1 years, SD = 2.8, n = 37) than males (10.3 years, SD = 3.5, n = 26) (t = 2.201, d.f. = 61, $P < 0.05$). Because the average age at death from all causes was the same for bulls and cows, this significant difference indicates that another cause of death must affect cows at a younger age than bulls to compensate for the age difference due to lion predation. However, I could not collect sufficient information for calves or juveniles, since for these categories I recorded only four cases of lion predation and one of leopard predation. Lack of data was caused by the high perishability of the skeletal remains of young animals (see also Caughley 1980) and the slim chance of finding such small remains. In addition, Malima (unpublished) recorded only three juveniles or buffalo calves in a total sample of 43 buffalo that were killed by lion between 1985 and 1988.

In the Serengeti young buffalo were underrepresented in a sample of lion kills (Schaller 1972, Table 50), but extensive observations on hunting lions in Botswana revealed that buffalo calves can form a large part of the lions' diet (McBride 1984). There is some indication that the juvenile mortality of females was higher than that of males, as in the population sample there were 67 subadult cows for 100 subadult bulls (χ^2 = 6.521, d.f. = 1, $P < 0.025$). As the sex ratio at birth was equal (32 female calves and 36 male calves were sexed within approximately 2 days of birth, which is not significantly different from an equal sex ratio: χ^2 = 0.243, d.f. = 1), this indicates that the unequal sex ratio for subadults was caused by differential juvenile mortality. As pointed out in Chapter 3, both subadult cows and subadult bulls stayed in the herd in which they were born, so the unequal subadult sex ratio cannot have been caused by migration. In 1991 the subadult sex ratio had changed in favour of cows; for 100 subadult bulls there were now 168 subadult cows (n = 436). Hence, differential mortality for the juvenile sexes is not certain.

There was no evidence that mother buffalo invested more in male offspring than in female offspring during the lactation period, which could have contributed to higher survival of male offspring, as has been observed in red deer (Clutton-Brock *et al.* 1982). I have taken as a very indirect measure of maternal effort (see for definition Barash 1980) the duration of the suckling bouts, and noted whether the mother terminated the suckling (often by lying down) or whether the calf did. Buffalo calves usually drink with their head between the hind legs of the mother, and this was often seen while mother and calf were on the march to keep up with the rest of the herd. Calves normally waited to suckle until their mother rose, and only rarely did the calf nudge or butt the mother to cause her to stand. The first suckling bouts by newborn calves were very short (mean 24.3 seconds, SD = 8.1, n = 3) and appeared to be terminated because the calf was still too feeble to drink. These suckling bouts I omitted from the analysis. Suckling frequency of individual calves could not be established. My impression is that a calf drank only twice a day: most suckling was observed between 10.00 and 11.00 and between 17.00 and 18.00, and never at night. The average suckling bout duration was about equivalent to the product of suckling bout duration and number of bouts per day in water buffalo calves (approximately 1600 seconds for very young calves to 1000 seconds for 1-week-old calves; Tulloch 1979). Hence, only one suckling bout in the Manyara buffalo was already as long as all suckling bouts of water buffalo calves in one whole day. Calves 3–4 weeks old had the longest suckling bouts (mean = 723 seconds, n = 14). This was longer than the suckling bouts of calves of between a few days and 2 weeks old (595 seconds, n = 22). At these ages it was normally the calf which terminated the bout (Figure 5.1A).

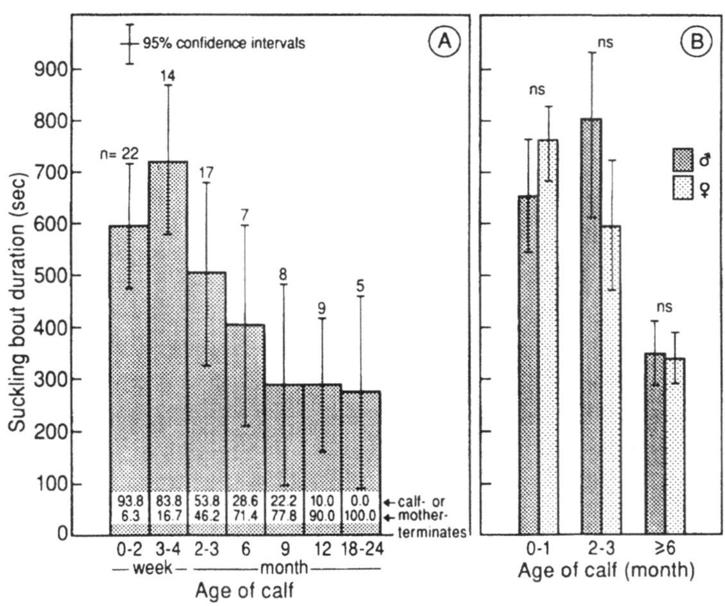

Figure 5.1 Suckling bout duration as a function of calf age. When calves are young, it is the calf that terminates suckling, but later the mother increasingly often terminates (A). There is no difference between suckling bout duration of male or female calves (B), which suggests that maternal effort is equal for the two sexes.

At an older age suckling decreased in length (F-test, $P < 0.05$) to only 276 seconds for the few calves of 1 year and over ($n = 5$) that were still sometimes suckling. After the age of 6 months it was nearly always the mother which terminated the bout. It is, however, unclear whether she terminated because of a 'mother–offspring conflict' (that is the calf 'demanding' more milk than the mother is 'willing' to give). This is because of my observations that at the age of 2–3 months, when the mother terminated equally often as the calf, the cases in which the calf terminated were even slightly shorter in duration (mean 604 seconds, $n = 6$) than when the mother terminated (mean 633 seconds, $n = 6$; NS, median test). A number of studies suggest a difference in maternal effort for sons or daughters (red deer, Clutton-Brock *et al.* 1982, p. 165 ff.; African elephant, Lee and Moss 1986). A difference in maternal suckling bout duration for male or female buffalo offspring was not significant for three discerned periods (t-test, Figure 5.1B). The decrease in suckling bout duration during the lactation period points to a decrease in costs to the mother, but as I could not collect data on milk flow or suckling frequency this conclusion is tentative.

5.4 ANNUAL MORTALITY

To compare the rate of predation in Manyara with that in other areas, I made an estimate of the annual lion predation pressure. I could do this only through indirect inference. In the absence of rinderpest 88.4% of the deaths were due to lions (Table 5.3). By analogy I assumed that this also applied to deaths of unknown cause. Thus $179 \times 0.884 = 158$ buffalo were killed by lions between 1 January 1980 and 1 February 1985 (49 months). I took a period of 49 months, because, although I started collecting the remains of dead buffalo on 15 November 1981 and continued until 1 February 1985, I assumed that the most disintegrated skulls had been lying in the field since 1 January 1980. This assumption was based on observations on the decaying rate of skulls kept outside in the Ndala Research Camp. I could collect the skulls of almost all buffalo carcasses; some exceptions were due to the illegal removal of skulls by tourists and some to the behaviour of hyena. The number of buffalo in the part of the park searched for buffalo remains, namely between Simba River and Maji Moto, was on average approximately 1250 during this period. Hence, the annual predation pressure was 158 buffalo/49 months $\times$ 12 months/1250 buffalo = 3.1% per year. This figure very much depends on the assumption that all buffalo remains were actually found, which is not very likely.

This predation pressure figure can be checked against the one calculated from the estimated annual consumption by lions in the park (Schaller 1972, also Kruuk and Turner 1967). I assumed, after many discussions with the park rangers, that there were 18 adult lion equivalents (following the definition of Hilborn and Sinclair 1979) in Manyara between 1981 and 1985. This accorded with Malima's observations on lion numbers between 1985 and 1988. I also assumed that these lions ate 5.8 kg per meat per day (Schaller 1972, p. 275 ff., Eloff 1984), and that the consumption equals 67% of the weight of the carcass (Schaller 1972, p. 276). This was confirmed by Malima's (unpublished) data on consumption by lion in Manyara – I calculated from his data that on average about 75% of the flesh of a buffalo was eaten from a sample of 13 cows and 21 bulls. Feeding time on bull carcasses, which are heavier, was slightly longer than on those of cows (1.8 days and 1.5 days respectively; $t_{32} = 1.353$, $P < 0.1$, one-sided). The Manyara lions would then have eaten $18 \times 5.8 \times 365 = 38\,106$ kg per year, and would have killed 57 159 kg per year. As 82.3% of the lions' consumption came from buffalo (Table 5.1), this would have represented 47 042 kg of buffalo. This represents 112 buffalo killed per year. This calculation is for the whole buffalo population (about 2000 head) of the Manyara area, including the part of the park not searched for buffalo remains. Following this line of reasoning, the calculated annual predation pressure equalled 5.6%. This figure was undoubtedly closer to the truth than the 3.1% mentioned above, and in my other calculations I have accepted 5.6% as the annual mortality estimate. This implies that, of every two dead buffalo, only one was found by me and the rangers, but of course calf mortality has be considered too. The predation pressure on the Manyara buffalo was thus twice as high as in the Serengeti, where it was approximately 2.5% per year (section 5.2).

Even if the predation pressure was quite high, this does not imply that predation should be viewed as a regulatory factor. Many studies have investigated the effect of predation on population regulation, and a number of authors have come to the conclusion that, even in populations in which many animals die a violent death, predation is not to be viewed as a population regulatory mechanism (Field 1968, Hirst 1969, Pienaar 1969b, Grimsdell 1969, Hornocker 1970, Spalding and Lesowski 1971, Schaller 1972, Peterson 1977, Eloff 1984, Mills 1984, but see the argument to the contrary put forward by for example, Bergerud *et al.* 1983a, b). I think that in the Manyara buffalo social herd size in combination with social inequality has to be viewed as a population regulatory factor but that predation is not.

Taking into account the fact that the predation risk for herd animals was approximately one-third that for adult bulls (section 5.3), and given the composition of the population (Table 5.5), this means that the annual predation pressure on herd animals was about 4.4%, while on adult bulls this pressure equalled 12.9%. This difference in predation risk for the two sexes is, I think, enough to explain the very unequal sex ratio of adult buffalo in Manyara, which was, as was shown, one adult bull for 4.5 adult cows.

The importance of predation pressure on different categories of buffalo might be assessed by an intuitive approach by comparing the risk to which certain classes of buffalo are exposed with the risks faced by western people in daily life. The predation risk for bachelor bulls (0.135% per day) is similar to the mortality risk of certain sport activities in humans. The risk of being killed by a lion for herd-living buffalo (0.012% per day) is very similar to the overall risk of dying for all ungulates in the Serengeti (0.02% per day) (Houston 1979), and herd-living buffalo run a risk twice as high as the risk of death or hospitalization faced by a driver or passenger in a car (0.006% per car-hour; Central Bureau for Statistics of The Netherlands). In Chapter 7 I will investigate whether there are indications that buffalo took these risks into account in their daily decisions of where to forage.

5.5 RINDERPEST

One of the most catastrophic diseases in wild ruminant populations in Africa has been rinderpest. It also had a great effect on the Manyara buffalo. Most authors state that it was accidentally introduced to Somalia from India in 1889 and again in 1903 (Von Ostertag and Kulenkampff 1941). Others think that it was introduced into Egypt from Asia even earlier and spread from Egypt, because many cattle thought to have died from rinderpest were seen in Masailand as early as 1884 (Thomson 1885 cited in Mack 1970). The disease spread like wildfire both in wild ruminants (Table 5.6) and in domestic stock, and it reached Malawi in 1892 and the Zambesi River in 1893.

In 1896 it jumped the river and reached the Cape in 1887 (Mack 1970). Apart from Thomson's 1884 observation, its first occurrence in northern Tanzania and southern Kenya can be dated through oral history.

> ... there was an eclipse of the sun and it took place at about five o'clock in the afternoon... It was then that the rinderpest attacked the cattle. The epidemic finished the Masai cattle...
>
> (Marieni ole Kertella in interview in Hanley, 1971, p. 285)

This eclipse took place on 22 December 1889 (M. Drummen, personal communication). In Masailand, the Engaruka Basin 'was littered with the remains of cattle and people – so many and so close together that the vultures had forgotten how to fly' (Branagan and Hammond 1965, p. 225).

Plowright (1982) classified rinderpest as the most lethal and potentially dangerous infectious disease which affects wild and domestic Artiodactyla. Iraqw oral history recalls that among the ruminants buffalo and impala were particularly badly affected in the rift valley (Table 5.7). In the Serengeti ecosystem the buffalo was nearly wiped out (Sinclair

Table 5.6 Sensitivity to rinderpest of different African ungulates

	Symptoms	*Remarks*
Extremely sensitive		
Buffalo	Peracute	
Warthog	Peracute	
Eland	Peracute	
Greater kudu[a]	Peracute?	
Lesser kudu[a]	Peracute?	
Very sensitive		
Bushbuck	Acute	
Bush pig	Acute	
Giraffe	Acute	
Sitatunga	Acute	
Bongo	Acute	
Wildebeest	Acute	
Giant forest hog	Acute	
Uganda kob[b]	Subacute	
Moderately sensitive		
Reedbuck	Subacute	
Topi	Subacute	
Blesbok	?	
Bontebok	?	
Gemsbok	?	
Roan antelope	Subacute	
Sable antelope	Subacute	
Oribi	Subacute	Very variable[c]
Impala	Subacute/acute	Very variable[c]
Springbok	?	
Not very sensitive		
Waterbuck	Subacute/acute	Variable[d]
Duiker	Acute	Very variable[c]
Oryx	Subacute	Variable[d]
Grant's gazelle	Subacute?	Very variable[c]
Dik-dik	Acute	
Hartebeest	Subacute/acute	
Hardly sensitive		
Thomson's gazelle	Subacute	
Hippopotamus	Subacute	
Gerenouk	Subacute	

Sensitivity according to Pastoret *et al.* (1988) with additional notes from Plowright (1982). Classification according Scott (1981, p. 23): 'Most peracute and many acute reactions are fatal, death following prostration and occurring 1–2 weeks after the illness is evident; The signs of subacute reactions vary – most animals undergoing subacute reactions survive'.
[a]Pastoret *et al.* and Scott only mention kudu; Plowright gives greater and lesser kudu.
[b]Plowright.
[c]Plowright mentions 'species whose natural susceptibility is reported very variable'.
[d]Normally resistant but affected by the 1945 epidemic in Kenya (Plowright).

Table 5.7 Major epidemics of the rift valley area of north Tanzania, as recorded through oral history of the Iraqw; number of generations is counted by genealogy of chiefs

Year of epidemic	*Number of generations*	*Animal species affected*	*Number dying*	*Area*
1770	10	Elephant, buffalo, impala	'Many'	Rift valley
1790	9	Buffalo, wild pigs	'Many'	Rift valley
1880	5–6	Impala, buffalo, elephant	'Many'	Rift valley
1907	4	Impala, buffalo, elephant	'Many'	Rift valley
1911	4	Buffalo, wild pigs	'Many'	Rift valley
1939	2	Elephant, impala, buffalo	'Many'	Rift valley
1941	2	Wild pigs	?	Mbulu Plateau, Manyara, Mangola
1973–74	1	Buffalo, wild pigs	'Many'	Ngorongoro Conservation Area

Note: wild pigs can include warthog and bush pig.

1977), and in the Kruger National Park the buffalo population was as low as only 20 individuals in only one herd (Stevenson-Hamilton 1911) in 1902, while it had been common before the catastrophe. It recovered to some 10 000 head in the early 1960s (Pienaar 1963) and to 35 000 afterwards. In many parts of Africa the interaction between rinderpest virus and wild and domestic populations of ruminants has been the principal factor that has kept those populations at very reduced levels for many decades (Sinclair 1979). There is, however, good indication that buffalo recovered quickly in other areas, since Radclyffe Dugmore (1910, p. 14) noted that 'The buffalo which was nearly wiped out by the frightful ravages of the rinderpest only a few years ago, are on the increase and... bid fair to hold more than their own... At present they are by no means as rare as some people imagine'. In 1921 he observed that:

> Next to the giraffe, the animal that has increased most noticeably is the buffalo... The increase in their numbers is due not alone to the protection afforded by the game laws, but to their recovery from the effects of rinderpest, which about forty years ago threatened their complete extermination. Being fairly prolific, and having few enemies save humans, the scattered bands of less than half a dozen have now grown to be good-sized herds, numbering in some cases as many as two hundred and even more. The only other animal which seems to have gained in numbers is the eland...
>
> (Radclyffe Dugmore 1925, p. 257)

Cattle numbers also recovered quickly. In 1907 Patterson observed that 'I myself have noticed how greatly these flock and herds have multiplied during the last decade [1897–1907]. Where ten years ago only a few cattle, sheep and goats were to be seen, now there are thousands' (Patterson 1909, p. 179). In 1991 the effects of this rinderpest epidemic were still visible in the vegetation structure of Manyara, manifested by the presence of even-aged stands of mature *Acacia tortilis* trees dating back to about 1890 (Prins and Van der Jeugd 1993).

Rinderpest was a recurring phenomenon in the first 60 years of this century (Sinclair 1979). Northern Tanzania was badly affected by rinderpest first in 1890 and again in 1913–17 (when the First World War caused extensive cattle movements in East Africa), in 1927 (spreading again from Kenya), and in 1945 (spreading from Ethiopia through Kenya into Tanzania) (Plowright 1982). Branagan and Hammond (1965) report various outbreaks, namely 1917–21, 1928–29, 1937–40, 1945–47, 1950–52 and 1959, and Scott (1981) states that outbreaks of rinderpest occurred in 1913–21, 1931, 1937–41 and 1949. Generally, however, the periods more or less coincide.

Although buffalo recovered, they were still rather rare in many places during the first half of the twentieth century. This is also suggested by

Watermeyer and Elliott (1943), who stated that 'buffalo are common in swampy patches [in Manyara] **which can be excelled by few places** in the opportunity it offers for viewing buffalo in the open' (my emphasis). However, it is also possible that generally the buffalo's behaviour was affected so that they were generally little seen at that time because of a shift to nocturnal activity (Chapter 7). There are several reports that the buffalo's behaviour changed during a rinderpest epidemic. For example, Thomas and Reid (1944, p. 10) stated that 'the herd instinct is... very strong and breaking up with scattering of individuals, as is described when severe mortality occurs, was not seen or heard of in this [outbreak]', while Plowright (1982, pp. 8–9) observed that 'buffalo... were especially dangerous [as agent of transmission], as affected herds tended to fragment and move more quickly and further than usual in search of water with which to slake their thirst, enhanced by diarrhoea and dry seasons. Buffalo also became unusually aggressive'. This is also described by Scott (1981, p. 23), who wrote that 'buffalo become dangerously aggressive and stampede in search of water to allay thirst... [but] other buffalo stand immobile, backs arched, in the same spot for days, dying where they stand'.

Rinderpest was such a virulent disease, and so difficult to control, that extreme measures were taken. To protect the cattle areas of southern Africa against the disease from East Africa, belts of immune cattle were maintained and reinforced by the construction of a game-proof fence stretching over a distance of 270 km between Lake Nyasa and Lake Tanganyika (the border area between Tanzania and Zambia) (Thomas and Reid 1944). The next step was the creation of game-free zones, 40 km wide at either side of the fence (Plowright 1982). Ultimately, this might have been a blessing in disguise, since among veterinarians support was now growing for the idea of game sanctuaries, an idea that originally was put forward by white hunters. For example, Thomas and Reid, who worked on the game fence construction, postulated that:

> there is ample room in Africa for game in plenty as well as for livestock, but their interest need not and must not be allowed to clash. In other words it is becoming more and more evident that wild animals will have to be segregated in sanctuaries as much as for their own protection as for that of livestock and agriculture.
>
> (Thomas and Reid, 1944, p. 19)

This viewpoint was, I think, much more enlightened than what Wilde wrote in 1953 (p. 286): 'If rinderpest is to be eradicated... some means of control of game must be evolved. Methods of attaining this end must be considered and the first which springs to mind is complete eradication [of game]. This strikes one as being drastic and sure to evoke considerable opposition from the civilised world' but even he admitted that 'as alternative, some policy of segregation may be possible... [which depends] on the availability of lands suitable for game preservation but unsuitable for stock'. In the end, however, rinderpest was only controlled by means of an enormous vaccination campaign in the 1950s and 1960s, when literally millions of cattle were vaccinated. Only then were the reservoirs of non-immune animals too small to maintain the virus (Branagan and Hammond 1965, Plowright 1982).

Rinderpest is not the only disease that was spread or introduced by the advance of colonialism (Ford 1971, Kjekshus 1977, Johnson 1989, Giblin 1990, Maddox 1990). Another example is east coast fever in southern Africa. This tick-borne disease caused by *Theileria* protozoans was apparently endemic in Tanzania and Kenya and perhaps at some places along the coast in Mozambique. When Cecil Rhodes imported 1000 cattle from Australia through the port of Beira in 1901, there was a problem with railway transport and the cattle had to be grazed for a few days near Beira. There they picked up the disease and, although most died before they came into what is now Harare, the disease was transferred to the local cattle in Zimbabwe. In 1903 and 1904 it quickly spread into the Transvaal and Natal (Von Ostertag and Kulenkampff 1941), where it caused very high mortality among cattle in 1904 and again in 1918 (Stevenson Hamilton 1974, pp. 106 and 159 ff.). However, buffalo were apparently resistant because there was no indication that they became diseased at the time (Von Ostertag and Kulenkampff 1941, p. 191). Sinclair (1977, p. 243) reported a very high incidence of *Theileria* exposure in buffalo but observed that generally speaking buffalo are not affected (*op. cit.*, p. 245).

Rinderpest is caused by an RNA virus (morbillivirus subgroup of the paramyxoviruses; Robertson 1976). Viruses of this group cause measles, polio and influenza in humans, but also foot-and-mouth disease (Robertson 1976, Reanny 1982). Transmission requires close contact between sick and healthy animals. The virus is excreted by the host in expired air in rather large airborne droplets, nasal and oral secretions, and in the faeces. The infected droplets are inhaled by the next animal and the virus penetrates the mucosa of the upper respiratory tract. Because the droplets are rather large, windborne dispersion over long distances appears to be rare. Incubation times in cattle range from 3 to 15 days. Death is caused mainly by rapid dehydration after severe diarrhoea, which explains the stampeding behaviour described in diseased buffalo. If the animal survives, life-long immunity is normally acquired (Robertson 1976, Scott 1981).

The means of transmission of related virus 'species' by 'healthy' carriers shedding viruses in their faeces (Prins and Weyerhaeuser 1987) makes it likely that buffalo can acquire rinderpest easily through healthy carriers travelling from area to area. These carriers could be adult buffalo bulls, which, at least in the Masai ecosystem encompassing Manyara and Tarangire, were not permanent members of any herd (Chapter 4). Among the large game species of East Africa, buffalo in particular have been incriminated as the vector of disease spread, although cattle transport (legal and illegal) and cattle rustling (Branagan and Hammond 1965, Plowright 1982) may have been much more important.

5.6 THE RINDERPEST OUTBREAK OF 1959

In the Manyara area there was a rinderpest outbreak in 1950, since Branagan and Hammond (1965, p. 238) stated that:

> in 1950, game carried rinderpest to isolated groups of cattle in the Magalarasy riverine system, and buffalo, from Lake Manyara, took the infection up the Rift Valley Escarpment to initiate an outbreak among Mbulu cattle. In 1952, infected buffalo and wild pig at Oldeani passed the disease to zebu and grade cattle...

Whether the 1950 outbreak killed many buffalo in Manyara is not known. In 1959 there was again an outbreak of rinderpest, but this time it was a very serious one and many buffalo died.

> Rumours of the disease among buffalo and eland herds in the Ngorongoro highlands were persistent from September 1958 onwards. Despite thorough search, no definite evidence of disease was found but, in January 1959, rinderpest appeared in young cattle below the rift valley escarpment (at Engaruka) [some 50 km north of Manyara] and in the immediate vicinity in the line of descent from the Ngorongoro highlands followed by buffalo and eland some two weeks previously. A chain of small outbreaks among young cattle then appeared along the trails of wildebeest migrating north and west from this point. While these outbreaks among calves were very mild indeed, what was presumably the original infection continued southwards with buffalo herds migrating *[sic]* in the thick forest at the foot of the rift valley escarpment. It was not possible to observe these buffalo until they emerged at Lake Manyara, when the infection was seen to have attained great virulence...
>
> (Branagan and Hammond 1965, pp. 239–240)

Data on this epidemic among the Manyara buffalo were fortunately collected by A.M. Morgan-Davies, erstwhile park and honorary game ranger of Manyara. I calculated the apparent mortality since the beginning of the outbreak as logit $(X_t) = \ln X_t / (1 - X_t)$. In this equation X_t is the fraction of dead animals of the original population (Armitage 1971,

Zadoks 1972). The apparent infection rate is defined as:

$$r = (1 / t_2 - t_1) \times [(\ln X_2 / 1 - X_2) - (\ln X_2 / 1 - X_1)]$$

in which t is time and X is as defined above (Zadoks 1972). More sophisticated methods of analysing epidemics are available (Anderson and May 1979, 1985, May and Anderson 1979; see also Wangersky 1978), but they require a number of parameters for which I did not have the necessary data.

On 17 August 1959 it was reported that rinderpest had broken out among buffalo in the Kitete area, some 30 km north of the present park boundary, and on 31 August 1959 the first five buffalo died from this disease in the present park (Warden's Report, August 1959). The number of victims eventually totalled 310 (Warden's Report, January 1960), some 14% of the total number of buffalo at that time. Branagan and Hammond (1965) reported that 1000 buffalo died during this very virulent rinderpest epidemic of 1959, and Sinclair (1977, p. 253) estimated that some 50% of the Manyara buffalo had died. The Warden's Reports, however, were very precise about describing the epidemic and give a much lower figure. I quote from the November 1959 report:

> [1] ...the only species of game affected has been buffalo. During the epidemic two wildebeest were found dead, a circumstance which can reasonably be attributed to other factors.
>
> [2] ...the estimated total loss has been 300 head.
>
> [3] ...the animals most affected were those belonging to the Mto-wa-Mkindu Herd [that is the present-day Central Herd], which lost most of their yearling stock. About 100 head or 17% of the estimated 600 died.
>
> [4] ...the next most affected was the Mto-wa-Mbu Herd [Northern Herd] and of the estimated 700 head about 50 or seven per cent died.
>
> [5] ...the balance of 150 deaths were from the remaining 1700 spread throughout the reserve.

I have accepted the estimates of the warden as the most reliable ones, since he was daily in the area and a very keen observer, but note that the reserve was larger than the present-day park. Apparently, the disease spread neither to other wild ruminants (Warden's Report, December 1959) nor to cattle (Branagan and Hammond 1965, p. 240). The total population of buffalo at that time is estimated at 2000 [975–1000 animals in the present-day park, and 600–700 animals in a herd which used the swamps near the northern end of Manyara in the dry season only (Chapter 3) plus the 310 which died during the rinderpest epidemic].

From this information I could reconstruct the course of this epidemic (Figure 5.2A).

However, a major problem here is although the size of the original population of buffalo was known, I did not know the number of the initially susceptible animals in this population. In contrast to the impala, of which (nearly) the whole population appeared to be susceptible to anthrax during the anthrax outbreak of 1984 (Prins and Weyerhaeuser 1987), it is mentioned in the Warden's Report of February 1960 that the 'yearling stock' of buffalo was affected specifically. This indicates that the adult buffalo had acquired immunity against rinderpest, perhaps during the 1950 epidemic. The calculated apparent infection rate varies depend-

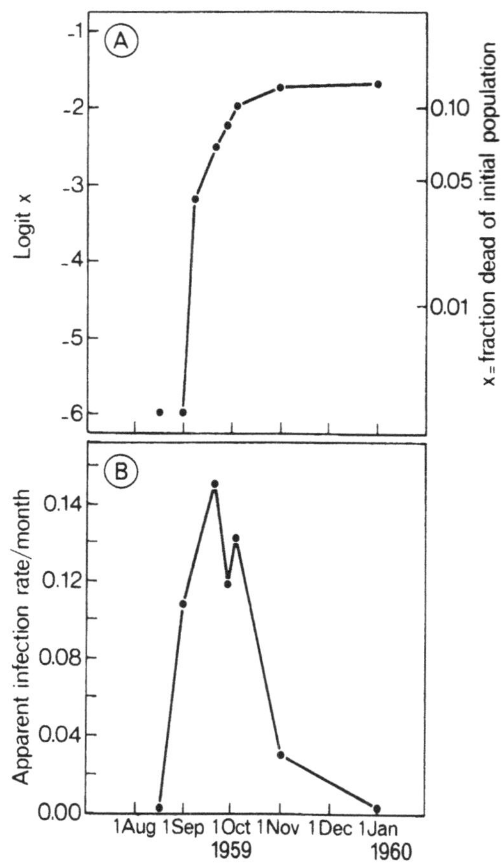

Figure 5.2 The apparent mortality of buffalo (A) during the rinderpest epidemic of 1959, and the apparent infection rate (B) during this epidemic.

ing on the size of the initially susceptible population, but the form of the curve of the apparent infection rate (Figure 5.2B) is not much affected by the size of this initial population estimate (Prins and Weyerhaeuser 1987). The rinderpest epidemic spread rapidly, and the highest apparent infection rate was reached only 1 month after the epidemic started. One month after this peak the disease was nearly over (Figure 5.2).

Since 1960 rinderpest has been suspected several times in Manyara but has not been confirmed (Warden's Reports 1960 to 1974). In 1982 there was a rather serious outbreak of this disease in the northern Serengeti. It spread from there through the Ngorongoro area as far as Karatu (30 km from Manyara), but it did not reach the Manyara buffalo. Apart from parasitic diseases with a low frequency (Anonymous 1966; section 5.8), and some cases of pneumonia which I observed in 1982 and 1983, the population of buffalo was free of major disease between 1960 and 1991. At the end of October 1991 there was a minor anthrax epidemic among the buffalo which killed about 10 adult buffalo. Anthrax will be discussed in the next section.

5.7 CATASTROPHIC DISEASES

Catastrophes have occurred a number of times over the last 200 years, and most likely before that time too. Based on interviews with Mzee Massay Ama Sangka (born 1930), Mzee Gerard Gwaha (born 1942) and Mzee Shaban Swalleh (born 1932), Table 5.7 gives major disease events as remembered for the Manyara area; years of outbreaks in the eighteenth and nineteenth centuries were based on the sequence of tribal chiefs, taking an average tenure of 22 years per chief (average for this century). The dying-off of large game in 1770 and 1790 might have been coincident with a massive drought that afflicted much of the rift valley region at that time (Lamphear 1988). Independent evidence for this catastrophe for buffalo, impala and elephant at the end of the 18th century was the dating of bush encroachment of *Acacia tortilis* in Manyara to about 1780 (Prins and Van der Jeugd 1993). Some of these catastrophes may have been rinderpest (such as the 1939 and 1941 epidemics of Table 5.7), others may have been anthrax. The dying-off of elephant, as remembered for the epidemic of 1880 for example, is not likely to have been caused by rinderpest, since elephant are not susceptible to this disease even though there are reports to the contrary (D'Costa 1936 cited in Scott 1981). At the end of the last century there was an upsurge in elephant hunting for ivory for the European market, which event might have been confused with the rinderpest. Sale of ivory was frequently one of the few means to restore wealth so that cattle could be bought after the destruction by rinderpest (Sobani 1988, Johnson 1989).

That catastrophes occurred is clear, and, as Johnson (1989) observed for the Upper Nile region, a succession of catastrophes is a constant fact of life in many African ecosystems. But how these catastrophes occurred, and under what circumstances, is less clear from oral history or monthly reports. The epidemic years of the twentieth century do not show a correlation with recorded famines or disease outbreaks from other areas in eastern Africa (cf. Johnson 1989, Giblin 1990, Maddox 1990).

In 1983 I was fortunate to witness the onset and course of an anthrax outbreak in the impala population of Manyara (Prins and Weyerhaeuser 1987). In contrast to rinderpest, anthrax is an indigenous disease in Africa (Von Ostertag and Kulenkampff 1941, Pienaar 1960, 1961) and must have been acting in African ecosystems for many generations. Immunity against anthrax can be acquired if the animal survives the disease, but is lost again, in cattle at least, after 12 months (Robertson 1976). This is in contrast to immunity against rinderpest, which is maintained for life. Although rinderpest and anthrax epidemics cause different responses in the affected populations, and are very different types of disease, the ultimate effect of these diseases on population numbers is similar. Manyara offers excellent conditions for anthrax outbreaks because all the conditions necessary for these outbreaks are met: high rainfall in the wet season, ephemeral swampy areas, alkaline soils and a dusty dry season (Chapter 1; Prins and Weyerhaeuser 1987). Impala have been present in Manyara in large numbers for scores of years and probably even centuries. Iraqw oral history tells that impala were present in the area at the end of the eighteenth century (Table 5.7). The oldest written reference mentions that impala occurred in large herds (Watermeyer and Elliott 1943). The first reliable game censuses in 1959 resulted in an estimate of 1500–2000 impala (Warden's Report, October 1960). That year also provides the first record of an anthrax outbreak in Manyara; only impala and some zebra died from it. Another anthrax outbreak took place in 1977 and large numbers of impala died during this epidemic. The small anthrax outbreak of November 1991, when a number of buffalo were killed, did not noticeably affect the impala. The 1984 anthrax epidemic caused a mortality of about 92% (Prins and Weyerhaeuser 1987).

The population of impala started increasing again in December 1984. At the beginning of 1985 there were about 150 impala in the park, and in 1991 this had increased to over 1200. Thus the annual rate of increase, calculated on the basis of compound interest (Anonymous 1989), was 41%. This is about the maximum rate of increase that can be expected in impala, since the maximum growth rate (r_{max}) of a population is related to body weight (45 kg in impala) according the formula (Ndhlovu and Balakrishnan 1991) $r_{max} = 1.5 \times W^{-0.36}$, which for impala is 38.1% per year. This observed growth rate is very much higher than that of impala in Lake Nakuru National Park (Kenya), where, even in the absence of predators, the population increased from 255 impala in 1970 to 4083 in 1993 (J.K. Kairu, personal communication); this represents an annual growth rate of 'only' 12.9%. This indicates that a mammal population can recover extremely fast after a catastrophe.

There seems to be a seasonal pattern in the occurrence of disease outbreaks. Sinclair (1979) mentions that in the Serengeti rinderpest tends to take its toll at the end of the dry season. The same applies to the Manyara rinderpest epidemic of 1959, and Branagan and Hammond (1965) remarked that, indirectly, drought was a most important disseminating factor. Anthrax also occurred at the end of the dry season in the Kruger National Park (Pienaar 1961, 1967), and in the Manyara anthrax outbreaks of 1977, 1983 and 1991. In domestic stock anthrax outbreaks at the end of the dry season are also the normal pattern (Von Ostertag and Kulenkampff 1941). This has several ecological explanations (Prins and Weyerhaeuser 1987), of which loss of physical condition is most important. This is most pronounced at the end of the dry season and the beginning of the wet season (Figure 3.6). The two catastrophic diseases, rinderpest and anthrax, both show a strong link with the time of year when buffalo (or other herbivores) are in their worst condition. This has important implications for understanding the social organization of buffalo, since, as I showed Chapters 3 and 4, condition is also closely associated with reproduction. Both mortality and reproduction, two important parameters of fitness, are thus related to condition.

5.8 RISK OF DISEASE AND SOCIAL INEQUALITY

In Chapter 2 I pointed out that dietary protein is very important for both nutrient and energy acquisition. This is directly linked to a ruminant's ability to maintain or improve its physical condition (Chapter 3). There is also a relation between protein nutrition and resistance against diseases through the potential host animal's ability to form antibodies against biological disease agents. This explains to a large extent why many diseases take hold in a population when the animals' physical condition is at its lowest. Moreover, intestinal parasites negatively affect the host's ability to utilize protein effectively, as has been demonstrated for sheep and cattle by several authors (Stewart 1932 cited in Venkatachalam and Patwardhan 1953, Parkins *et al.* 1973, Roseby 1973, 1974, 1977, Sykes and Coop 1977, Van de Nieuwenhuizen 1979, Holmes 1986), and also in humans (Stephenson *et al.* 1980, Ash *et al.* 1984). Hence, an animal in poor physical condition has an increased chance of becoming infected by parasites. They consequently cause a lowered efficiency of utilization of protein through which the host's immune responses can be impaired and which lowers the host's rate of nutrient accumulation. This then results in an increased chance of continuation of the infestation and a continuation of poor physical condition. This led Sinclair (1977, p. 265 ff.) to formulate a model of population regulation through resource limitation, in which mortality is dependent on the quality of the resource and on the interaction between the host's physical condition and the effects of disease.

In natural populations most animals are infected with parasites, but normally these infections do not result in pathogenic symptoms or mortality. Basson *et al.* (1970) found infections in almost all African buffalo they examined, but only a small fraction were seriously affected. Also, Sinclair (1977) reports high frequencies of immunity to an array of diseases in the buffalo population of the Serengeti, indicating that many animals have been exposed to these diseases but survived. However, the number of diseases that can cause mortality in buffalo under natural circumstances is considerable, for example rinderpest, anthrax, Allerton-type herpes, theileriasis, foot-and-mouth disease, trypanosomiasis, and so on (see Sinclair 1977, p. 240 ff., for an extensive review; see also Mohan and Gotts 1970).

I have shown already that the Manyara buffalo experienced rinderpest and anthrax, but together with Henrich Bruggemann and Floris Deodatus I also investigated their parasite status. We concentrated on the Northern and Central Herds from April to August 1983 to find out more about individual disease risks because it was impossible to study this in relation to rinderpest or anthrax. Eggs of intestinal parasites are often found in dung, and the parameter eggs per gram (EPG) gives an indication of parasite load. We took dung samples of individually classified buffalo (sex, age, condition) but also sampled a series of dung pats of which we only knew the herd of the defecating buffalo. In these samples we determined the EPG following the procedure of Thienpont *et al.* (1970). Most samples (90%) proved to be from adult individuals. We could distinguish six groups of helminth eggs, namely those of Cestoda, Trematoda (within this category were *Fasciola, Schistosoma* and *Paraphistomum*), *Trichuris* (Nematoda), *Toxocara* (Nematoda), *Nematodirus* (Nematoda) and strongylid-type (Nematoda) eggs such as *Haemonchus*, and *Trichostrongylus* (see Soulsby 1968, Thienpont *et al.* 1970 for details). Because the incidence of eggs of *Cestoda* (3), *Trichuris* (7) and *Toxocara* (3) was very low in the total sample ($n = 59$) and the EPG values were very low (< 10), we distinguished in the statistical analyses only trematodes, strongylids and *Nematodirus*. Because egg excretion is not linearly dependent on infection intensity, and the sample size was small, we used logarithmic transformed EPG values in statistical tests, and tested two-sided.

Differences in parasite load between the herds can be shown by comparing the proportion of infected buffalo per herd (without considering condition) and the average EPG per herd. Both variables were higher in the Northern Herd than in the Central Herd for the groups of parasites, but only in the case of trematodes was this significantly higher ($P < 0.001$) for the Northern Herd than for the Central Herd. Even within the same condition class, the average trematode EPG was still significantly higher ($P < 0.035$) in the Northern Herd than in the Central Herd. This implies that, regardless of condition, either buffalo in the Northern Herd were more exposed to trematode infestation than those in the Central Herd or that the level of trematode infection had little effect on the condition of the buffalo. It is highly unlikely that the identified trematodes did not cause condition deterioration in buffalo as *Schistosoma* is considered very detrimental and *Paraphistomum* mildly so (Soulsby 1968, Basson *et al.* 1970). The most likely explanation for a higher infection rate in buffalo of the Northern Herd was that they spent much time in the *Typha* swamp, where water snails, the intermediary hosts of trematodes, occur (cf. Basson *et al.* 1970). The Central Herd buffalo, on the other hand, hardly ever ventured into these swamps. Most of the rivers of Manyara were sampled for host snails by Beesley (1965) but, except for Bagayo River, without results.

The relation between condition and parasite load could be investigated by calculating the proportion of infected buffalo and the average EPG per condition category (Table 5.8).

Logarithmic transformed EPG values of animals in a very poor and poor condition (combined) was significantly higher than those of animals in a good and excellent condition (combined) for strongylids ($P < 0.02$), *Nematodirus* ($P < 0.05$), and for trematodes ($P < 0.02$), and we concluded

Table 5.8 Percentage of infected buffalo per condition category and average logarithmic transformed parasite load (eggs per gram faeces) for each condition category (standard deviations in parentheses)

	Condition category				
Parasites	*1 (n=1) (very poor)*	*2 (n=5)*	*3 (n=20)*	*4 (n=9)*	*5 (n=1) (excellent)*
Percentage of buffalo infected					
Strongylidae	100	80	90	67	0
Nematodirus	100	80	35	44	0
Trematoda	100	100	58	67	0
Transformed EPG values					
Strongylidae	3.2	1.7 (0.9)	1.9 (1.5)	0.8 (0.7)	0
Nematodirus	0.15	0.32 (0.27)	0.20 (0.32)	0.10 (0.13)	0
Trematoda	4.9	2.2 (1.9)	1.0 (1.3)	0.8 (1.0)	0

that buffalo in better condition had fewer parasites than others. Intestinal parasites can impair the utilization of dietary protein by the host, and our results on the wild buffalo of Manyara are in line with the reported results on cattle and sheep.

The crude protein concentration of faecal samples was not dependent on sex (Northern Herd: $t_8 = 0.133$, NS; Central Herd: $t_{25} = 0.793$, NS) so it was not necessary to distinguish between the dung of bulls and cows. The faecal crude protein concentration of the two herds was different. In the post-rain season the faecal protein for the Northern Herd was 10.0% (SD 3.4, n = 10), which significantly differs from that for the Central Herd (7.5%, SD 1.2, n = 27) ($t_{35} = 3.364$, $P < 0.01$). We concluded that a difference in faecal crude protein concentration between buffalo was a reflection of a difference in nitrogen retention and not a difference in dietary crude protein, since their diets were the same (Scarce 1986).

The striking differences in condition between the two herds (Table 5.9) thus demands explanation.

Table 5.9 Condition changes of the Northern and Central Herds

	Condition					
	1	*2*	*3*	*4*	*5*	n
Northern herd						
16 May–15 June	0	7	51	40	2	733
16 June–15 Aug	2	30	55	13	0	1537
Central herd						
16 May–15 June	0	10	55	34	1	2956
16 June–15 Aug	0	12	59	28	0	829

n, number of buffalo.

It is likely that the difference in faecal crude protein concentration between the two herds can be ascribed to an interaction between the condition of the buffalo and faecal crude protein concentration through the negative effect of parasites on protein utilization. Indeed, buffalo in a very poor and poor condition (combined) had a faecal crude protein concentration of 10.4% (SD 4.0, n = 6), buffalo in a moderate condition had a faecal crude protein concentration of 8.0% (SD 1.8, n = 20) and buffalo in a good and excellent condition (combined) had a faecal crude protein concentration of 7.4% (SD 1.2, n = 11) ($F_{2,35} = 4.271$, $P < 0.01$; Scheffé multiple comparison at $P < 0.05$: poor ≠ moderate = good). Thus, buffalo in a poor condition had a higher parasite load, and were

Plate 1 Lake Manyara National Park lies in the Great African Rift Valley of northern Tanzania. The lake itself is about 960 metres above sea level; the Marang Plateau (background to the right) rises about 500 metres above the lake.

Plate 2 The lowlands of Manyara are characterized by grasslands, where herbivores such as wildebeest (here in the Endabash area) abound, and by gallery forest (background).

Plate 3 At the southern end of Manyara the Endabash River has formed a large alluvial fan in the Rift Valley. Open grasslands line the river, which in the park flows only during the rainy season. Further away dense woodlands occur. This landscape is ideal for buffalo.

Plate 4 Manyara has a number of permanent fresh water rivers, such as the Simba River, where hippopotamuses rest during the daytime. In the background is Mount Essimingore, an extinct volcano east of Lake Manyara.

Plate 5 Although the lake is highly alkaline and saline the number of pelicans is very high in years when fish abound after a good rainy season. In the background are buffalo of the Northern Herd and beyond these are some of the approximately one million lesser flamingos on the lake.

Plate 6 Fresh water is of extreme importance to buffalo since they have to drink nearly every day.

Plate 7 During the dry season fresh water can get very scarce in the central part of the park. Animals like these elephant then drink from springs at the foot of the escarpment.

Plate 8 Buffalo form dense herds and graze in very close proximity on the dense swards of *Cynodon dactylon*.

Plate 9 The largest buffalo herd of Manyara, the Central Herd, numbered about 1000 individuals. During the daytime they often could be found on the alkaline grasslands along the lake.

Plate 10 Identification of individual buffalo is possible as many adults, like this adult bull, have distinctive scar patterns on their ears. Other scars and horns are also used for identification.

Plate 11 Buffalo cows do not have bosses as the adult bulls do and the bases of their horns are covered by skin with hair on it.

Plate 12 The Manyara buffalo spend much of their time in woodland where, during the wet season, they make use of the numerous wallows.

Plate 13 Most buffalo in Manyara die due to predation by lions. Most lion attacks start from the ecotone between woodland and grassland.

Plate 14 Zebra are the preferred prey of the Manyara lions although most of the food they eat is from buffalo they have killed.

Plate 15 Facilitation of the sward by grazing buffalo appears to be of minor importance for the Manyara zebra although it is possible that they prefer to graze closely with buffalo as an anti-predator adaptation.

Plate 16 A typical communal decision procedure of the Manyara buffalo before the night falls. Most buffalo are still resting and a few are standing. The cow to the right shows a typical 'voting stance'.

therefore less able to utilize protein than buffalo in a better condition. Note that this analysis was conducted within a short period of time and differences are small. The pattern of diet quality over the year and condition changes (Chapters 2 and 3) are not affected by these conclusions. Most disease–host systems are characterized by an intermediate level of virulence of the disease and a moderate level of resistance of the host against this disease, with the result that the quality of the diet, and especially the level of protein in the diet, often tips the balance towards infection (Scrimshaw *et al.* 1968, Chandra 1976, 1977, 1979, Neumann 1977, Ash *et al.* 1984). The Manyara results show that buffalo in a poor physical condition had a high parasite load, which resulted in impaired protein retention and consequent further deterioration of body condition. Thus Sinclair's (1977, p. 265 ff.) model of (parasitic) diseases as amplifiers of the effects of undernutrition on mortality appears to be a good one, although I think too little emphasis is played on the interaction between social status, food acquisition and physical condition (Chapter 3).

An equivocal relationship between social status and parasite load in animals cannot be substantiated, as two contrasting mechanisms appear to play a role. First, there is generally a positive relation between social standing and intake of preferred food (Banks *et al.* 1979, Appleby 1980, Kikkawa 1980a, b, Baker *et al.* 1981, Ens and Goss-Custard 1984, Teunissen *et al.* 1985). If these preferred resources are associated with higher concentrations of intermediary hosts or parasites, then high status leads to increased risk of parasite infection. This explains the positive relation between status and parasite load in yellow baboon males and females (Hausfater and Watson 1976) or in reindeer calves (Halvorsen 1986). If, on the other hand, a good condition is related to high status, and if this good condition confers an increased protection against infection, then a negative correlation between social standing and chance of parasite infection can be expected. Such a negative correlation was found in bowerbirds (Borgia 1986). This also appears to be the case in the Manyara buffalo. In buffalo the negative relation between parasite load and social standing is enforced through the relation between social standing, the place in the herd and food intake (Chapter 3). Buffalo in the rear of a herd have a higher chance of being infected by nematodes as they are forced to graze closer to old faecal pats than animals in the front of these herds; grass close to these pats has a higher density of nematodes than grass further from these pats (unpublished results). This implies that social inequality results in a number of individual buffalo to having poorer condition, a higher chance of exposure to parasites, a lower protein retention and, most likely of all, a higher chance of reduced health and future condition. This will undoubtedly affect longevity and total reproductive output. It also strongly suggests that, if a major disease takes hold in a population, animals with a low social position will have the highest chance of succumbing to the disease because they have the worst condition. Very little is yet known about the fate of individuals during an epidemic, but I believe that this will be an exciting field of study. At the level of populations more is known already, so I will now turn to population dynamics.

5.9 THE MANYARA CENSUSES

Numerous census counts of large herbivores in East African ecosystems have been conducted, but repeated censuses over a long period are available in only a few areas. Yet the study of population dynamics requires repeated censuses over a long period (for example, Sinclair *et al.* 1985, Borner *et al.* 1987). However, in areas where repeated censuses have been carried out, the technique employed has nearly always involved sample counts with associated large confidence limits.

The data on Manyara, however, are ideal for the study of population fluctuations as most figures are 'total' counts. Manyara is unique in East Africa as the density of large herbivores is very high and poaching was nearly absent between 1959 and 1986. However, between 1987 and 1991 poaching became rampant with the black rhinoceros population being totally wiped out and elephants reduced by some 90% (Prins *et al.* 1994). Before this, Manyara had the highest mammalian biomass density in any natural ecosystem, similar to the system of Queen Elizabeth National Park, Uganda (which, like Manyara, was dominated by elephant and buffalo; Delany and Happold 1979) before severe poaching reduced mammal numbers drastically there in the 1970s.

I obtained estimates from a variety of sources (see Table 5.10 footnotes) of the number of large herbivores (that is heavier than 20 kg) within the present park boundaries between 1959 and 1984 (see for details Prins and Douglas-Hamilton 1990).

I used mean-square successive difference testing to detect whether or not a time series of census results of different herbivore species was randomly distributed around the mean (Wonnacott and Wonnacott 1977). I then used the constancy parameter s, defined as the standard deviation of the base 10 logarithm of population estimates, to assess the 'stability' of the population; when s is smaller than approximately 0.3 the population is considered stable (Ostfeld 1988). It is, however, better to speak of constancy (that is lack of change in a numerical parameter) than of stability, as the latter implies that the population returns to a certain state after disturbance (Orians 1975). Most species in Manyara showed a high level of constancy in their numbers, but most species also showed non-random fluctuations (Table 5.11).

Fluctuations in the herbivore numbers can be caused by a change in

Table 5.10 Numbers of large herbivores in Manyara between 1959 and 1984, and park size

	1959	*1961*	*1965*	*1967*	*1970*	*1975*	*1976*	*1981*	*1984*
Elephant	380 [a]	421	421	300	427	386	453	483	373
Buffalo	1750 [c]	1507	1507	1500	1548	1823	2032	1800	2097
Hippopotamus	150 [b]	150 [b]	150 [b]	150 [b]	150 [b]	151	118	150	179
Burchell's zebra	180 [c]	115	115	100	75	156	157 [d]	162	255
Impala	600 [c]	451	285	380	451	985	1071 [d]	1500 [e]	150 [e]
Black rhinocerous	30 [c]	23 [f]	20 [f]	20 [f]	23 [f]	35 [f]	20 [f]	20 [f]	25 [f]
Giraffe	60 [a]	60	31	50	60	60	60	72	66
Wildebeest	250 [c]	250 [a]	7 [a]	2 [a]	0	80 [a]	80	169	675
Waterbuck	20 [a]	20 [a]	20	20 [a]	20	20 [a]	20 [d]	16	17
Bushbuck	20 [a]	20 [a]	20	20 [d]	20	50 [b]	50 [b]	50 [b]	50 [g]
Reedbuck	220 [a]	220 [a]	220	220 [a]	20	20 [a]	20 [a]	20 [d]	10 [g]
Baboon	550 [c]	450 [a]	400	400 [a]	400	450 [a]	450	500 [g]	500 [g]
Warthog	0 [c]	2 [c]	6 [c]	6 [a]	35 [c]	50 [d]	50 [c]	65	95
Size of park (km^2)	100	100	86	86	84	98	100	100	100

Sources: 1961, Morgan-Davies cited in Watson and Turner (1965); 1965, Watson and Turner (1965); 1967, Vesey-FitzGerald (1969); 1970, Douglas-Hamilton (1972); 1975, Mwalyosi (1977); 1976, Douglas-Hamilton unpublished; 1981, Douglas-Hamilton unpublished; 1984, Douglas-Hamilton unpublished.
Owing to lake level changes, the real size of the area which can be used by herbivores fluctuates. The park size figures refer to the area covered by vegetation, and are based on aerial photographs, Warden's Reports and information from park rangers.

[a] Unpublished estimates based on counts of other years and inferences from Warden's Reports.
[b] Unpublished estimates; the published counts are certainly too low (oral information from park rangers) (1961, 10; 1965, 10; 1967, 20; 1970, 10).
[c] Unpublished estimates mentioned in Warden's Reports.
[d] Interpolation.
[e] Estimates based on road sampling (Prins and Weyerhaeuser 1987).
[f] These data are published rough estimates and are very uncertain.
[g] Estimates on personal information from park rangers and personal observations.

Table 5.11 Fluctuations in population numbers and constancy of the populations over time for the large herbivores

Population fluctuation	*Random (i.e. no tracking)*	*Not random (i.e. tracking)*
Population levels with high degree of constancy: $s<0.3$	Elephant ($z_0=1.029$, NS; $s=0.06$) Hippopotamus ($z_0=-1.415$, NS; $s=0.05$) Rhinoceros ($z_0=0.197$, NS; $s=0.09$)	Buffalo ($z_0=-4.222$, $P<0.0001$; $s=0.06$) Zebra ($z_0=-3.745$, $P<0.0001$; $s=0.16$) Giraffe ($z_0=-2.101$, $P<0.05$; $s=0.11$) Waterbuck ($z_0=-3.846$, $P<0.0001$; $s=0.04$) Bushbuck ($z_0=22.562$, $P<0.0001$; $s=0.21$) Baboon ($z_0=-4.125$, $P<0.0001$; $s=0.05$)
Population levels with low degree of constancy: $s>0.3$	Impala ($z_0=-1.617$, NS; $s=0.31$)	Wildebeest ($z_0=5.344$, $P<0.0001$; $s=1.01$) Reedbuck ($z_0=-5.936$, $P<0.0001$; $s=0.59$) Warthog ($z_0=-5.962$, $P<0.0001$; $s=0.72$)

The largest herbivores appear to be constant and with random fluctuations; giraffe is a borderline case.

the carrying capacity for the species involved over time as result of fluctuations in rainfall, as was demonstrated for the nearby Serengeti (Sinclair 1979). Rainfall fluctuations are large in Manyara, with 1960–61 as the driest meteorological year on record (only 214 mm) and 1967–68 as the wettest (1275 mm) (Chapter 1). However, no correlation could be demonstrated between herbivore consumption and rainfall. Correlations between herbivore consumption (section 5.10) and annual rainfall of the census year, between consumption and the cumulative rainfall of the census year and the preceding year, and between consumption and the cumulative rainfall of the census year and the two preceding years, all resulted in slopes not significantly deviating from zero (*t*-tests, two-sided; see Prins and Douglas-Hamilton 1990). In the Serengeti it appears that dry season rainfall is particularly important in explaining the increase in herbivore carrying capacity (Sinclair 1979, p. 93 ff.; Sinclair *et al.* 1985). However, in Manyara it appears that if anything the opposite happened, as dry season rainfall decreased slightly from 1959 through 1984, although this is not significant (Prins and Douglas-Hamilton 1990). In association with dry season rainfall there is a significant negative correlation with total herbivore pressure, which suggests that Manyara functions as a dry season drought refuge for a number of herbivore species.

5.10 FUNCTIONAL STABILITY

I calculated the herbivores' daily food intake (dry matter consumption) to be 2.5% of their live body weight (Chapter 2), which agrees well with the published data for the food intake of East African herbivores (Delany and Happold 1979). Some of the herbivores are considered real grazers, and others real browsers, but many species are 'mixed feeders' (Hofmann 1973). As I was interested in quantifying the consumption of grass and browse in Manyara, the consumption by 'mixed feeders' had to be divided into 'grass consumption' and 'browse consumption'. In the analysis I distinguished five groups of herbivores ('functional groups'), namely buffalo, 'elephant-as-grazer', 'elephant-as-browser', 'other grazers' (that is hippopotamus, Burchell's zebra, wildebeest, impala, reedbuck, waterbuck, warthog, baboon and bushbuck) and 'other browsers' (that is giraffe, black rhinoceros, baboon, impala, bushbuck, waterbuck, reedbuck and warthog); the categories are weighted according to body weight and the proportions grass or browse in the diet (Table 1.7). The calculated estimates of grazing and browsing pressures (Table 5.12) are gross indices of consumption per unit land, corrected for variation in dry land area due to changes in the level of Lake Manyara (Chapter 1) as calculated from aerial photographs (Table 5.10).

Table 5.12 Grazing and browsing pressure (kg dry matter ha^{-1} $year^{-1}$) by large herbivores over time

	1959	*1961*	*1965*	*1967*	*1970*	*1975*	*1976*	*1981*	*1984*
Grazers									
Buffalo	519	447	520	517	547	552	603	534	622
Elephant	413	457	532	379	552	428	492	524	405
Hippopotamus	137	137	159	159	163	135	108	137	163
Zebra	41	26	31	27	20	36	36	37	58
Wildebeest	51	51	2	0	0	17	16	35	139
Impala	26	20	14	19	23	44	46	65	7
Reedbuck	8	8	9	5	1	1	1	0	0
Waterbuck	3	3	3	3	3	3	3	2	2
Warthog	0	0	0	0	2	2	2	3	4
Baboon	2	2	2	2	2	2	2	2	2
Bushbuck	0	0	0	0	0	0	0	0	0
Subtotal grazers	1205	1150	1274	1117	1318	1230	1318	1347	1405
Browsers									
Elephant	177	195	228	162	237	183	211	225	174
Giraffe	42	42	25	41	50	43	42	50	46
Rhino	27	21	21	21	25	33	18	18	23
Baboon	7	6	6	6	6	6	6	6	6
Impala	1	1	1	1	1	2	2	3	0
Bushbuck	1	1	1	1	1	3	2	2	2
Waterbuck	1	1	1	1	1	1	1	1	1
Reedbuck	0	0	0	0	0	0	0	0	0
Warthog	0	0	0	0	0	0	0	0	0
Subtotal browsers	259	292	285	237	328	274	284	310	256
Total	1464	1459	1559	1354	1646	1504	1602	1657	1661

Consumption is calculated after correcting for lake level fluctuations (Table 5.10), under the assumption that the food intake per day equals 2.5% of the body weight.

The results clearly show that the Manyara system is dominated by buffalo and elephant, as these account for 35 and 43%, respectively, of the total consumption. The 11 other large herbivores together consume on average only 22%. Most of the food consumed in the system consists of grass (82%). Browse consumption is much less important.

The regressions of consumption over time of four of the five functional groups do not significantly deviate from a zero slope between 1959 and 1984 (t-tests, two-sided, $P > 0.1$; Figure 5.3).

Total consumption, however, increased over the study period (t-test, two-sided, $P < 0.025$). This was mainly caused by an increase in buffalo grazing pressure associated with a significant increase in the absolute number of buffalo in the park since 1959 (Figure 5.4A).

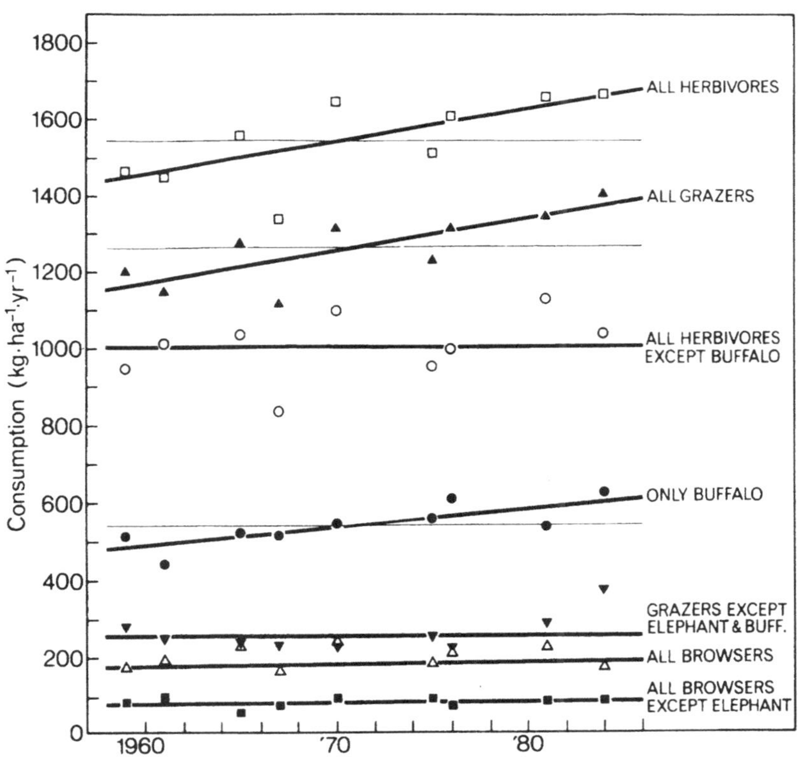

Figure 5.3 Consumption (kg dry matter) per hectare over time of different functional groups of herbivores in Manyara. Mean consumption pressure (horizontal lines) is given too. See the text for herbivore species which form the different functional groups.

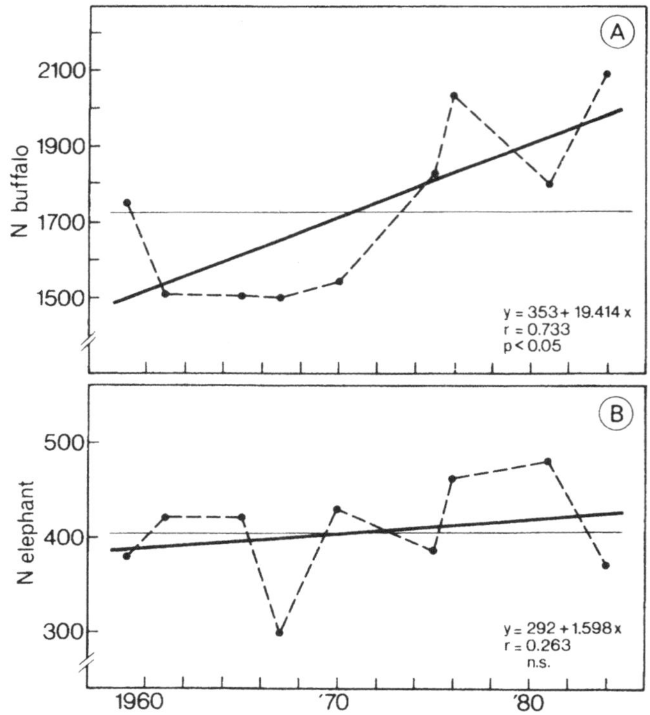

Figure 5.4 The number of buffalo (A) and elephant (B) in Manyara over time.

This increase was undoubtedly a result of the absence of rinderpest since 1959. Note that, apart from buffalo, warthog also increased in numbers; warthog also is extremely sensitive to rinderpest (Table 5.6). The other herbivore species remained stable over the study period (for example elephant; Figure 5.4B), although their numbers show strong fluctuations (Table 5.10) or even decreased significantly (reedbuck, $r = -0.890$, $n = 9$, $P < 0.01$). This reduction was caused by the disappearance of swamp vegetation as a consequence of very high lake levels around 1967. The elephant fluctuations cannot be caused solely by mortality and natality as the fluctuations are too large but must have arisen from extensive interchange with the surrounding Masai Ecosystem (Prins 1987; compare Croze *et al.* 1981). This explains why the constant elephant population does not show tracking in its population numbers – fluctuations are random (Table 5.11); in contrast buffalo hardly leave the area

(Chapter 3). On the basis of the present census results I conclude that consumption (and thus pressure on the vegetation) of the different groups of herbivores fluctuates around a more or less stable equilibrium except for consumption by buffalo. The fluctuations of the different grazers was larger than that of the grazers in totality, and the same was true for the browsers (Table 5.13), which in turn were both larger than those of all herbivores (grazers and browsers) combined.

Table 5.13 The herbivore species assemblage exerts a more stable consumption pressure on the vegetation than the individual species, since the coefficients of variation of their combined consumption is much less than the coefficients of variation of the individual species

	Number of individuals		*Consumption = pressure on vegetation (kg ha^{-1} year^{-1})*	
	Mean	*Coefficient of variation*	*Mean*	*Coefficient of variation*
Grazers				
Buffalo	1729	13.3	540.1	9.5
Elephant	405	13.1	464.7	13.5
Hippopotamus	150	10.2	144.8	12.6
Zebra	146	36.4	34.7	31.7
Wildebeest	168	127.5	34.6	127.1
Impala	653	67.4	29.3	62.6
Reedbuck	96	103.9	3.5	103.0
Waterbuck	19	8.1	3.0	13.8
Warthog	34	97.4	1.5	93.5
Baboon	456	11.6	1.7	7.1
Bushbuck	33	47.4	0.2	24.8
Subtotal grazers			1262.7	7.5
Browsers				
Elephant	405	13.1	199.1	13.5
Giraffe	58	20.1	42.3	17.4
Rhinoceros	24	22.0	23.1	20.7
Baboon	456	11.6	6.1	5.9
Impala	653	67.4	1.6	62.3
Bushbuck	33	47.4	1.7	44.7
Waterbuck	19	8.1	0.8	11.7
Reedbuck	96	103.9	0.2	108.1
Warthog	34	97.4	0.1	104.5
Subtotal browsers			280.5	10.0
Total herbivores			1545.1	6.9

The most important result of the censuses is that the numbers of the different herbivores, and the related consumption figures, show large fluctuations but that the different species compensate for each other's effects on the vegetation. Hence, this community may be functionally stable since total plant biomass consumption does not change with changes in herbivore community structure (Rejmánek 1992).

5.11 COMPENSATION

The main conclusion is that total food consumption within the Manyara system has varied much less through the years than the species composition of the herbivore assemblage. The total amount of food on offer will have depended largely on the area under grass (only 18% of the total consumption was browse), and variations in the extent of Lake Manyara are thus seen as the main source of variation in animal numbers. This is clearly illustrated by the local extinction of wildebeest when the lake level was very high at the end of the 1960s when this population was trapped between the lake, the escarpment and the groundwater forest; the population was exterminated by lions (Warden's Reports 1959–73). Note that in Table 5.11 it is shown that wildebeest had a low degree of constancy but that the fluctuations were not random. Likewise, the reedbuck population nearly became extinct because the high alkalinity of the water caused the *Typha* swamps, the favoured habitat, to disappear when the lake flooded these swamps. Other causes of population fluctuations were trapping of baboons for export purposes at the end of the 1950s (Warden's Reports 1959–73), the death of large numbers of impala, the death of elephant from pneumonia and the death of many buffalo in the rinderpest epidemic of 1959. These effects on baboon, elephant or buffalo populations were not so large that their constancy parameter was affected, but in the case of buffalo and baboon the recovery of the population caused tracking in the data set (Table 5.11). It is striking that, independent of these large fluctuations in the population size of the different species (see the large coefficients of variation in Table 5.13), total consumption by browsers or grazers showed much less fluctuation and total consumption by all herbivores combined even less, suggesting that the different herbivores were compensating for each others fluctuations. The reduction of the variance is not a mathematical artefact but a consequence of a system composed of different independently fluctuating elements or different compensating elements. The suggestion that the different animal populations compensate for each other's fluctuations is borne out by the results presented in Table 5.14.

Does this also mean that the maximum carrying capacity of the system had been reached and that carrying capacity has to be viewed at the

Table 5.14 Correlations between consumption of different functional groups of herbivores in the period 1959 to 1984

Grass consumption by buffalo (%)		=	52.9 – 0.597 × grass consumption by elephant (%)
	R^2	=	45.2%, t_7=2.401, $P<0.05$
Grass consumption by 'other grazers' (%)		=	40.5 – 0.795 × grass consumption by elephant (%)
	R^2	=	70.2%, t_7=4.063, $P<0.01$
Grass consumption by 'other grazers' (%)		=	8.3 + 0.240 × grass consumption by buffalo (%)
	R^2	=	5.1%, t_7=0.611 NS
Browse consumption by 'other browsers' (%)		=	7.4 – 0.061 × grass consumption by buffalo (%)
	R^2	=	3.8%, t_7=0.526 NS
Browse consumption by 'other browsers' (%)		=	4.5 + 0.050 × grass consumption by 'other grazers' (%)
	R^2	=	2.9%, t_7=0.459 NS
Browse consumption by 'other browsers' (%)		=	7.3 – 0.157 × browse consumption by elephant (%)
	R^2	=	5.9%, t_7=0.660 NS

Consumption is expressed as the percentage of total consumption by large herbivores in the years of census to remove the effects of fluctuations in total consumption. There appears to be a competitive interaction between elephant and buffalo and between elephant and the other grazers.

level of the herbivore assemblage in total instead of at the level of the different component species? The interpretation favoured by me is that the Manyara system was 'full' and that the carrying capacity for the herbivore species assemblage in total was reached **and** maintained at a sustainable level. An independent indication that the maximum carrying capacity for all herbivores combined has been reached is the finding that approximately 85% of the annual grass production in Manyara is consumed by these large mammals (Chapters 1 and 2). Competitive interaction appears to play a role in explaining the fluctuations of the component species. There were strong negative correlations between grass consumption by elephant and that by all other grazers ($r = -0.84$, $P < 0.01$) and between elephant and buffalo alone ($r = -0.67$, $P < 0.02$), and a non-significant negative relation between elephant browse consumption and that by all other browsers combined ($r = -0.24$, NS) (Table 5.14). The increase in buffalo over time, however, cannot be explained by a decrease in elephant over time because elephant numbers did not change significantly (Figure 5.4B). Past outbreaks of rinderpest are the most likely explanation of changes in buffalo numbers. Buffalo recovery was slow, I think, because of competitive interaction in a system that remained full as far as the other herbivores were concerned. From a scientific point an interesting series of events then started in 1987, namely the excessive poaching of elephants: the major competitor was nearly removed, and one would expect a clear effect of this on buffalo population numbers. Yet before I show this effect, I will first discuss recruitment.

5.12 CONDITION AND REPRODUCTION

I explained in Chapter 2 that the best feeding opportunities for buffalo start during the short rains (around mid-November) and that the quality of the diet of the buffalo remains high until June when the dry season begins. During the dry season, food quality is generally still above maintenance requirements but is no longer good enough for fattening or milk production. This seasonal pattern (Figure 2.2) is reflected in the pattern of condition loss and gain over the year (Figure 3.6). The buffalo cows' condition is at its nadir in December and at its zenith in May. Grimsdell (1969) also reported peak condition for buffalo cows at the end of the rainy season/beginning of the dry season and lowest condition at the end of the dry season. He found no significant variation in the condition index for buffalo bulls, and likewise in the pooled data for the Manyara bulls (combining both bachelor bulls and herd bulls; Chapter 4) there is no profound variation (Figure 5.5). Fryxell (1987) also observed a greater decline of body condition in females than in males of the white-eared kob in Sudan during the dry season. Grimsdell (1969) suggests that, in relation to nutritional conditions, seasonal changes occur in buffalo cows as regards ovulation rate, intensity of oestrus and success of conception (see also Field and Blankenship 1973, Spinage 1973, Tulloch 1979, Lochmiller *et al.* 1986). The results of my study make this very likely (Figure 5.5): the peak of the conceptions occurs when the cows' condition improves.

For 133 calves, I determined the date of birth within a margin of approximately 14 days. Births were unevenly distributed over the year ($\chi^2 = 40.7$, d.f. = 11, $P < 0.001$) and, especially in the late dry season, significantly fewer births than the monthly average were observed (Figure 5.6), while during the long rains there was a strong tendency for more births than the monthly average.

In this respect, buffalo in which conception is favoured by good nutritional conditions show good timing because, by having a gestation period

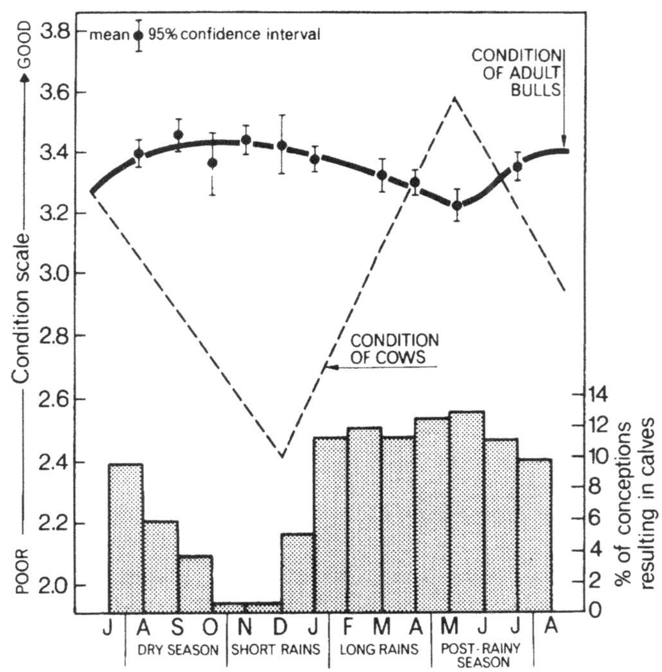

Figure 5.5 Condition of adult cows (from Figure 3.6) and adult bulls in relation to season, and the conception rate (based on calving dates minus 340 days to account for average pregnancy duration; Figure 3.6). At peak conception buffalo bulls have their lowest average condition.

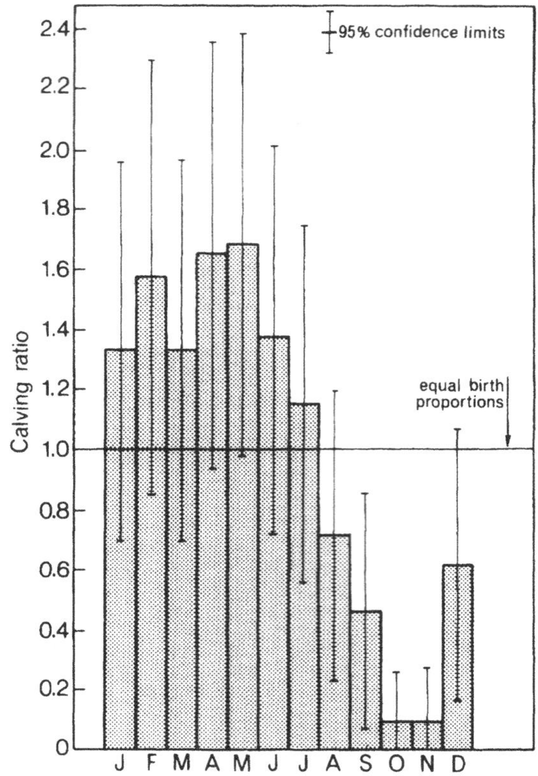

Figure 5.6 The ratio of calf births over the year as related to a hypothetical equal monthly distribution of births (horizontal line). The calving peak in April and May is nearly significant, and during the dry season (September to November) only a few calves are born.

of nearly 1 year (340 days), they also show calving at the best time of the year.

I could determine the calving interval of only six individual adult cows. The average interval was 879 days and the mode 558 days, which is much longer than the gestation interval of about 340 days (Sinclair 1977, p. 166). I had additional information on 42 individually known cows in which I am certain that the calving interval was longer than the gestation period: for these cows the calving interval exceeded the modal value between first sighting without calf and parturition (mode 666 days, mean = 644.5 days, 95% confidence limits 52.6 days, n = 42). These data strongly suggest a calving interval of at least 2 years in Manyara.

Grimsdell (1969) proposed that the reproductive rate in buffalo is positively related to their own population density and to that of other large mammals, that is until very high densities are reached, because a high grazing pressure appears to enhance the availability of good-quality food. However, the Manyara data refute this hypothesis. In Aswa (Uganda), an area with a low buffalo density, the calving interval is 25 months (Grimsdell 1969); both in the Serengeti and in Kruger National Park, with reasonably high buffalo densities, it is 15 months (Sinclair 1977, p. 169); and in Queen Elizabeth National Park (Uganda), an area with a high density, it is 18 months (Grimsdell 1969). However, in Manyara, with 18–20 buffalo km^{-2}, which is a very high density (and also 6 elephant km^{-2}), the calving interval is at least 24 months. Nevertheless, the quality of the diet of Manyara buffalo is better than that of the Serengeti buffalo and there is only little indication of intake limitation (Chapter 2). Hence, Grimsdell's proposition has to be discarded as an explanation for the low calving rate of the Manyara buffalo, and instead social factors and herd structure appear to play a role (section 5.13).

The data on the calving interval of the Manyara buffalo resulted in an estimated pregnancy rate of 0.523, and a production of 0.56 calves per female per year (see Sinclair 1977, p. 168, for calculations). However, given the average number of calves in the herd (6–12%), and considering the early calf mortality of 41.6% ($n = 24$), even this estimate of pregnancy rate is too high, and a more realistic calving interval would be 3 years for most females that live in the large herds. As an average cow has her first calf when she is approximately 6 years and dies when she is 11.1 years (section 5.2), an average cow in Manyara can be expected to produce only two calves during her lifetime. Based on the formula for the calculation of compound interest, annual calf mortality was 42.9% in 1983 and 55.3% in 1991 (which was a drought year). The rather high calf mortality combined with the low fecundity results in a very low population growth of only about 2% per year between 1959 and 1984 (Figure 5.4). However, it may be misleading to speak of an average cow because of the effect of social inequality. This became very evident after the elephant poaching.

5.13 HERD SIZE AND GROWTH RATE: THE EFFECTS OF POACHING

The interpretation of the long-term observations of population fluctuations of different herbivore species suggests that the carrying capacity of the Manyara system can only be viewed at the level of all the herbivores combined and not at that of the individual species, as in this system independent factors (such as epidemics and rainfall events) cause population numbers of the separate species to fluctuate (that is regulation by extrinsic factors; Caughley and Krebs 1983, Borner *et al.* 1987, *pace* Sinclair *et al.*

1985). However, because the different species are affected differently, their numbers can expand when competition is reduced, such as when other species are affected negatively. This mechanism appears to cause overall stability of the system under natural conditions.

In 1987 elephant poaching began, which resulted in a decrease in the elephant population by some 90% (Prins *et al.* 1994). Elephant is the major competitor of buffalo (Chapter 6) and, at least on a very long time scale, a negative interaction between elephant and buffalo is made likely by the observation that African buffalo increased in numbers in the fossil record after the extinction of *Elephas recki* in southern Africa (Klein 1988); bison also expanded their range or increased in numbers with the extinction of *Mamuthus* (Guthrie 1990). If there is a negative interaction between elephant and buffalo numbers then one would expect the buffalo to have increased after 1987. However, total buffalo numbers in Manyara remained nearly constant between 1985 and 1991. Two factors can explain this **apparent** lack of response by the buffalo population. First, there has been poaching of buffalo in some herds within Manyara (Chapter 3), and, secondly, the largest herd of Manyara even had a zero growth rate. However, closer inspection of the data reveals a clear density-dependent relation between herd growth rate and herd size for the period 1975–85 (Figure 5.7), which implies that density dependency is negative (and not positive as Grimsdell (1969) proposed) when it comes to individual herds.

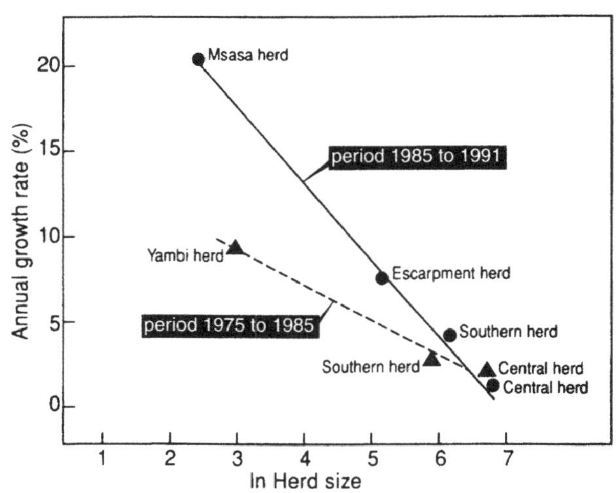

Figure 5.7 Herd growth rate is negatively related to herd size, which indicates density-dependent population regulation. The two lines with different slopes show that competition with elephant plays a major role too, because when the elephant population of Manyara was seriously reduced due to poaching (upper line) herd growth rate increased. This effect is especially noticeable in the small herds.

This is confirmed by the observation that the average calving interval for the very large Central Herd is 2–3 years (see above) while for the very small Msasa Herd, which lives within the home range of the Central Herd, is only 12–14 months. When poaching started and the elephant population was severely reduced, the annual growth rate of the different herds increased: this effect was most pronounced in the small herds, in which social inequality and intraspecific competition did not mask the interspecific effects.

So, for the greater part of the last decades the buffalo of Manyara have been living in a stable assemblage of large herbivores. Yet this overall stability was only possible because of their competitive ability compared with other major consumers in the Manyara ecosystem. In the next chapter their competitive interactions with the other herbivores is discussed.

6

Competition for food

6.1 THE FOUR MAJOR GRAZERS

In the previous chapter I presented evidence that the Manyara ecosystem is so full of herbivores that it seems that herbivores of any one species can only increase in numbers if individuals of another species decrease in numbers. This strongly suggests that there is competition for food between the different herbivore species. The main grass consumers in the Manyara system are buffalo, elephant, wildebeest and zebra. The data on the feeding behaviour of buffalo were provided in Chapter 2. In this chapter I first explore first the feeding behaviour of the buffalo's competitors and I then present evidence for food competition.

The major competitors of the buffalo are elephant, zebra and wildebeest, which are all more or less sedentary in Manyara. They utilize the same resources to a large extent, face the same environmental changes, but differ in their feeding strategies. Buffalo are ruminants which are dependent on water and which are classified as roughage grazers (Chapter 2). Wildebeest are also ruminants, but they are classified as water-dependent, selective grazers of fresh grass (Hofmann 1973, p. 195 ff., Hofmann and Stewart 1972, Van Soest 1982, Van Wijngaarden 1985). The sedentary population in Manyara is small (100–150 individuals) and subject to fluctuations (Chapter 5). In the dry season, a varying number of migrant wildebeest (up to 10 000) are attracted to the northern section of the park, where water and food are abundant. This appears to have been the case for at least the last 70 years (Arbuthnot 1954, Lamprey 1963, Warden's Reports 1959–66) but probably much longer. The Burchell's zebra is a hindgut fermenter which is classified as a water-dependent, non-selective roughage grazer (Van Soest 1982, Van Wijngaarden 1985). Manyara has a semisedentary population of approximately 200 animals (Chapter 5), but in the dry season some few hundred migrant animals may also be found in the area. Manyara zebra once moved from the park to the Mbulu plateau and back (Warden's Reports 1959–73), but this has come to an end over the last 10 years. The African elephant is also a hindgut fermenter, and is classified as a water-dependent, intermediate bulk feeder that is not very selective, preferring grass to browse (Van Soest 1982, Van Wijngaarden 1985). Between 1958 and 1988 the Manyara population was fairly stable between 400 and 500 individuals (Chapter 5) but has greatly reduced since then (Prins *et al.* 1994).

We collected data on the activity of all four species, as explained in section 2.3. In total, about 8000 road counts were done, which resulted in more than 10 000 group observations. In total, 480 000 buffalo individuals were scanned (that is group sizes times number of group observations), plus 160 000 wildebeest, 22 000 zebra and 10 000 elephant. Data on wildebeest and zebra were collected from May to September 1983 and from February to January 1985, but for elephant only in the latter period. Thus, for all four species data were collected during the most important periods in the yearly cycle (Figure 1.3), that is the short rains season (November to February), the long rains season (February to May), the post-rains season (May to August) and the long dry season (August to November). Night observations were few, mainly because of the herbivores spending the night in habitats with poor visibility for the observers. Data on night activity do not allow for a detailed quantitative analysis throughout the year for zebra or wildebeest but, based on the data available and on a few extensive records, qualitative conclusions can be drawn. Data on elephant and buffalo are considered adequate for quantitative conclusions. The same type of data collected between April and November 1991 did not give rise to a different interpretation of the time budgets of the herbivores. However, as a result of poaching, elephant had become so rare that sufficient data regarding their activity could no longer be collected.

6.2 WILDEBEEST AND ZEBRA GRAZING ACTIVITY

From wildebeest in Manyara, we have only collected quantitative data on grazing activity as we rarely saw them ruminate during the day. Night observations (with and without moon) showed that wildebeest contracted into dense groups far out on the alkaline grasslands, where they were never seen grazing, though they did ruminate. Much of the day is spent grazing, we suspect that Manyara wildebeest graze mainly during the day and ruminate mainly during the night. This activity pattern is different from that of buffalo, but must be quite similar to that of domestic cattle in East Africa, which are kept in kraals without food during the night. Serengeti wildebeest show a different activity pattern during the night, as revealed by observations on moonlit nights. After dusk they have a marked rumination period until approximately 22.00 hours, after which time a varying part of the herd begins to graze. Grazing reaches a peak around midnight, after which more and more wildebeest start ruminating again. After 03.00 hours a fairly synchronized grazing period occurs (D. Kreulen, personal communication). The daytime pattern of grazing activity of the Manyara wildebeest is rather constant throughout the year (Beekman and Prins 1989): most of their day is taken up with grazing and moving to new forage. During the daytime period, Manyara wildebeest spend on average 73.2% of the time grazing (with a range of 60–88%) (Figure 6.1).

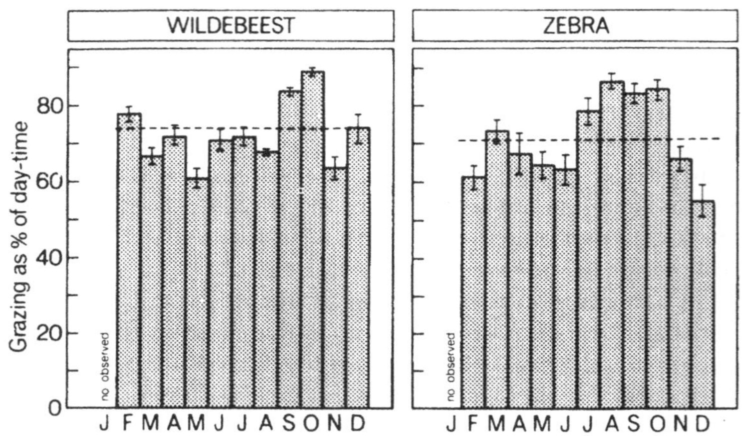

Figure 6.1 Grazing time as proportion of daylight hours over time for wildebeest (left) and zebra (right) over the year. Horizontal broken lines give the means, and the vertical bars 99.9% confidence limits. November to January is the short rains season, February to April the long rains season, May to July the post-rains season and August to October the long dry season.

There is no clear pattern over the year, except that more time is spent grazing at the end of the dry season than during the rest of the year. This increase can be explained by the deterioration of the food conditions. Wildebeest are typically selective grazers, preferring green leaves, but these become scarce during the dry season (Chapter 2).

The daytime grazing pattern of zebra is quite different from that of buffalo but rather like that of wildebeest. Zebra spend most of the day grazing, but this pattern is not constant through the year (Beekman and Prins 1989). They have an afternoon resting period when food conditions are good, but they seem to use this time for grazing when food is less abundant. Night activity could rarely be observed in zebra. The few night-time sightings of zebra in the woodlands did not result in good observations on activity as they immediately grouped together and were vigilant on approach of the car. In the adjacent Ngorongoro Crater zebra have sleeping grounds to which they move during the night, and on which they graze to a certain extent (40–60% as compared with 70–80% during the day) (Klingel 1967). It is quite possible that in Manyara zebra have sleeping grounds far out on the alkaline grasslands, but we never found these.

A comparison of our results for daytime grazing with results of other studies leads me to suppose that night-time grazing will be of the same order of magnitude as found by Klingel (1967). Tyler (1972) observed that horses graze at night in both summer and winter, during summer even more at night than during the day. Doreau *et al.* (1980) found that ponies graze for 23% of the night. Other studies on Burchell's zebra report approximately 65% of grazing time during a 24-hour period (Grogan 1978, Gakahu 1984) (this is also the average of Klingel's 1967 values). Horses have to spend at least 60% of the time grazing to meet intake requirements (P. Duncan, personal communication). On average over the year, Manyara zebra spend 70.9% of the day grazing, with a range of 54–86% (Figure 6.1). In Manyara, the pattern of grazing over the year shows two peaks. A minor peak occurs during the long rains season, when food is abundant, the other during the long dry season when food is scarce. The peak in the long rains season may reflect the build-up of body reserves for the dry season (although zebra always look fat), while the peak in the long dry season is thought to reflect compensation for the deterioration of food conditions.

6.3 ELEPHANT FEEDING ACTIVITY

Unlike the other major grazers, the elephant is an intermediate feeder, consuming both grass and browse. In addition to nutritional differences between these types of food, their spatial distribution is different, which has consequences for the rate of food intake, and hence for the allocation of feeding time in the time budget. Grasses form dense mats of vegetation, while leaves, twigs and fruits are more widely spaced.

Elephant in Manyara feed almost continuously over the 24-hour cycle (Figure 6.2). This is possible because they are hindgut fermenters like zebra, and thus there is no need to interrupt feeding bouts by ruminating the ingested food. Elephant were regularly observed feeding in the *Acacia* woodlands during the night and were less easily disturbed than other species by approach of the vehicle. Elephants feed almost all hours of the day, and this is independent of the time of year (Beekman and Prins 1989) and, although they rest for a short period during the night and often have a midday wallowing break, 'their life must virtually be one continuous meal' (Buss 1961). Over the year the Manyara elephant spend on average 76.4% (range of 73–83%) of the day feeding, which is remarkably constant compared with the range observed in other species.

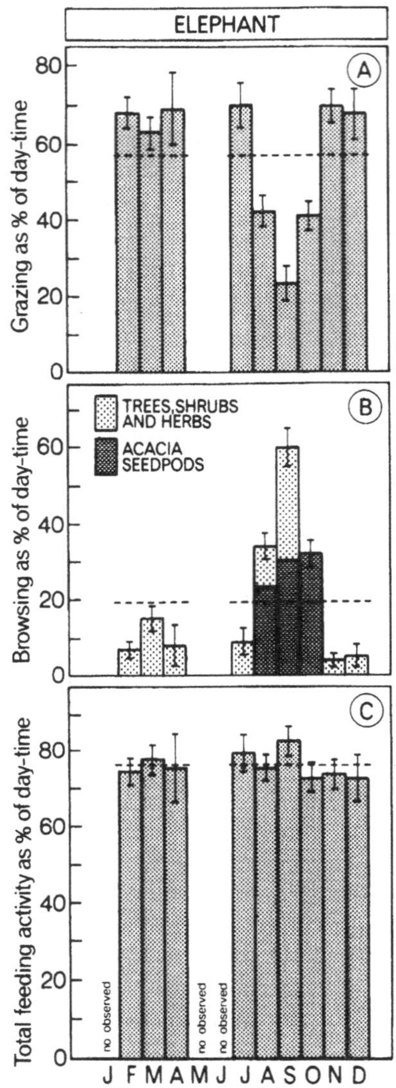

Figure 6.2 Elephant grazing time (A), browsing time (B) and total feeding time (C) as proportions of daylight hours for different months of the year. Horizontal broken lines give the means, and the vertical bars 99.9% confidence limits. November to January is the short rains season, February to April the long rains season, May to July the post-rains season and August to October the long dry season.

In Uganda, elephant showed 74.2% feeding activity over the 24-hour cycle (Wyatt and Eltringham 1974), while in Zimbabwe only 50–60% was observed by Guy (1975). The difference between the two sets of observations might stem from the better feeding conditions in Zimbabwe compared with Uganda (Guy 1975).

The monthly variation in total feeding time in Manyara is small (Figure 6.2). We have no reason to believe that feeding time is different in those months for which we have no data. Feeding patterns alter between the wet and the dry season. In the wet season 67% of the time is spent grazing (*Cynodon plectostachyus*, *Cynodon dactylon* and *Chloris gayana*) and 8% browsing, while in the dry season grazing time drops to a minimum of 23% and browsing increases to 60% in September. A similar season-dependent shift in diet was reported by Field (1971), Wyatt and Eltringham (1974) and Barnes (1983). Elephant in Manyara spend half their browsing time in the dry season collecting protein-rich *Acacia* seed pods from the ground. These seed pods are abundant from mid-August to mid-October. Shrubs (namely *Capparis* spp., *Cordia* spp., *Maearua triphylla* and *Salvadora persica*) are also very heavily browsed and appear seriously ravaged by the end of the dry season, while the alkaline grasslands are little visited by the elephant, even when there is almost no grass left in the woodlands. The quality of browse, in terms of crude protein and fibre content, did not drop as much as the quality of grasses in the dry season (Chapter 1), which is probably the cause of this diet shift in elephant.

6.4 SIMILARITIES AND DIFFERENCES IN ADAPTATION BETWEEN THE FOUR SPECIES

Grass (roughage)-eating ruminants, because of the selective delay of food in the rumen, digest their high-fibre food more efficiently than non-ruminant herbivores. However, these ruminants are constrained to diets with a crude protein concentration of more than 7–8% (Chapter 2), while large non-ruminant herbivores can utilize lower quality diets (Van Soest 1982). Hindgut fermenters can compensate for lower digestion efficiency by increasing food intake, resulting in retention times approximately half as long as in ruminants of the same size (that is the throughput rate is higher as particle flow is about 2.5 times faster in horses than in cattle), and the energy and nutrient assimilation achieved by hindgut fermenters on low-quality diets is higher than that achieved by ruminants (Bell 1971, Rittenhouse 1986). It is thought that hindgut fermenters have a selective advantage when food quality is a limiting factor, whereas ruminants perform better when the quantity of food is in limited supply (Janis 1976).

The Manyara zebra appears to fit in this scheme very well, as they increase their grazing time to compensate for reduced food quality in the dry season (Figure 6.1). Unlike the other species, zebra do not show any change in external body features indicative of condition loss (as described for buffalo in Chapters 3 and 4), nor is there any evidence for condition loss in the literature. However, it appears likely that zebra are time limited in the dry season since they have to work harder to obtain their nutrients (Figure 6.1), and their idling time is minimal then (Tables 2.4 and 6.1).

Similarly, ponies in the New Forest in England spend much more time feeding in winter than in summer (up to 85%, Pratt *et al.* 1986). We would expect high night-time grazing percentages to spread the increased workload, but we have not observed this. The reason for this absence may be the high risk of predation by lions, to which zebra appear to be especially vulnerable (Chapter 5).

Elephant are also hindgut fermenters but they do not fit the above-mentioned scheme. First, elephant lose body condition in the dry season in East Africa (Malpas 1977). Second, it appears that, instead of increasing the throughput rate of their food by increasing their total feeding time (Figure 6.2), elephant switch in the dry season from mainly grasses to mainly browse and *Acacia* seed pods. Apparently, the low intake rate of the small seed pods is better than grazing on the available short grasses. Feeding from a long sward, an elephant gathers a thick bunch of grass with its trunk then pulls it over one of the tusks to break it. The amount of material collected in this way is relatively large. If the sward height is below 20–30 cm, they grab the grass with their trunk and kick it loose carefully with one of their forefeet. When using the foot-kick grazing technique, food intake rate is undoubtedly much lower than when grazing in the 'normal' fashion. Both the change of grazing technique with decreasing leaf table height and the diet shift to browse suggest that elephant try to avoid a reduced food intake rate during the dry season. Yet, I still favour the interpretation that the elephant's intake rate is reduced during the dry season. Although elephant, throughout the year, spend large proportions of the time feeding (Figure 6.2), their idling time is very low during the (late) dry season (Tables 2.4 and 6.1). It is very likely that they are time limited, especially during the dry season. Hence, they are then unable to compensate by feeding longer, which forces them to shift to another diet. This diet shift, however, cannot prevent condition loss.

In zebra compensation for deteriorating food quality is achieved mainly through higher intake rate, which speeds up the throughput rate in the gut. Thus zebra obtain more energy and nutrients per unit of time. This option is impossible for grass-eating ruminants such as buffalo and wildebeest, first because they are constrained by the minimum protein concentration in the diet to make fermentation in the rumen possible and secondly because of the retention of large and coarse food particles in the rumen. Yet I think that buffalo partly do adopt this strategy (Chapter 2). The normal expected adaptation of ruminants is to increase their selectiveness in the dry season in order to keep fermentation going, and wildebeest express this option. Wildebeest do not increase their grazing time until September, while buffalo do so as early as July (Figures 6.1 and 2.5B). The possible cause of this difference might be that wildebeest can graze the sward down shorter than buffalo (to about 2 and 10 cm, respectively), owing to the wildebeest's grazing technique of cutting the grass with the incisors rather than plucking it with the tongue as buffalo mostly do. Wildebeest also have a narrower incisor arcade than buffalo so that they can more easily select within the sward. Thus, when primary production has ceased in the dry season and the standing crop is grazed down by other herbivores, wildebeest can still utilize these short vegetations for a prolonged period. As the dry season progresses, wildebeest also appear to have difficulty in obtaining good-quality forage and have to increase their selectiveness and thus their grazing time (Figure 6.1); idling time is at its lowest during the (late) dry season (Table 6.1). In domestic cattle, long grazing times are indicative of difficulties in satisfying nutritive requirements (Stobbs 1970, Penning 1986) and one would expect that wildebeest would compensate for the high workload during the day by grazing at night. The absence of night-time grazing in Manyara pleads against the idea that the end of the dry season is a difficult period for wildebeest. However, this absence may be caused by the high predation risk for wildebeest in Manyara, where lion density is high (Chapter 5). So, if wildebeest are time limited then this appears to be due to lion predation pressure.

The four herbivores, including buffalo, show a remarkable correspondence in the average proportion of time they devote to food collection and moving through their home range (buffalo 68.3% + 6.8% = 75.1%; wildebeest 73.2% + 4.0% = 77.2%; zebra 70.9% + 5.1% = 76.0%; and elephant 76.4% + 10.4% = 86.8%). All four species appear to be adapted to the same environment, but their 'solutions' to the times of shortages and abundance appear to be different. The different feeding strategies of zebra, wildebeest, elephant and buffalo in Manyara are summarized in Table 6.1. It appears likely that buffalo are not time limited but are constrained by other factors (Chapter 2), zebra and wildebeest are possibly time limited in the dry season (perhaps because of predation risk) and elephant are quite definitely time limited in the dry season. So these four herbivore species have different adaptations to handling their food. These adaptations are partly reflected in differences in digestive morphology and physiology and partly in the allocation of time to different activities. To a large extent these four species depend on the same food resources in Manyara; what evidence is there that they minimize competition?

Table 6.1 Overview of the feeding strategies of the four major herbivores in Manyara

	Short rains season (Nov–Jan)	*Long rains season (Feb–Apr)*	*Post-rainy season (May–Jul)*	*Late dry season (Aug–Oct)*
Green biomass	High	Very high	Decreasing	Low
Quality of food	High	Mixed	Mixed	Low
Strategy of herbivore				
Buffalo				
Average grazing time = 35%	Grazing time low (35%), not very selective	Increase in grazing time to 40%, becoming more selective	Grazing time low (25%)	Grazing time high (50%), very selective
Average rumination time = 30%	Rumination time low (30%) because fibre content of food is low; it increases from 25 to 45%	Rumination time high (35%) because fibre content of food is high; it decreases from 45 to 30% during this time	Rumination time low (30%)	Rumination time very low (20%); presumably increased rate of throughput
Average moving time = 5%	Moving around is low (5%)	Moving around is low (5%)	Much moving (10%)	Little moving (5%)
Average idling time = 30%	Inferred intake is high	Inferred intake is somewhat reduced	Inferred intake is reduced, especially because of social unrest	Inferred intake is low
	Idling time is high (30%)	Idling time is low (20%)	Idling time is high (35%)	Idling time is moderate (25%)
Wildebeest				
Average grazing time during the day = 75%	Grazing time during the day is high (70%); selective	Grazing time during the day slightly lowered (65%); selective	Grazing time during the day slightly lowered (65%); selective	Grazing time during the day is very high (80%); very selective

Rumination at night	Rumination at night – because of lions?	Rumination at night – because of lions?	Rumination at night – because of lions?	Rumination at night – because of lions?
Average moving time during the day = 5%	Moving around is low (5%)	Moving around is low (5%)	Much moving (10%)	Little moving (5%)
Average idling time during the day = 25%	Idling time moderate (25%)	Idling much (30%)	Idling moderate (25%)	Little idling (15%)
Zebra				
Average grazing time during the day = 70%	Grazing time during the day is relatively low (60%); not very selective	Grazing time during the day increases (65%); increased selectivity	Grazing time during the day increases (70%); increased selectivity	Grazing time during the day very high (85%); very selective
Average moving time during the day = 5%	Little grazing at night – because of lions?	Little grazing at night – because of lions?	Little grazing at night – because of lions?	Little grazing at night – because of lions?
Average idling time during the day = 25%	Little moving (5%)	Little moving (5%)	Much moving (10%)	Little moving (5%)
	Idling time high (35%)	Idling time high (30%)	Idling time low (20%)	Idling very low (10%); time limited?
Elephant				
Average feeding time = 75%	Very high feeding time (75%); feeds mainly on Gramineae	Very high feeding time (75%); feeds mainly on Gramineae	Very high feeding time (75%); feeds mainly on Gramineae	Very high feeding time (75%); switches to browse and seed pods
Average moving time = 10%	Much moving (10%)	Little moving (5%)	Little moving?	Very much moving (15%)
Average idling time = 15%	Little idling (15%)	Moderately idling (20%)	Moderately idling?	Idling very low (10%); time limitation?

The amount of habitat overlap between species pairs made up from buffalo, elephant, zebra and wildebeest was calculated using the formula of Pianka (1973; see also Leuthold 1978).

$$\alpha_{jk} = \Sigma(p_{ik} \times p_{ik}) / \sqrt{[\Sigma(p_{ij})^2 \times \Sigma(p_{ik})^2]}$$

where α_{jk} is the mutual habitat overlap between the j-th and the k-th species, and p_{ij} and p_{ik} = proportions of the i-th resource utilized by the j-th and k-th species, respectively. Overlap in habitat occupancy between the four study animals in Manyara over the entire year is shown in Table 6.2.

Habitat overlap between zebra and wildebeest was the greatest of the four investigated species pairs, that between buffalo and elephant was the smallest. For all four species pairs, the average degree of habitat overlap in the dry season was not significantly different from that during the rainy season (calculated after arcsine transformation and paired t-tests). Because of this lack of significance the data from the wet season were combined with those of the dry season.

Table 6.2 Habitat overlap between pairs of herbivores in Manyara

Month	*Habitat overlap between*			
	Buffalo and elephant	*Buffalo and zebra*	*Buffalo and wildebeest*	*Zebra and wildebeest*
Jan	–	0.50	–	–
Feb	0.52	0.74	0.88	0.98
Mar	0.65	0.77	0.95	0.87
Apr	0.49	0.60	0.96	0.58
May	–	0.75	0.73	0.98
June	–	0.74	0.74	0.73
July	0.66	0.66	0.74	0.86
Aug	0.42	0.60	0.81	0.90
Sep	0.23	0.74	0.77	0.88
Oct	0.56	0.80	0.81	0.91
Nov	0.44	0.88	0.95	0.90
Dec	0.44	0.68	0.80	0.77
Rainy season (Nov–Jun inclusive)	0.51	0.71	0.86	0.83
Dry season (Jul–Oct inclusive)	0.47	0.70	0.78	0.89
Between seasons	t_7=0.520, NS	t_{10}=0.172, NS	t_9=1.679, NS	t_9=0.496, NS
Yearly average	0.49	0.71	0.83	0.85

The degree of habitat overlap is higher when the value is closer to 1.0.

To determine whether the four herbivore species influence each other through their presence or absence at particular times, we analysed 7788 observations that encompassed date, time, map coordinates, vegetation type, herbivore species, number of animals, and an activity score. This score was the percentages of time animals spent standing, moving, lying, wallowing, grazing, browsing and ruminating (where appropriate for the species). An important part of the data set consists of observations on the **absence** of herbivores (De Boer and Prins 1990). Because measuring interactions such as competition is so difficult (Wiens 1989) we concentrated our efforts on watching particular patches of vegetation where the Manyara herbivores grazed frequently. We defined visits of individuals to certain patches as events, and then calculated whether or not two events were randomly distributed and mutually independent. Our selection of patches where we studied the interactions between the different herbivores was based on the following conditions.

- A patch had to be regularly visited by the four species because the size of species-specific and patch-specific data sets should be large enough for statistical analysis.
- Each patch had to have a homogeneous monospecific vegetation because the diet of the herbivore species in a patch should be as similar as possible.
- A patch had to have a very good visibility to the observer, even at large distances, as the chances of overlooking herbivores had to be negligible.

The *Cynodon dactylon* and *Sporobolus spicatus* grasslands, as defined and mapped by Loth and Prins (1986), were ideally suited for our investigation. These particular *Cynodon* grasslands cover the deltas of several perennial streams. The *Sporobolus* grasslands are situated along the lake shore, more or less in between the deltas. All these patches, because of their small size (all less than 50 ha), their relatively short vegetation (shorter than 20 cm), the absence of trees and bushes and their location at the lake shore, were very easy to observe. Because we could not measure food consumption by the different groups of animals during their visits to these patches, we estimated it. Indeed, food consumption depends on the number of animals in a patch, on their activity, on the time spent in the patch and on their body weight. Hence we transformed all observations into estimates of grazing pressure by means of the formula:

Grazing pressure (kg dry matter day^{-1} $patch^{-1}$) = (N animals in an observation) × (% animals grazing) × (0.025 W) × A/B .

In this formula W = average weight (kg) of the individuals of a species, A is the average residence time in a patch, B is the total time feeding per day and the factor 0.025 W is a good estimator of total daily dry matter intake (Chapter 2). Average weight was taken as 325 kg for buffalo, 1700 kg for elephant, 250 kg for zebra and 225 kg for wildebeest (Chapter 5). Because the total daily dry matter intake is consumed during the total daily period that the animals are feeding, the formula holds if it is accepted that the consumption rate was constant over the total foraging period. On the basis of many days of observation we have taken patch residence time to be 3 hours in case of buffalo, 2 hours in case of wildebeest and zebra and only 1 hour in case of elephant. We knew from Manyara that the average daily foraging time was 16.4 hours for buffalo, 18.3 hours for elephant, 9.9 hours for zebra and 10.2 hours for wildebeest.

For the four species the grazing pressure figures were tabulated in a patch-specific calendar. All observations pertaining to the absence of herbivores were also tabulated, as well as the days on which no observations were collected (missing values). Each day on which a patch was visited was defined as day 0. We then took the grazing pressures on the 10 days preceding any day 0 and on the 10 days following that day 0 as individual cases. By calculating the average grazing pressure over time for all these days (that is for all days 1, all days 2, etc.) we obtained an average pattern of the grazing pressure for each patch and for all four herbivore species.

6.5 ZEBRA AND WILDEBEEST: MINOR COMPETITORS WITH BUFFALO

Buffalo and zebra were observed more often than expected on the same day on the same *Cynodon* patch (Table 6.3), but not so on the *Sporobolus* patches; the same pattern was observed in case of buffalo and wildebeest.

Absence or presence of buffalo did not affect group size of zebra or wildebeest, nor did zebra or wildebeest influence group size of buffalo (Table 6.4).

It is possible to investigate whether the grazing pressure by one species influences that by the other species on the same day. This we did through scoring the visits as higher than the median grazing pressure by one of the two species, lower than the median, or as absenteeism. The results of the foraging consumption by buffalo and zebra and by buffalo and wildebeest on a *Cynodon* patch on the same day are shown in Table 6.5.

If within the zebra–buffalo pair one of the two species was absent, grazing pressure by the other species was less than expected; on the same day a high grazing pressure by one species occurred when the grazing pressure from the other species was also high. This occurred more often than expected. A strong association between buffalo and wildebeest was also clear, but high grazing pressures of both buffalo and wildebeest did not occur more often than expected. The observed patterns of association between buffalo and zebra or between buffalo and wildebeest could be indicative of mutual benefits from each other's presence on *Cynodon* patches. An inhibiting effect, which would point at competition, is not apparent from the data in Table 6.4. The patterns of visits to *Sporobolus*

Table 6.3 Contingency of occurring together of pairs of herbivores on the same patch at the same time

		Buffalo–elephant	*Buffalo–zebra*	*Buffalo–wildebeest*	*Elephant–zebra*	*Elephant–wildebeest*	*Zebra–wildebeest*
Cynodon dactylon							
A3 vegetation	N	400	688	610	400	400	609
	χ^2	2.83	44.36	17.44	21.04	10.73	109.68
	P	NS	0.0001	0.0001	0.0001	0.002	0.0001
Sporobolus spicatus							
A2 vegetation	N	324	572	504	324	324	504
	χ^2	0.22	2.71	1.19	0.88	1.58	63.34
	P	NS	NS	NS	NS	NS	0.0001

N is the total number of observations.

Table 6.4 The influence of presence or absence of different large herbivores on the group size of buffalo, elephant, zebra and wildebeest

	Group size of			
	Buffalo	*Elephant*	*Zebra*	*Wildebeest*
On the *Cynodon dactylon* (A3) vegetation				
Buffalo				
Present	–	10	25	53
Absent		11	20	54
		$t=0.399$, $n=70$	$t=1.199$, $n=130$	$t=0.054$, $n=150$
		NS	NS	NS
Elephant				
Present	406	–	14	53
Absent	350		25	45
	$t=0.604$, $n=105$		$t=2.158$, $n=93$	$t=0.510$, $n=118$
	NS		$P<0.05$*	NS
Zebra				
Present	442	10	–	61
Absent	363	10		46
	$t=1.661$, $n=188$	$t=0.046$, $n=70$		$t=1.296$, $n=150$
	NS	NS		NS
Wildebeest				
Present	343	9	28	–
Absent	416	12	16	
	$t=1.199$, $n=171$	$t=0.899$, $n=70$	$t=2.814$, $n=123$	
	NS	NS	$P<0.01$**	

On the *Sporobolus spicatus* (A2) vegetation				
Buffalo				
Present	–	4	13	19
Absent		4	17	38
		$t=0.735$, $n=44$	$t=0.784$, $n=90$	$t=0.852$, $n=79$
		NS	NS	NS
Elephant				
Present	351	–	13	20
Absent	253		13	38
	$t=0.615$, $n=52$		$t=0.025$, $n=64$	$t=0.691$, $n=72$
	NS		NS	NS
Zebra				
Present	246	5	–	38
Absent	342	4		29
	$t=0.967$, $n=99$	$t=0.798$, $n=44$		$t=0.462$, $n=77$
	NS	NS		NS
Wildebeest				
Present	247	5	13	–
Absent	356	3	17	
	$t=0.972$, $n=87$	$t=1.792$, $n=44$	$t=0.754$, $n=79$	
	NS	NS	NS	
Average group size on *C. dactylon*	391±30	10±2	23±2	54±6
Average group size on *S. spicatus*	321±40	4±1	16±2	33±9
Difference	$t_{285}=0.028$ NS	$t_{112}=2.269$ $P<0.05$*	$t_{218}=0.727$ NS	$t_{227}=0.179$ NS

Table 6.5 Pairwise comparisons of the interaction between the grazing pressure exerted in a patch by buffalo and the grazing pressure exerted in the same patch on the same day by another herbivore species

	Grazing pressure by											
	Elephant				*Zebra*				*Wildebeest*			
Grazing pressure by buffalo	*0*	*>0 to median*	*>median*	*Total*	*0*	*>0 to median*	*>median*	*Total*	*0*	*>0 to median*	*>median*	*Total*
On *Cynodon dactylon* patches												
0	259NS	22NS	25NS	306	455^{+***}	36^{-***}	31^{-***}	522	371^{+***}	48^{-**}	42^{-**}	461
	(254.0)	(26.0)	(26.0)		(424.1)	(48.5)	(49.3)		(349.9)	(59.0)	(52.1)	
>0 to median	37NS	6NS	4NS	47	52^{-***}	17^{+***}	14^{+*}	83	44^{-***}	15^{+*}	15^{+**}	74
	(39.0)	(4.0)	(4.0)		(67.4)	(7.7)	(7.8)		(56.2)	(9.5)	(8.4)	
>median	36NS	6NS	5NS	47	52^{-***}	11NS	20^{+***}	83	48^{-**}	15^{+*}	12NS	75
	(39.0)	(4.0)	(4.0)		(67.4)	(7.7)	(7.8)		(56.9)	(9.6)	(8.5)	
Total	332	34	34	400	559	64	65	688	463	78	69	610
Total χ^2 with d.f.=4			3.342:NS				55.603:P<0.0005				22.308:P<0.0005	
On *Sporobolus spicatus* patches												
0	241NS	18NS	15NS	274	415NS	38NS	31NS	484	364NS	30NS	29NS	423
	(241.0)	(17.8)	(15.2)		(411.3)	(31.5)	(31.3)		(361.8)	(30.2)	(31.0)	
>0 to median	24NS	0NS	1NS	25	34NS	7NS	3NS	44	33NS	3NS	4NS	40
	(22.0)	(1.6)	(1.4)		(37.4)	(3.8)	(2.9)		(34.2)	(2.9)	(2.9)	
>median	20NS	3NS	2NS	25	37NS	4NS	3NS	44	34NS	3NS	4NS	41
	(22.0)	1.6)	(1.4)		(37.4)	(3.8)	(2.9)		(35.0)	(2.9)	(3.0)	
Total	285	21	18	324	486	49	37	572	431	36	37	504
Total χ^2 with d.f.=4			3.543:NS				3.436:NS				0.941:NS	

Categories of grazing pressure are: (1) no grazing, (2) from zero to median, (3) above the median. Each cell contains the number of observations in each grazing pressure category for the two species observed on the same day; expected values are given in parentheses. Tested with Habermann's method of adjusted residuals (significant deviations: $*P < 0.05$, $**P < 0.01$, $***P < 0.001$). Superscripts: + indicates that the number of observations was greater than expected, and – less than expected. Where observed and expected values are not significantly different (NS), the grazing pressures of the two species are independent of each other.

patches by buffalo and zebra or by buffalo and wildebeest were independent of each other (Table 6.5).

The pattern over time of zebra grazing pressure in relation to that by buffalo on any one day (day 0) is not presented here. An analysis analogous to that presented in Table 6.6 on the interaction between elephant and buffalo showed that in the periods before or after a day with a high grazing pressure by buffalo there was no effect of buffalo grazing on that by zebra (or vice versa).

Also, in the periods before or after a day with a high grazing pressure by buffalo there was no effect of buffalo grazing on that by wildebeest (or vice versa). The grazing pressure distribution of zebra and buffalo or of wildebeest and buffalo on the *Sporobolus* patches over time was as on the *Cynodon* patches: there were no effects. Thus, neither buffalo and zebra nor buffalo and wildebeest seem to influence each other's grazing pressures, and neither mutual nor one-sided effects could be detected at a level of analysis of several days. However, at the level of a single day there was a significant positive interaction between their grazing pressures on the same *Cynodon* patch, which is indicative of some form of symbiosis, perhaps through mutual benefits in detecting predators better (section 6.6).

Although the interaction between buffalo on the one hand and zebra or wildebeest on the other is minimal in Manyara, the interaction between zebra and wildebeest is generally thought to be important through a process called facilitation. This is associated with a grazing succession of several species which causes a positive effect on the food intake of each species. For instance, zebra, wildebeest and Thomson's gazelle can all benefit from a grazing succession (Gwynne and Bell 1968, Bell 1970, McNaughton 1976, Maddock 1979): zebra have a relatively high proportion of grass stalk in their diet and 'open' the vegetation (by grazing first in the succession) for selective leaf eaters such as wildebeest. However, the mentioned publications reveal the difficulty of interpreting correlative data. The assumption that one species benefits from the grazing by another species earlier in the succession is not supported by the population dynamics of the species involved (Sinclair and Norton Griffiths 1982, Stelfox *et al.* 1985). In Manyara zebra and wildebeest were seen together on the same day on the same patch more frequently than expected (Table 6.3), but there was no evidence for facilitation in the Manyara ecosystem (De Boer and Prins 1990). The positive correlation between the annual grass consumption of the two species between 1959 and 1985 ($t_7 = 3.53$, $P < 0.01$, $R^2 = 64\%$; Figure 6.3) is probably caused by the disappearance of suitable habitat for both species in the years 1965–75, which resulted from a higher lake level in those years and the concurrent flooding of the low-lying grasslands (Chapter 1; see also Table 5.10).

In Manyara most swards are maintained under quite heavy grazing by the numerous herbivores (Chapter 2) and, perhaps as a result, grasses

Table 6.6 The effect of grazing pressure by buffalo or elephant in the days preceding a visit on the grazing pressure by the other species

	Numbers of days preceding day 0											
	10	*9*	*8*	*7*	*6*	*5*	*4*	*3*	*2*	*1*	*0*	
Grazing pressure on *C. dactylon* (A3) by												Elephant grazing on *C. dactylon* (A3) on day 0
Buffalo	1804	894	248	925	4	93	251	204	243	163	196	Higher than 31
Buffalo	402	506	58	347	848	127	457	780	347	378	362	Lower than 31
Significance of cumulative difference (from day 0)	NS	NS	NS	NS	<0.025	<0.025	<0.025	<0.1	<0.025	<0.05	–	
Grazing pressure on *C. dactylon* (A3) by												Buffalo grazing on *C. dactylon* (A3) on day 0
Elephant	32.9	1.2	0.7	11.9	4.7	15.8	10.3	5.5	5.1	8.6	4.4	Higher than 600
Elephant	4.3	0.7	8.1	14.6	10.3	4.8	6.1	4.6	7.2	11.7	21.7	Lower than 600
Significance of cumulative difference (from day 0)	NS	NS	NS	NS	NS	NS	NS	NS	NS	NS	–	
Grazing pressure on *S. spicatus* (A2) by												Elephant grazing on *S. spicatus* (A2) on day 0
Buffalo	0	4.3	229.0	0.9	0	5.1	2.9	3.5	0	0	9.2	Higher than 9
Buffalo	142.0	53.5	0	164.5	1.5	13.8	163.6	40.7	63.8	14.0	33.0	Lower than 9
Significance of cumulative difference (from day 0)	0.01	0.025	0.05	0.05	0.1	0.1	0.1	NS	NS	NS	–	

Grazing pressure on *S. spicatus* (A2) by												Buffalo grazing on *S. spicatus* (A2) on day 0
Elephant	0.9	7.1	0.3	0	0.3	1.3	0	0	0.5	0.5	0.2	Higher than 329
Elephant	0	0.5	0.3	1.9	0.6	1.4	0.2	0.3	0.5	0.5	1.5	Lower than 329
Significance of cumulative difference (from day 0)	NS	NS	0.05	0.05	0.1	NS	NS	NS	NS	NS	–	

The values for 'higher than' or 'lower than' represent the median grazing pressures for buffalo and elephant.

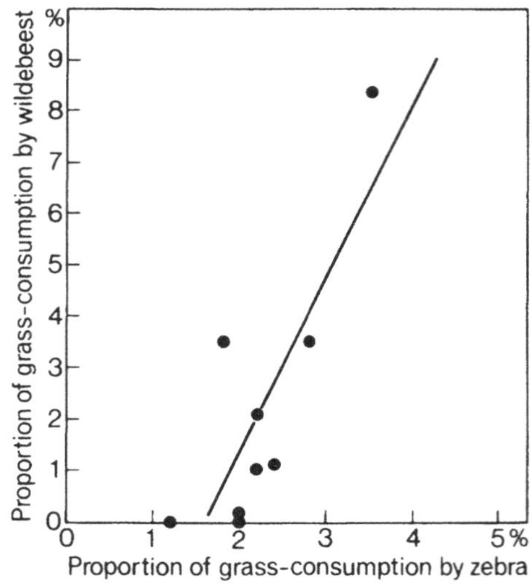

Figure 6.3 The relation between grass consumption by wildebeest and that by zebra as proportions of total grass consumption by all large herbivores in Manyara for the period between 1959 and 1984. The strong positive relation can be explained by the near-identical effects of lake level changes on both populations (see text).

rarely produce flowering stalks. This could explain the absence of observable facilitation by zebra in Manyara.

6.6 ELEPHANT: THE MAJOR COMPETITOR

Competition between two animal species is only possible when the following three conditions are met.

1. There must be an overlap in habitats visited by the two species.
2. The diet of the two species in this overlapping habitat must be overlapping too.
3. The food supply must be limited (which could be revealed when an increase in the population of one of the two animal species causes a reduction in numbers of the population of the other species) (see also Wiens 1989).

Thus, the following hypothesis was derived: grazing individuals of one herbivore species should react to the grazing pressure exerted by individuals of the competing species through having their intake reduced; this will be most apparent at relatively high grazing pressures. I have already shown that there was no evidence for a competitive relation between buffalo on the one hand and wildebeest or zebra on the other, but the relation between elephant and buffalo has not yet received attention.

If we look at the conditions necessary for interspecific competition to occur, namely overlap in habitat occupancy, diet overlap and a limited food supply (which could become evident through a limit on the total numbers of herbivores in the Manyara system) then the following becomes evident. Overlap in habitat occupancy was least between buffalo and elephant (Table 6.2). Diet analyses in Manyara are only available for buffalo (Chapter 1), but the literature on diet of the present study animals in other areas is extensive (Stewart and Stewart 1970, Jarman 1971, Field 1972, Hofmann 1973, Jarman and Sinclair 1979, Hansen *et al.* 1985). On the basis of these studies it is nearly certain that diet overlap occurs and that interspecific competition is possible. Moreover, the potential for differences in diet is almost eliminated because of our methods: we used only the data set of observations on the *Cynodon* and *Sporobolus* vegetation types. Both vegetation types consist of homogeneous monospecific swards, so that the herbivore species consume the same plant species when in the patch. Lastly, Manyara was until the end of the 1980s characterized by one of the world's highest densities of mammalian herbivores, namely 180 kg ha^{-1} (Drent and Prins 1987) facilitated by the high level of soil fertility and the high level of groundwater (Chapter 1). Total grass consumption was as high as approximately 80% of the annual production (Chapter 2) and total herbivore pressure was rather constant (Chapter 5), which suggests that the quantity of vegetation operates as a limiting factor. On the basis of these considerations I think that the necessary conditions for interspecific competition between the four large herbivores are fulfilled in Manyara, and thus that the association pattern between individuals of these four species could be influenced by interspecific competition for food. The interpretation that Manyara has reached its maximum sustainable herbivore density (Chapter 5) is additional support for the contention that interspecific competition plays a role in this area.

No evidence was found for competition between buffalo and zebra on the one hand and buffalo and wildebeest on the other. Neither zebra nor wildebeest seemed to alter their food intake in response to the (high) grazing pressures by buffalo (section 6.5). However, our data showed a clear negative influence of the grazing pressures by buffalo on that of elephant. Visits of buffalo and elephant, as observed through scoring their presence or absence, to the same patches on the same day were randomly distributed both on the *Cynodon dactylon* patches and on the *Sporobolus spicatus* patches (Table 6.3). Hence, the occurrence of an observation of both buffalo and elephant on the same patch is expected to be the product of the species-specific frequencies of visits to the patches. Group sizes of buffalo or of elephant were not affected by the presence or absence of individuals of the other species on either *Cynodon* patches or *Sporobolus* patches (Table 6.4).

The grazing pressure by one of the two species was independent of the grazing pressure by the other species on both vegetations (Table 6.5); the data in this table are not significantly different from results based on the expectation that days with a higher or lower than average grazing pressure by buffalo and elephant, respectively, were independent of each other. This indicates that interspecific competition between elephant and buffalo cannot be demonstrated at the level of a single day. However, on both types of vegetations, elephant reacted to rates of consumption by buffalo in the preceding days. On the *Cynodon dactylon* patches a low grazing pressure by buffalo during the preceding 6 days plus day 0 is followed by a high grazing pressure (that is higher than the median of all visits) by elephant on day 0. This buffalo grazing pressure was significantly lower than when followed by a low grazing pressure by elephant on day 0; this was more or less the same on the *Sporobolus* patches (paired *t*-tests; Table 6.6). This indicates that elephant reacted to the effect of the grazing pressure by buffalo, because the size of the elephants' grazing pressure was dependent on that by the buffalo in the foregoing days.

In comparison with the *Cynodon* patches, the reaction of buffalo to the grazing pressure by elephant on the *Sporobolus* patches was less clear. Over a 7- or 8-day period a significant distinction can be made: the elephant grazing pressure followed by a high buffalo grazing pressure on day 0 was lower than that followed by a low buffalo grazing pressure on day 0. Thus, on a time scale of several days, a high grazing pressure by individual elephant in a specific patch on a given day takes place only in the absence of a high grazing pressure by individual buffalo on the days preceding that day. These negative interactions prove the likely existence of interspecific competition for food between elephant and buffalo.

At the population level there is also evidence for competition between these two species: the annual grass consumption by these two species was negatively correlated ($t_7 = -2.43$, $P < 0.05$, $R^2 = 46\%$; Figure 6.4).

This confirms the conclusion about interspecific competition between buffalo and elephant based on annual herd growth rate before and after the poaching of elephant (Chapter 5). The finding that there is no association between elephant and buffalo visits to the same patch on the same day (Table 6.3), or between frequency of elephant and buffalo visits (Table 6.5), could be taken as evidence for scramble competition whenever the quantity of food on offer is high enough again to warrant a visit. Elephant and buffalo, thus, appear to be major competitors for food.

With the high density of buffalo in Manyara, buffalo appear to have the selective edge in the scramble competition between the two species. However, they appear to pay a (small?) price in having suboptimal regrazing intervals on the *Cynodon dactylon* patches from which they obtain a

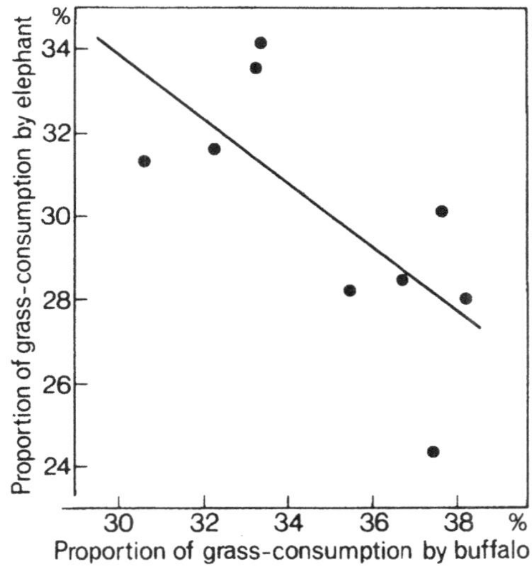

Figure 6.4 The relation between grass consumption by buffalo and that by elephant as proportions of total grass consumption by all large herbivores in Manyara for the period between 1959 and 1984. The negative relation between the two species is indicative of competition (see also Figure 5.7).

substantial part of their food intake. They seem to come too early to get the maximal harvest of the regrowth in these patches, perhaps in 'anticipation' of harvesting by elephant of the same regrowth (Chapter 7). I think buffalo can compete with elephant because they live in very large herds – as a group they can enjoy a despotic relationship with the other species (section 7.6) even in the days, before 1987, when the elephant density in Manyara was extremely high (some 6 elephants per km^2). Elephants can reduce this competition (in natural selection terms) through exploiting, apart from grass, other food sources, such as browse. The extinct *Elephas recki* was more a typical grazer than the still extant African elephant (Klein 1988), which might explain the expansion of African buffalo after the extinction of *Elephas recki* (Klein 1988). Also, the African elephant became less rare in East Africa after the extinction of *Elephas recki* (Haynes 1991, p. 8), so this species appears to have occupied a part of the niche that has been filled since its extinction by both the buffalo and the elephant. However, in the long term African elephants might have a positive effect on the population size of buffalo if they push the competitive balance between grass and woody species towards a vegetation that is dominated by grass (cf. Myers 1973, Caughley 1976, Norton-Griffiths 1979, Pellew 1983).

7

Patch selection: predators and grazing by 'rule of thumb'

7.1 ZONES OF DANGER

'There are many interrelationships between the habitat type normally occupied by a species and... [its] antipredator behavior... The choice of habitat is perhaps one of the most basic behaviorally mediated processes in an animal's life' (Leuthold 1977, p. 40; see also, for example, Hornocker 1970, Bergerud *et al.* 1983a, b), but proof for this assertion is scanty. Few studies have dealt with the problem of how the effect of predation risk must be considered when deciding where to graze or where to go. In this chapter I want to assess the influence of risk of predation on the habitat choice of buffalo. Manyara has a very diverse vegetation or, in the terminology of many zoologists, very many different habitats. The different vegetation communities and landscape ecological units were referred to in Chapter 1 (see for details Loth and Prins 1986). We used two different vegetation typologies to describe vegetation. One was based on the floristic composition of a unit and one was based on vegetation structure. In this chapter I use the term 'vegetation type' for groups of plant species that occur together and 'vegetation structure' for formations such as grassland, woodland or thicket. I used these different categories to investigate whether specific vegetation types or vegetation structures were more dangerous for buffalo, or more favourable for lions if these vegetations facilitated hunting success by lions.

I recorded all dead buffalo that I found, to nearest 50 m, on the 1:25 000 vegetation map of Manyara (Loth and Prins 1986) (Figure 7.1), registering both vegetation type and vegetation structure.

The distribution of lion kills over the vegetation types was the same as that of the buffalo which died from an unknown cause ($\chi^2 = 8.196$, d.f. = 8, NS). Hence I used the data on all carcasses to investigate the effect of vegetation type on predation. Table 7.1 shows that these buffalo carcasses had an unequal distribution over the different vegetation types.

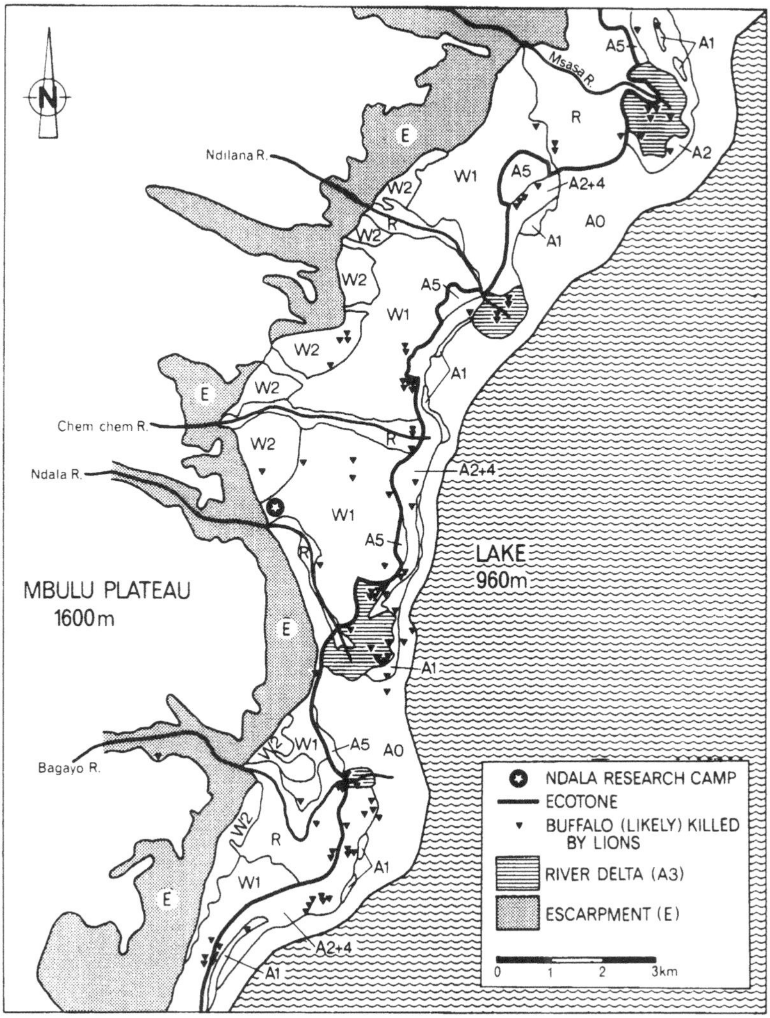

Figure 7.1 Distribution of carcasses of buffalo killed by lions in the central part of Manyara between 1982 and 1985. The thick line from north to south indicates the position of the ecotone between open grasslands and woodlands. Landscape ecological units: A1, areas with upwelling groundwater dominated by the sedge *Cyperus laevigatus*, A2, lake shore grasslands dominated by the grass *Sporobolus spicatus*, A3, riverine grasslands dominated by the grass *Cynodon dactylon*, A4, beach walls dominated by the grass *Dactyloctenium aegyptium*, A5, *Sporobolus consimilis* grasslands, R, riverine bushlands, W1, *Acacia tortilis* woodlands with many grass species such as *Chloris virgata*, W2, *Acacia tortilis* woodland with much less trees. No kills were found on the escarpment.

Table 7.1 Distribution of buffalo carcasses according to vegetation type

Vegetation *area (ha)*	*A0* 752	*A1* 97	*A2* 319	*A3* 365	*G1+2* 424	*A4* 31	*A5* 95	*G3+4+5* 979	*W* 708	*R* 666	*E* 948	*Total* 5348
Carcasses of herd buffalo	0	4	16	6	5	4	0	0	18	19	1	73
Preference index (±95% confidence limits)	0.00* ±0.82	3.04 ±6.37	3.70* ±1.94	1.21 ±1.69	0.87 ±1.46	9.52 ±19.92	0.00 ±6.50	0.00* ±0.63	1.88* ±0.87	2.10* ±0.93	0.08* ±0.65	χ^2_9=93.15 P<0.001
Carcasses of adult bulls	4	10	12	11	3	3	0	0	5	9	1	58
Preference index (±95% confidence limits)	0.49 ±0.92	9.57* ±7.14	3.49* ±2.17	2.80 ±1.90	0.66 ±1.63	8.98 ±22.35	0.00 ±7.29	0.00* ±0.71	0.66 0.98	1.25 ±1.04	0.10* ±0.73	χ^2_9=135.62 P<0.001
Average time budget (%) of herd buffalo	12.5	2.5	5.0	15.0	5.0	1.0	0.5	0.5	25.0	27.5	5.5	100
Carcasses of herd buffalo	0	4	16	6	5	4	0	0	18	19	1	73
Preference index (±95% confidence limits)	0.00* ±0.13	2.19 ±4.59	4.38* ±2.29	0.55 ±0.76	1.37 ±2.29	5.48 ±11.47	0.00 ±22.94	0.00 ±22.94	0.99 ±0.46	0.95 ±0.42	0.25 ±2.09	χ^2_9=62.99 P<0.001
Average time budget (%) of adult bulls	16.3	8.8	3.8	13.8	2.5	0.8	0.3	0.5	22.5	25.5	5.3	100
Carcasses of adult bulls	4	10	12	11	3	3	0	0	5	9	1	58
Preference index (±95% confidence limits)	0.42 ±0.79	1.96 ±1.46	5.45* ±3.39	1.37 ±0.93	2.07 ±5.15	6.27 ±16.09	0.00 ±42.89	0.00 ±25.47	0.38* ±0.57	0.61 ±0.50	0.33 ±2.43	χ^2_7=64.71 P<0.001

The distribution of buffalo carcasses, both of herd animals and of adult bulls, strongly deviates from the distribution expected on the basis of the areas of different vegetation types. This distribution also strongly deviates from the distribution as expected on the basis of the time budgets of living buffalo in the different vegetation units. An asterisk indicates that a preference index with its 95% confidence interval deviates from unity. A preference index value lower than 1 indicates overrepresentation.

In Chapter 5 I showed that adult bulls run a much higher predation risk than those buffalo that live in herds (that is adult cows, subadult cows and subadult bulls). Thus Table 7.1 also presents information on the different distributions of carcasses of adult bulls and buffalo of the other sex/age categories . The distribution of carcasses was very unequal ($P < 0.001$): some vegetation types contained many more dead buffalo were than in others.

What caused the unequal distribution of carcasses? The first issue that springs to mind when such a question is posed is whether this was an effect of the time budget of the buffalo. Although the time budgets in the different vegetation types were not known exactly, as night observations were relatively few, I could make an estimate on the basis of (a) the day time budget, (b) investigation of spoors and (c) information from the park rangers. The estimated time budget is given in Table 7.1. On the basis of the time budget for herd-living buffalo, it appears that the *Sporobolus spicatus* alkaline grasslands (A2) were risky. This was probably also the case for the *S. spicatus–Dactyloctenium aegyptium* ecotone (A4), but data were rather limited for this entity. The bare mudflats (A0) close to the lake were relatively safe. The distribution of dead adult bulls showed the same: the A2 (and possibly the A4) community appeared dangerous, while the A0 mudflats were apparently relatively safe. The last, of course, depends on the time budget of bulls, and for this purpose the modal bull was chosen.

Just as there was a relationship between vegetation **types** and predation risk, there was one between vegetation **structure** and risk. Again, because there was no significant difference between the distribution of buffalo killed by lions and that of buffalo which died from an unknown cause ($\chi^2 = 1.834$, d.f. = 5, NS), I have combined all data on dead buffalo. The mudflats were relatively safe, whereas grasslands appeared especially dangerous. Contrary to what one would expect from the literature (cf. Kruuk and Turner 1967, Schaller 1972, Sinclair 1977, Mills 1984 and others), dense bush and thicket did not appear especially dangerous, while it can be concluded that woodlands and bush [taken together in Table 7.2 as (wooded) 'bush'] were relatively safe. However, I was not able to construct an estimated time budget of buffalo for the different vegetation structures, although the interpretation was confirmed by the results presented in Figures 7.1 and 7.2. From the data presented here, I concluded that, although herd-living buffalo and adult bulls had a (slightly) different occupancy pattern in Manyara, one, and possibly two, area(s) were particularly dangerous for buffalo, namely the *Sporobolus spicatus* A2 grassland and probably the *S. spicatus–Dactyloctenium aegyptium* A4 bushed grassland. Other grassland types (*Cynodon dactylon* A3 grassland on riverine deltas, *C. dactylon* G1 forest glades and mixed *C. dactylon–Chloris gayana* G2 grasslands) were as dangerous as the average (Table 7.1). The mudflats adjacent to the lake appeared very safe for herd-living buffalo but not particularly so for adult bulls.

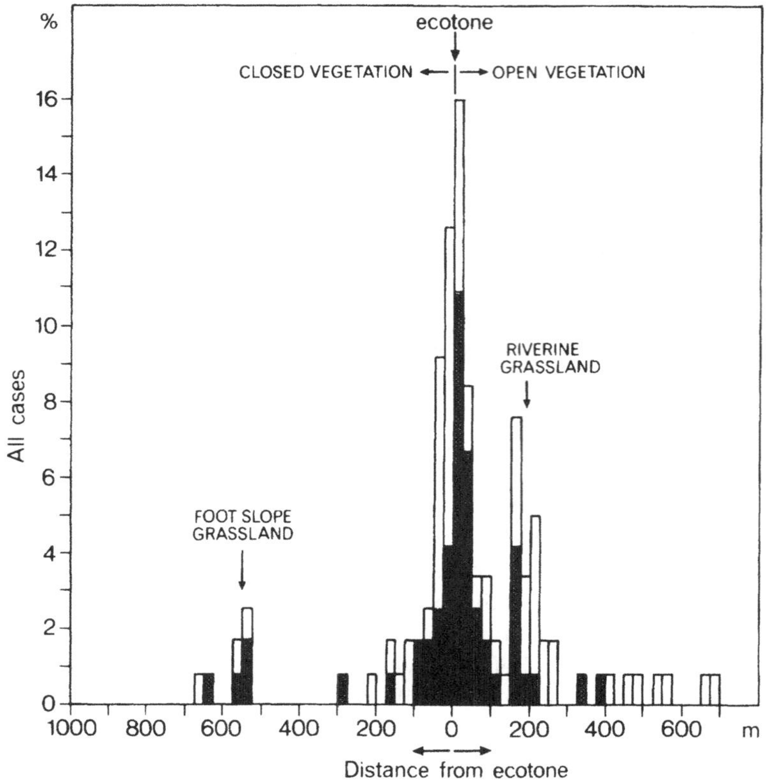

Figure 7.2 The distribution of buffalo probably and certainly killed by lion (n = 67; dark bars) and those possibly killed by lions (n = 92; open bars) in relation to the distance from the ecotone between open grasslands and woodlands (Figure 7.1). The small peak at the left indicates a second, minor ecotone between foot slope grasslands and woodlands. The peak at the right indicates the riverine delta grasslands where lions can easily hide in the many gullies.

Why should areas characterized by *Sporobolus spicatus* be dangerous for buffalo? The character species is a short perennial grass generally shorter than 15 cm. One does not associate such an area with preferential hunting areas of lions. This grass can occur together with the short perennial sedge *Cyperus laevigatus* (about 20 cm tall), the short annual grass *D. aegyptium* (about 30 cm) and the medium high herb *Indigofera costata* (about 35 cm). From the point of view of the hunting technique of lions, namely first stalking and then a short sprint to catch the prey

Table 7.2 The distribution of buffalo carcasses strongly deviates from the distribution as expected on the basis of the acreages of different vegetation structure units

	Vegetation structure						
	Mudflats	*Grassland*	*(Wooded) bush grassland*	*(Wooded) bush*	*(Wooded) dense bush*	*(Wooded) thicket*	*Total (Forest excluded)*
Area (ha)	*752*	*876*	*548*	*732*	*706*	*106*	*3720*
All carcasses	4	61	28	6	24	3	126
Preference index	0.16*	2.06*	1.51	0.24*	1.00	0.84	$\chi^2_5=75.13$
(±95% confidence limits)	±0.43	±0.37	±0.59	±0.44	±0.46	±3.06	$P<0.001$
Carcasses of herd buffalo	0	28	20	3	15	2	68
Preference index	0.00*	1.75*	2.00*	0.22*	1.16	1.03	$\chi^2_5=41.06$
(±95% confidence limits)	±0.59	±0.50	±0.81	±0.60	±0.63	±4.17	$P<0.001$
Carcasses of adult bulls	4	33	8	3	9	1	58
Preference index	0.34*	2.42*	0.94	0.26*	0.82	0.61	$\chi^2_5=39.34$
(±95% confidence limits)	±0.64	±0.55	±0.87	±0.65	±0.68	±4.52	$P<0.001$

An asterisk indicates that the preference index with its 95% confidence interval deviates from unity. Grasslands appear dangerous (all indices higher than 1), but mudflats and bush appear safe.

unawares (Schaller 1972, Elliott *et al.* 1977, Van Orsdol 1984, Sunquist and Sunquist 1989), these vegetations do not seem to be ideal, and my observation on the distribution of carcasses calls for an explanation.

Vegetation classifications are based on the idea of dividing the natural plant cover into discrete units. Phytosociologically defined units are thought to reflect ecological conditions such as soil fertility or water conditions. Structurally defined units have to reflect the way human observers experience spatial structure in the natural plant cover. I think that the first typology is very useful to understand, for example, the grazing of herbivores. The second typology can be used to clarify the behaviour of browsers, or thermoregulatory behaviour, or the incidence of tsetse flies.

However, to understand the relationship between lions and buffalo one has to look at the vegetation in an integrative instead of in a divisional way. Lions try to stalk their prey unobserved, so they need something to hide behind (Elliott *et al.* 1977, Van Orsdol 1984). In Manyara, buffalo spent most of the day on open grasslands, and especially on bare mudflats. Here they are difficult to approach by predators, were not bothered by tsetse flies and are exposed to the lake breeze. However, at night buffalo move preferentially into the woodlands to graze (section 7.3). In my opinion, the best tactic for lions is to observe buffalo from behind bushes in the ecotone between grasslands and woodlands or between grasslands and thickets. In this ecotone the lions can move freely without being noticed by buffalo and can look for the best position to start their attack when buffalo move in their direction. The A2 and A4 vegetation types happen to be always on or adjacent to the ecotone mentioned, more so than any other vegetation type (Figure 7.1; also see map in Loth and Prins 1986). The frequency distribution of carcasses in relation to the distance from the ecotone (Figure 7.2) shows clearly that most buffalo fell victim to lions just at the grassland side of the ecotone. Most kills took place within 100 m of the ecotone. Most carcasses for which the cause of death was unknown were found there too, indicating that these buffalo were probably killed by lions too (corroborating the conclusion drawn previously in Chapter 5). Figure 7.2 also shows that, generally speaking, buffalo were just as safe from lion predation when they were well into the structurally closed vegetation as when they were on the nearly completely open grasslands. This would appear at first sight to confirm observations from the Serengeti, where Schaller (1972), studied the interaction between large predators and large prey and stated that 'prey is particularly cautious about entering thickets. Herds may mill by a riverine forest for an hour or more before one individual leads the rest into or away from it... At the slightest hint of danger, animals rush away from dense vegetation into the open, then halt and look around'

(Schaller 1972, p. 231). However, I have never observed this behaviour during 7 years of observation in Manyara. Buffalo enter these ecotones without any visible apprehension. I will come to this again later in the present chapter (section 7.3).

As stated, lions appear to have a strong tendency to attack their potential prey from cover. However, I never fail to be amazed at how little cover is needed to hide a lion. To investigate whether kills were randomly distributed through the terrain or whether they showed a relation with the distance to cover on an even finer scale than was used in the analysis of the relation with vegetation type or vegetation structure, I measured the distances between a certain kill and the closest 'potential cover' for lions (n = 58). This I defined as any natural object behind which a lion could hide, which I based on both personal judgement and that of the rangers: grass taller than 50 cm, dead trees thicker than 50 cm, gullies deeper than 50 cm, dense bushes or termite mounts (compare Elliott *et al.* 1977, Van Orsdol 1984).

To determine whether buffalo ran more risk close to these terrain features, I used 'potential cover', as this could be investigated relatively easily, whereas actual cover used by lions before the kills was nearly always unknown, simply because nobody had seen the kill take place. I divided the data set into a group containing kills on open grassland and mudflats (n = 32), and one with kills in structurally closed vegetations (n = 26). The kills were not distributed according to a Poisson distribution with regard to distance (chi-square test, $P < 0.005$) but a $\log_{10}$ transformation of distances gave a near-normal distribution. Differences regarding the distributions were not sex or age dependent; in structurally closed vegetation, however, the distance between potential cover and the kills (8.0 m, SD = 2.9 m) was significantly shorter than in open vegetation (76.3 m, SD = 5.0 m). Figure 7.3 shows that the chance of being killed by lions declined rapidly with distance from potential cover: the regression formulae indicate that the chance of being killed increased quadratically when approaching potential cover. This was true for both open and closed vegetation, and applied equally to adult bulls and herd-living buffalo. The difference between the distribution of kills in closed vegetation and open vegetation was probably due to the structure of the habitat. The results corroborate the finding of Van Orsdol (1984) in Uganda that the charge distances after stalking on successful lion hunts was much shorter than on unsuccessful hunts (on average 19 and 33 m, respectively; calculated from his Table 6; see also Elliott *et al.* 1977 for charge distances in the Ngorongoro Crater).

From the observations I concluded that the ecotone between open grassland and closed vegetation was apparently the riskiest part of the park, and most kills took place within 100 m of this ecotone (Figure 7.2). This is in agreement with the hunting technique of the lions and the flee-

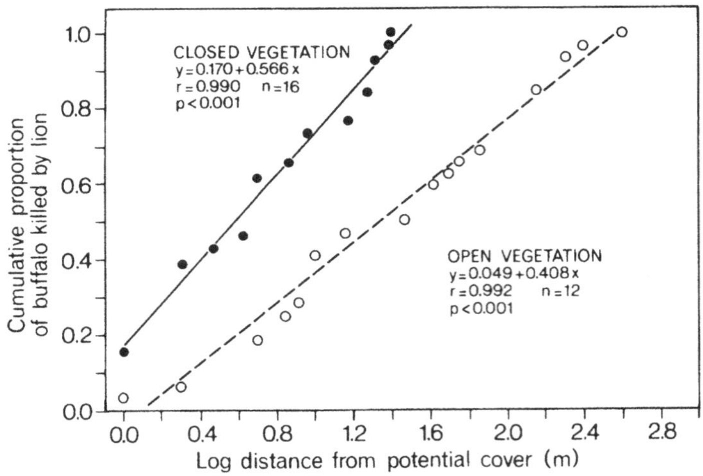

Figure 7.3 The cumulative proportion of the number of buffalo killed by lions in relation to the distance from potential cover. With increasing distance from cover the number of buffalo killed decreases. Sample size *n* stands for the number of classes and not the number of kills.

ing capabilities of the buffalo. The buffalo's full speed is nearly 60 km h^{-1}, and that of the lion 48–59 km h^{-1}. In most instances lions can capture an animal only if they approach it so closely that it has no time to attain full running speed before being grabbed (Schaller 1972; see also Elliott *et al.* 1977). This appears to be confirmed by my observations in Manyara that most kills take place very close to (potential) cover. But, contrary to the expectation based on these results, and on the observations of others (for example Schaller 1972, Leuthold 1977), the Manyara buffalo entered the thickets and woodlands in the evening without showing uneasiness, and passed the dangerous ecotone again in the morning seemingly quite undisturbed.

7.2 VIGILANCE AND PREDATOR DETECTION

Living in large groups may be advantageous for foraging by improving food quality (Prins *et al.* 1980, McNaughton 1984), but it may also afford protection against predators (Bertram 1978) through direct defensive action or by decreasing the probability that a particular individual will be killed (Hamilton 1971). This was the case with the Manyara buffalo, adult bulls that spent much of their lives outside the herd (Chapter 4) being exposed to a risk of predation that was about three times higher than that of adult cows (Chapter 5). Experimental evidence suggests that in groups vigilance behaviour results in earlier detection of predators (Powell 1974, Ydenberg and Dill 1986), and it makes economic sense for each individual to spend less time on vigilance as group size increases (Bertram 1980, Elgar and Caterall 1981). However, vigilance does not continue to decrease above a certain group size (approximately ten individuals), beyond which competitive interactions for food or mates increase vigilance (Alados 1985).

To discover more about potential predator detection in buffalo, Glenn Iason and I studied vigilance of buffalo. During daylight hours (07.00 to 19.00 hours) we used telescopes to observe the animals, but at night (19.00 to 07.00 hours) we conducted observations with the aid of a nightscope image intensifier and on very dark nights we illuminated the area by infra-red light from a laser when the buffalo were in the woodlands. The mean duration of watches was 4.7 hours. Every 5 minutes, five buffalo were randomly selected and each was classified according to sex and age. We recorded activity in the categories standing, walking, lying, ruminating, grazing and moving. The most striking aspect of the circadian pattern of buffalo activity was the extent of activity during the night: from about 22.00 to 01.00 hours there was a slight tendency for the animals to be lying (Figure 7.4).

The activity pattern of the buffalo changed over the year (Chapter 3) but for the present analysis I confine myself to the results of this observation period only. The proportions of the six activities recorded in Table 7.3 differed significantly between day and night (G = 697.69, d.f. = 5, P < 0.001), and much of this difference is attributable to the differences in the number of animals lying (daytime 53.7%, night-time 8.1%, G = 656.76, d.f. = 1, P < 0.001).

More animals were ruminating during the day than during the night. Rumination during the night was mainly while standing, 60.5%, as compared with 28.0% of rumination during the day. Activity changes in the herd were synchronous (Figure 7.4), and each evening around sunset the herds moved from the open grasslands to the structurally closed vegetations.

When we recorded activity, we also included vigilance, which we defined as standing with the boss of the horns or top of the head above the highest point of the shoulder. Normally the ears of a vigilant buffalo are extended sideways and it appears to be gazing in a particular direction when it adopts this posture. Our definition required that an animal stood; hence vigilance was expressed as percentage of standing animals. Ungulates seem to be aware of their surroundings while grazing (Underwood 1982, see also Ydenberg and Dill 1986), and I acknowledge

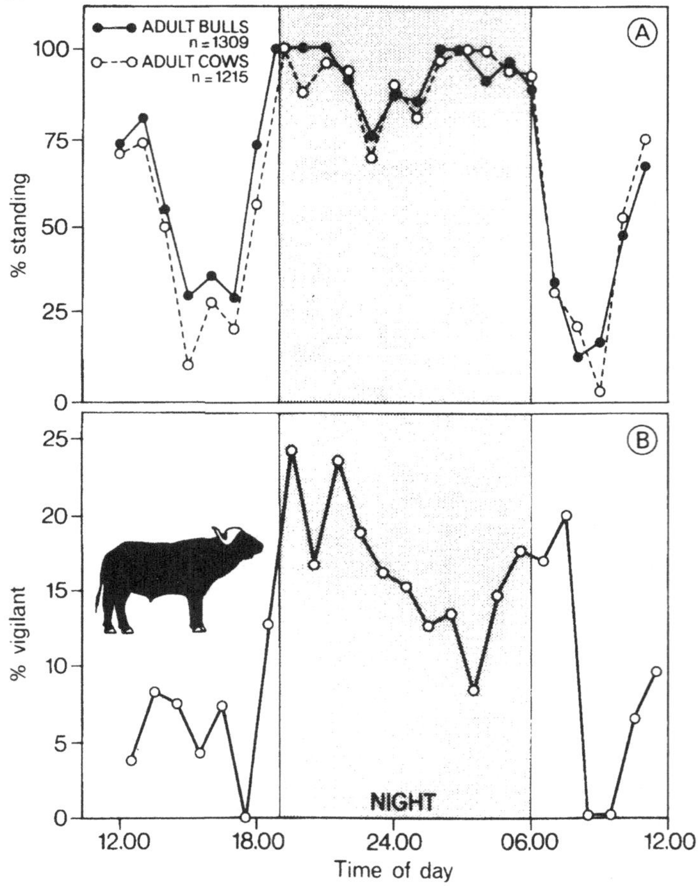

Figure 7.4 Proportion of adult male and adult female buffalo standing over the 24-hour cycle (A), and the proportion of vigilant individuals (B). The inset figure represents a vigilant bull.

this, but we assumed that during vigilance behaviour the awareness was enhanced. We also recorded whether each animal was central or peripheral to the herd. The percentage of standing buffalo that were vigilant at night was greater than during the day (Figure 7.5: night 15.4%, day 9.8%, $G = 13.539$, d.f. $= 1$, $P < 0.001$). In both the day and night periods, those animals at the periphery of the herd spent more time vigilant than those in the centre.

During the night, of the central animals 11.6% were vigilant and of the peripheral ones 18.7% ($G = 8.725$, d.f. $= 1$, $P < 0.01$). During the day,

Table 7.3 The percentage of the main activity types for each age–sex class of buffalo by day (07.00–19.00 hours) and by night (20.00–06.00 hours)

	Moving	*Grazing*	*Inactive standing*	*Ruminating standing*	*Ruminating lying*	*Inactive lying*	*No. of observations*
Night							
Subadult females	24.5	51.0	10.8	7.8	3.9	0	102
Adult females	23.8	46.8	9.8	9.5	6.7	2.0	357
Subadult males	24.4	30.7	18.9	14.2	6.3	1.6	127
Adult males	22.2	34.2	20.4	8.6	6.4	1.5	406
Day							
Subadult females	9.3	15.3	6.9	6.5	31.5	30.1	216
Adult females	10.4	20.9	7.0	6.0	30.8	24.5	766
Subadult males	10.6	16.2	12.3	4.3	22.5	32.8	302
Adult males	11.2	21.1	10.2	7.5	18.8	28.7	805

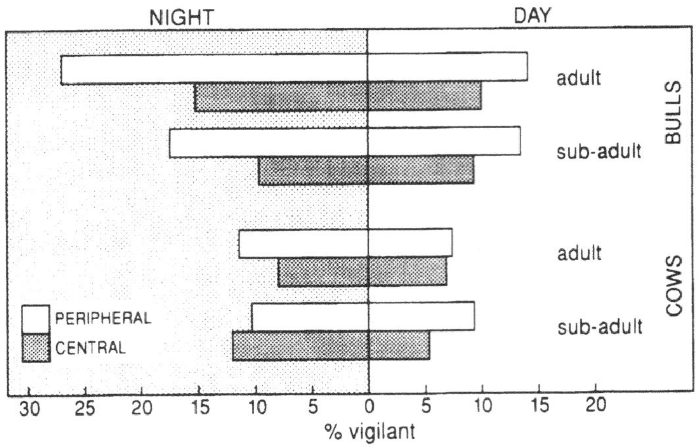

Figure 7.5 The percentage of standing animals which were vigilant broken down by age–sex category, central and peripheral to the herd, and for day and night.

7.6% of the central buffalo were vigilant and 11.9% of the peripheral ones were ($G = 5.029$, d.f. = 1, $P < 0.05$). These differences between central and peripheral buffalo, the peripheral ones being more vigilant, are in agreement with the findings of Underwood (1982). There was no difference between age–sex classes in proportions of animals vigilant during daytime, either for peripheral animals ($G = 5.010$, d.f. = 3, NS) or for centrally positioned animals ($G = 2.907$, d.f. = 3, NS). At night, however, there was a significant association between age–sex class and percentage of animals vigilant for peripheral buffalo ($G = 17.262$, d.f. = 3, $P < 0.001$) but not for central buffalo ($G = 4.515$, d.f. = 3, NS) (Figure 7.5). The association between vigilance and age–sex category in peripheral buffalo was due primarily to a sex difference: bulls 24.5% vigilant and cows only 11.1% ($G = 14.505$, d.f. = 1, $P < 0.001$). This is in agreement what would be expected if adult bulls ran a much higher risk of predation than adult cows (Chapter 5).

In section 7.1 I showed that risk of predation was essentially the same on the grasslands and mudflats as in the closed vegetation, but much higher on the ecotone between these two groups of vegetation structures (Figure 7.2). Hence, we expected vigilance to be the same in open areas as in closed vegetations. This prediction, however, was difficult to test owing to the confounding factors of the day–night dichotomy in the activity pattern of the buffalo: at night they were in the woodlands and much more vigilant than during the day, when they were on the grasslands or mudflats. However, peak vigilance was around dusk and dawn (Figure 7.4) when the herds crossed the ecotone between grasslands and woodlands.

Several herbivore species in the Serengeti are found in the neighbourhood of large herds of wildebeest more often than would be expected on the basis of a random distribution, and it has been suggested that this is due to a lower risk of predation in the neighbourhood of these herds (Sinclair 1985). In Manyara wildebeest are sedentary and their herds are small as compared with the migratory herds in the Serengeti (average group size was about 44 individuals; Table 6.4). Therefore it can be assumed that the Manyara wildebeest are more vulnerable to predators than those in the Serengeti. The most numerous herbivore in Manyara is buffalo, which constitutes the staple diet of lions. Lion predation of zebra and wildebeest is thought to be considerable, because lions prefer them to buffalo (Chapter 5). As predation is an important death cause in Manyara, one could hypothesize that individuals of species that are vulnerable to predation lower this risk by following or staying close to large herds or to animals which can protect themselves against predation. Large herds of buffalo sometimes chase lions, and are even known to kill them, so buffalo could be important to other herbivore species for protection against predators. The same applies to elephant, which I have seen chase lions. Moreover, because buffalo live in large herds with many vigilant individuals, they could be important for predator detection, from which individuals of other species, especially zebra and wildebeest could profit.

These observations and ideas lead to the following hypothesis: a relatively strong association is expected between individuals of species that are highly vulnerable to predation (wildebeest and zebra) and individuals of species that can defend themselves against predators (namely elephant and large herds of buffalo). Both zebra and wildebeest were seen more frequently in association with elephant on the *Cynodon* vegetation than would be expected on the basis of a random distribution of the individuals of these species, but this was not the case on *Sporobolus* patches (Table 6.3). It is remarkable that, on the *Cynodon* vegetation, the group size of zebra in the presence of elephant was significantly smaller than in their absence (group size of zebra was on average 14 and 25 individuals respectively). An effect on group size was not found for elephant itself or for wildebeest (Table 6.4). This effect of elephant on zebra could be caused by predator protection, but it is unclear why this mechanism would not work on the *Sporobolus* vegetation or, alternatively stated, why we did not detect such an effect on this latter vegetation.

Although the hypothesized mechanism of protection against predators (physically or through a better chance of detection) could be important in Manyara, its effects are difficult to detect through a correlative study, and most evidence for it is ambiguous. However, the observation that zebra and wildebeest occur together much more often than expected, could well be due to an increased chance of predator detection. On a daily basis, the visits of buffalo and elephant to the same patches occurred at random (Tables 6.3 to 6.5). Both species are well equipped to defend themselves against large predators such as lions, and would not benefit from an association. However, the positive association between buffalo and zebra and between buffalo and wildebeest, could be caused by individuals either increasing their chances of detecting predators or receiving protection against predators. These positive associations were only present on the *Cynodon* patches but not on the *Sporobolus* patches. In this chapter I have shown that the risk of predation for buffalo is higher on *Sporobolus* patches than on *Cynodon* patches. Thus, the positive association between buffalo and the other herbivore species on the *Cynodon* patches may be responsible for this relatively lower predation risk. However, the positive association of wildebeest and zebra with buffalo could also be caused through the food manipulation strategy of the buffalo on *Cynodon* patches (see section 7.4 on patch selection), from which association the buffalo would benefit through a lowered risk of predation but I believe that for buffalo predator detection or protection against predators through association with other species is not important in Manyara.

7.3 BUFFALO IGNORE PREDATION RISK WHEN DECIDING WHERE TO GRAZE

From the observations I concluded that buffalo, although they ran a measurable risk of being killed by lions (over 88% of the adult and subadult buffalo finished in a lion's belly: Chapter 5), did not regulate their behaviour in relation to this risk alone. Perhaps buffalo even ignore the risks all together. Leuthold (1977) expressed the opinion that ungulates should adjust the timing of their activity pattern relative to that of the predator. There was no evidence that the Manyara buffalo did this. They passed the dangerous ecotone at the most dangerous times of the day, predictably, when lions were most active. I think that it would have been safer to pass the ecotone randomly, providing synchrony within the herd is achieved. At first sight, the choice of the lake shore plains during daytime was probably due to the presence of more cooling air movements and fewer troublesome tsetse flies than in the woodlands. However, one observation pleads against this explanation. The Escarpment Herd always had its afternoon rest in the woodlands, while the other herds rested in the open. Even if the Central Herd and the Escarpment Herd were grazing within 100 m of each other in the morning, the Escarpment Herd would rest in a woodland and the Central Herd would recede to the lake shore. I noticed this pattern between 1981 and 1985, and in 1991 the herds still maintained this difference in behaviour. Vegetation characteristics such as productivity (Chapter 1) as well as feeding competition with other herbivores (Chapter 6) were also likely determinants of habitat choice. In January and February, during the wet season, grazings available within woodlands were only exploited at night, with concomitant risk of predation (see Van Orsdol 1984, who found that mortality of large prey was greater at night). One way in which this risk was perhaps reduced was by ruminating while standing at night rather than adopting a potentially vulnerable lying posture (Table 7.3). However, to minimize the risk, buffalo should have grazed randomly in the woodlands and should not, as they did, enter these woodlands every night predictably.

In the study period, buffalo were more vigilant at night than during the day (Figure 7.5). Possible reasons for this include: (1) predators were more active at this time and prey was less able to detect them (Van Orsdol 1984); (2) evening movements into areas of denser vegetation cover broke down communication between members of large herds, producing many units of fewer individuals (Underwood 1982), each spending more time on anti-predator vigilance; or (3) increased buffalo activity at night meant that intraspecific encounters and alerting behaviour were more likely. An observational study such as this cannot distinguish between these causes, which are not mutually exclusive. However, the spatial distribution of lion kills indicates that lions mainly killed in the ecotone corridor rather than in the woodlands, and vigilance peaked at the time the herds crossed the ecotone. Hence, night-time predator activity *per se* cannot be separated from the breakdown in communication.

How do the other data on vigilance bear on this problem? Peripheral buffalo were more vigilant than those within the herd. Unless interactions between buffalo were more prevalent at the periphery, causing higher vigilance, as suggested for birds (Murton 1971), this observation is consistent with the hypothesis that peripheral animals were more vulnerable to predation, and hence were more vigilant (Jennings and Evans 1980). Hirth and McCullough (1977) contend that dominant males of wapiti and pronghorn antelope force subordinate males to the periphery of the (fleeing) herd where, as they claim, these males are more vulnerable. Concerning anti-predator vigilance, there was no clear a priori prediction as to which sex of buffalo should have been most vigilant when in large mixed herds. Many studies, including the study on the Manyara buffalo, have shown a bias towards males in lion kills (Chapter 5). These observations might have resulted from males being less vigilant for predators than females. It has been suggested however, that males killed by lions were from small bachelor groups (Sinclair 1977; Chapter 5), and then we might expect the converse, that is that males might be more vigilant on returning to large herds because of a carry-over effect from their recent experience when in small bachelor groups. Our data that bulls spent more time vigilant than cows contradict the first and are not entirely consistent with the second of these possible explanations. In particular, if solely the carry-over effect were correct, then bulls in central positions would have been equally as vigilant as those at the periphery. Our data showed greater male vigilance of both adults and subadults in large herds in January to February, while subadult bulls have not been outside their herds, again making the carry-over effect explanation unlikely. This period, just after the short rains and before the long ones, is immediately before the main period of conceptions (Chapter 5) when males are seeking access to mates, and we speculate that interactions between males, or males checking for oestrus females, may have increased the levels of vigilance in bulls. A similar conclusion was reached regarding male vigilance behaviour in harbour seals (Renouf and Lawson 1986).

Although buffalo were the major prey item of lions in Manyara (Chapter 5), some important aspects of their activity cycle were contradictory to the generally predicted anti-predator strategy. Lions, especially those groups likely to include buffalo in their diet, hunt mainly at night (Van Orsdol 1984). However, buffalo not only grazed more at night, but they did so in thicker, closed vegetation types, although darkness and vegetation cover increases lion hunting success (Schaller 1972, Van Orsdol 1984). I doubt whether Kingdon's (1982, p. 64) idea that the 'nocturnal activity pattern can also be seen as simulation of the cool, dark conditions of the ancestral forests' is sufficient to explain this behaviour. It should be noted here that Radclyffe Dugmore observed at the beginning of the twentieth century that:

> during the daytime [the buffalo] usually stay in the dark forests... late in the afternoon they come out to feed, stay out in the open throughout the night and until the morning... by nine o'clock usually return to the seclusion of the forest. This is more or less the rule... but there are places where they may be seen feeding in the open or resting under scattered trees, as they used generally to do **before the great rinderpest epidemic** [emphasis added]... the few that survived became almost entirely nocturnal in their habits, keeping in the thick cover of papyrus or forest from sunrise to sunset. From observations during the past few years it seems as though they are gradually reverting to their former habits.
>
> (Radclyffe Dugmore 1925, pp. 11 and 12)

The causal link between the disease and the changed behaviour, though implicated, is unclear, but there might have been an effect of increased risk of predation at that time because ruminant densities were so depressed (Chapter 5) relative to lion density (which also explains the high incidence of human-eating behaviour at that time).

Other aspects of the Manyara buffalo's behaviour contradictory to the generally predicted anti-predator strategy were that they crossed the ecotone between closed vegetation and open areas, the major danger area, at the time of peak lion activity and at a predictable time, without showing clear uneasiness but with enhanced vigilance. Their vigilance did not appear to have been clearly geared to the detection of predators but, if they detected large predators, they even attacked them. Nevertheless, they fell victim to lions. Is there any evidence that buffalo basically ignored lions, and took their daily decisions regarding the choice of habitats without taking risk into account? I believe that the data as presented in Tables 7.4 and 7.5 strongly point in this direction and that buffalo made decisions regarding habitat choice on the basis of thermoregulation, fly avoidance and especially their food requirements, given the fact that they spent most of their lives in herds, which have probably evolved as a reaction to predation pressure among other factors.

In this respect individual buffalo behaved like individual people (for example Vlek and Stallen 1981): statistically it can be shown that individuals run certain risks attached to certain actions but in their daily lives they appear to ignore these risks entirely. With their large horns and sharp hooves buffalo can defend themselves against attacking lions and their behaviour reflects this (Schaller 1972, p. 232). Not only do buffalo watch lion pad past them a mere 20–30 m away, despite the fact that their flight distance is considerably lower than that of gazelle and zebra (*loc. cit.*), but I have several times observed buffalo going out of their way to attack or chase away lions without any provocation, a phenomenon which apparently is not rare (A. Geertsema and J. Hanby, personal communication; see also Mitchell *et al.* 1965, Mloszewski 1983).

My conclusion then is that buffalo basically ignored the risks of lion predation, not because they were invulnerable but because the advantages outweighed the risks associated with a particular behaviour. For bulls these advantages were mainly gaining physical condition as fast as possible (Chapter 4) and the trade-off between risk and being bachelor can be coined in terms of lifetime reproductive output. For cows the risk of predation was reasonably low, about 4% per year, provided they stayed in their herds (Chapter 5). This low risk, most likely, cannot be lowered further by a modification of the behaviour of the cows: it is a consequence of the behaviour and density of lions. It must be concluded then that for the cows living alone would be too risky in terms of lifetime reproductive success as compared with what they can achieve by staying in a herd. The explanation for this difference between the two sexes

Table 7.4 Expected and observed time spent by buffalo in different vegetation types

	Vegetation type (based on phytosociological analysis)											
	A0	*A1*	*A2*	*A3*	*G1+2*	*A4*	*A5*	*G3+4+5*	*W*	*R*	*E*	*Total*
Area (ha)	*752*	*97*	*319*	*365*	*424*	*31*	*95*	*979*	*708*	*666*	*948*	*5384*
Expected time budget for all buffalo	14.0	1.8	5.9	6.8	7.9	0.6	1.8	18.2	13.2	12.4	17.6	100%
Observed time budget for herd buffalo	12.5	2.5	5.0	15.0	5.0	1.0	0.5	0.5	25.0	27.5	5.5	100%
Preference index	0.89	1.39	0.85	2.21	0.63	1.67	0.28	0.03	1.89	2.22	0.31	
Observed time budget of all adult bulls	16.3	8.8	3.8	13.8	2.5	0.8	0.3	0.5	22.5	25.5	4.0	100%
Preference index	1.16	4.89	0.64	2.03	0.32	1.33	0.17	0.03	1.70	2.06	0.23	
Observed time budget of bachelor bulls only	20.0	15.0	2.5	12.5	0.5	0.5	0.0	0.5	20.0	23.5	5.0	100%
Preference index	1.43	8.33	0.42	1.84	0.06	0.83	0.00	0.03	1.52	1.90	0.28	

If buffalo used all vegetation types equally, then the amount of time spent in the different vegetation units would be proportional to the surface area of these units. However, a Pearson rank correlation test shows that this is not the case (for herd buffalo $r_s = 0.354$, for all adult bulls $r_s = 0.319$, and for bachelor bulls $r_s = 0.254$). The difference between preference for different vegetation units of herd buffalo and of adult bulls is small (Wilcoxon test, $N = 38$, $n = 10$, NS), and the same applies to the difference in preference of herd animals and bachelor bulls (Wilcoxon test, $N = 38$, $n = 10$, NS), or between difference in preference of all bulls and bachelors only (Wilcoxon test, $N = 38$, $n = 10$, NS).

Table 7.5 The relation between risk and vegetation type for different categories of buffalo (based on Table 7.1: S, safe; D, dangerous; –, neutral) compared with the deviation of the percentage time spent in the different vegetation units (based on the expectation that buffalo spend their time in relation to the acreage of the different vegetation units – from Table 7.4; more time than expected is defined as a preference index of 2.0 or higher and less than expected as preference index of 0.5 or lower)

	Vegetation type (based on phytosociological analysis)										
	A0	*A1*	*A2*	*A3*	*G1+2*	*A4*	*A5*	*G3+4+5*	*W*	*R*	*E*
Risk for herd-living buffalo (from Table 8.1)	S	–	D	–	–	D	–	–	–	–	–
Time spent in area compared with expectation based on acreage (from Table 8.5)	Equal	Equal	Equal	More	Equal	Equal	Less	Less	Equal	More	Equal
Risk for adult buffalo bulls (from Table 8.1)	–	D	D	–	–	D	–	–	S	–	–
Time spent in area compared with expectation based on acreage (from Table 8.5)	Equal	More	Less	Equal	Less	Equal	Less	Less	Equal	Equal	Less

If predation were an important factor in deciding where the buffalo would like to spend their time, one would expect more time to be spend in relatively safe areas and less time in relatively dangerous areas. This is not the case and factors other than predation risk appear to co-determine where buffalo spend their time in Manyara.

should be sought in the fact that a bull in prime condition can probably sire many offspring in a few days. For him, in terms of fitness, it pays to strive for a better condition than his competitors. The Manyara buffalo cows, however, with their average calving interval of some 3 years (Chapter 5), could not risk reducing their reproductive life by falling victim to a lion while living alone because they do not gain increased fecundity through increased condition when foraging alone. Many females pay for protection against lions by social inequality and its associated lowered reproductive success (Chapters 3 and 5). Breaking away and starting a new herd is one possibility but this needs the support of other females. One female alone apparently cannot break away because the increased risk would not be compensated for by higher fecundity. The key to fecundity is physical condition (Chapter 5), and this is determined by food intake and food quality (Chapters 2 and 3). Hence I concluded that group living (except for adult bulls), in addition to its anti-predator benefits, has benefits associated with food intake. In the remaining part of this chapter I will look again at food intake and quality in relation to the behaviour of group-living buffalo.

7.4 PATCH SELECTION AND REGRAZING

A picture of buffalo social organization emerges from which it becomes clear that buffalo are not just a random collection of individuals that move around to graze or to avoid being killed by lions. On the contrary, females and subadults move in highly organized social units which are very permanent and have a high fidelity to their home range (Chapter 3). The risk of predation for the Manyara buffalo was reasonably low, and they apparently ignored the risk of lion predation as long as they lived within the protection of their herd companions. The consequence of this type of social organization is severe social inequality in the large herds, rear animals suffering most. The question that has vexed me since I discovered this is: 'Why do these rear animals stay in the herd?', or, alternatively, 'Why do these animals not start a new herd?'. Part of the answer is that a particular female who leaves her herd will be exposed to a risk of predation at least equal to that of the bachelor bulls, that is her risk of predation will have increased by at least a factor of 3 (Chapter 5). In anthropomorphic terms, her problem is to convince other females to join her so as to reduce the risk. It would be very interesting to study the exact relation between herd size and predation risk for each individual, because such a study would find, I am sure, that at some size of the 'splinter herd' the risk for each individual would equal that of the risk in the original herd. At that particular size of the splinter herd the decision as to whether or not to leave the main herd would become a neutral one in risk terms. This would then explain why in the fusion–fission pattern that I described in Chapter 3 the group of 'notorious splitters' occurred in a herd fragment of at least 100 individuals. Yet, despite these considerations, I came to the conclusion that buffalo gain something extra by staying in a large herd even when suffering from social inequality. I hinted at the possibility of information sharing (Chapter 3; see also Chapter 8). Here I will examine the spatial pattern of resource exploitation by the buffalo of Manyara. This could lead to an understanding of the foraging decision rules of buffalo.

Buffalo relied to a large extent on the grass *Cynodon dactylon* to cover their nutritional needs but utilized other graminoids too. In most years, say nine out of ten, they nearly completely ignored browse species as sources of food. Only in exceptionally dry years were they known to browse to some extent (Chapters 1 and 2). In all other years the Manyara buffalo assembled a diet from *C. dactylon* and the other food plants with a quality that ranged between 6 and 13% crude protein. Such a diet was nearly always sufficient to cover maintenance requirements (Chapter 2). In Chapter 1 I showed that *C. dactylon* grows in Manyara nearly exclusively in monospecies stands in areas with groundwater within rooting depth. Hence, this grass remains green throughout most dry seasons. During grazing the buffalo strongly avoided ingesting dead plant material, that is dead leaves and dead stalks.

Buffalo graze in dense herds, and during grazing they feed closely together (Figure 3.11). Observations both in Manyara and in Arusha National Park (also northern Tanzania) led Vesey-FitzGerald (1974) to conclude that buffalo maintain grazing swards (see also McNaughton 1979, 1983). The idea was that grazing could increase the net primary production of the sward; there has been some controversy about the adequacy of the methods that have been used to study this 'grazing optimization' hypothesis (Belsky 1986, 1987, Belsky *et al.* 1993) but Prins *et al.* (1980) and Cargill and Jefferies (1984) indicate that production optimization is at least possible under certain circumstances. In the study of Prins *et al.* (1980) on brent geese grazing on saltmarshes in The Netherlands, geese showed a very clear revisitation interval of approximately 4.5 days: particular patches were grazed and regrazed again. This revisitation appears to ensure harvesting by the geese of the new plant growth. Buffalo in Manyara also show a clear revisitation interval and regrazed patches of *Cynodon dactylon* on average after 4.5 days (Table 7.6). On average, the annual harvest by all grazers combined was approximately 90% of the leaf production of this grass species (Table 1.8). From this, I would suggest that, if the revisitation of buffalo is geared towards some form of primary production maximization, severe defoliation of this grass would promote new leaf growth; green grass leaves form the preferred food of the buffalo. An additional reason to revisit patches might be that they collect information about the relative quality of the different patches (section 7.5).

Table 7.6 Grazing intervals and visit intervals to the same patches; during a visit buffalo can have grazed or just walked through

Sample date *Period*	*31 Aug* *(10 Aug–14 Sep)*	*5 Oct* *(22 Sep–18 Oct)*	*31 Oct* *(16 Oct–3 Nov)*	*5 Dec* *(20 Nov–19 Dec)*	*ANOVA*
Grazed patches					
Grazing interval (days)	8.2±6.0 n=9	5.3±1.5 n=7	8.4±1.5 n=5	6.4±3.1 n=12	$F_{3,29}$=1.21 NS Mean=6.96, n=33
Visit interval (days)	4.6±4.4 n=22	4.9±1.8 n=8	6.3±2.3 n=11	3.6±1.5 n=29	$F_{3,66}$=2.53 NS Mean=4.46, n=70
Not-grazed patches					
Visit interval (days)	6.8±7.1 n=17	4.2±1.6 n=35	5.7±2.0 n=15	3.2±1.2 n=29	$F_{3,92}$=4.97 P<0.01 Mean=4.57, n=96
Difference between visit intervals of grazed and not-grazed patches	t_{37}=1.18 NS	t_{41}=1.04 NS	t_{24}=0.73 NS	t_{56}=1.07 NS	t_{164}=0.41 NS

I have conducted one clipping experiment in Manyara on *Cynodon plectostachyus*, which is closely related to *C. dactylon*. In this experiment I removed well-quantified amounts of leaf from tillers during the late rainy season, and measured the regrowth (in cm new leaf per tiller). I thought that the late rainy season was the best time of the year to conduct this field experiment, since at this time of the year both a moderate water deficiency and nitrogen deficiency occur. Indeed, Georgiadis *et al.* (1989) showed that in the grass *Sporobolus kentrophyllus* of the short grassland plains of the Serengeti overcompensation of clipped plants occurred only when plant growth was limited by soil water availability (see also Chapin and McNaughton 1989). Although the data from the Manyara field experiment are few, there is a clear indication that the highest regrowth occurred when approximately 90% of the leaves were removed but when the stalk was damaged as little as possible (Figure 7.6).

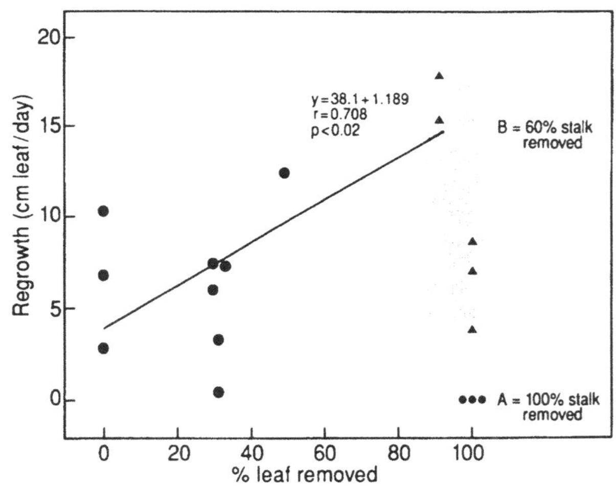

Figure 7.6 Regrowth in relation to clipping intensity in the grass *Cynodon plectostachyus*. Regrowth is more vigorous when proportionally more leaves have been removed, but if too much stalk is removed regrowth can fail totally.

This is similar to how buffalo graze from this species: they remove nearly all leaves by wrapping their tongue around a few tillers and then jerk their heads up. In this way, only the top part of the stem is harvested. This experiment has to be repeated on other species, but the present data hint at the possibility of regrowth maximization under severe grazing pressure as exerted by the Manyara herbivores and specifically buffalo.

Other experiments that I conducted on clipping *Cyperus laevigatus* (unpublished) indicate that severe defoliation increases the tannin concentration in the regrowth (cf. Bryant *et al.* 1989). Tannin can decrease the digestibility of the food (Owen-Smith *et al.* 1993). Vesey-FitzGerald's (1969, 1974) original ideas on the maintenance of grazing lawns by buffalo need more investigation and especially experimentation before the interaction between food plants and herbivores are really understood.

In the experiment on *C. plectostachyus* the maximum regrowth of leaves was about 15 cm of new leaf per tiller per day; four new leaves per tiller were formed over the 14 days that the experiment lasted (then termites cleared the experimental site). This very high regrowth rate suggests that the buffalo could return to grasses they have harvested after only a few days. The data in Table 7.6 show that on average buffalo returned after 4.5 days to the same patch. In a larger data set that I collected over 6 months, the revisitation interval by buffalo was 5.1 days but the median revisitation interval was shorter: only 3 days (Figure 7.7).

The data that I have suggest that this is actually too short a revisitation interval to maximize the harvestable regrowth. In contrast to the brent geese referred to, which ensured a maximum harvest with their grazing regime, I think that buffalo return too soon to previously grazed patches. My explanation is that because buffalo compete with elephant (Chapter 6)

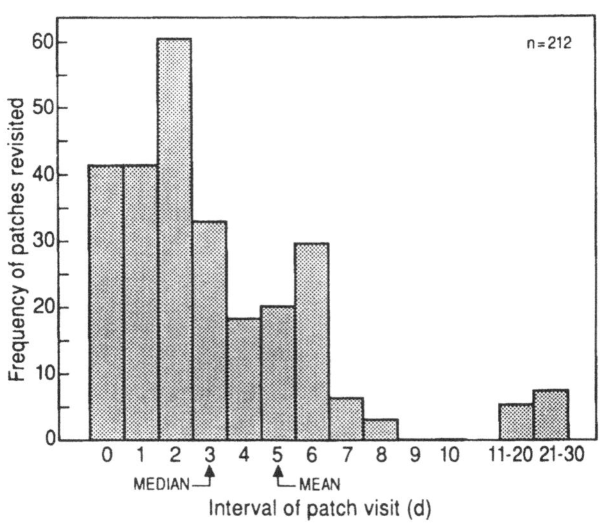

Figure 7.7 Number of patch revisitation intervals in relation to interval length (days) between visits of buffalo to individual patches. Patches are on average about 20 ha in area and are covered by the grass *Cynodon dactylon*. Patches are revisited too early (median is 3 days) relative to the optimal regrazing interval of about five days, possibly due to competition with elephant.

they have to regraze a patch before elephant can start harvesting the grass. This will need more research, but if this is the case, then competition by elephant would force the buffalo to adopt a suboptimal harvesting strategy although other herbivores could benefit from their intense grazing.

Large herds of buffalo have a foraging strategy characterized by a cyclic return time on the *C. dactylon* grass vegetation. Because of the high grazing pressure of these herds, they appear to manipulate the vegetation at patch level. By choosing the optimal return time, it is possible that buffalo could optimize the quality of the vegetation regrowth in relation to their food requirements. Smaller herds or other species are, because of their substantially lower grazing pressure, not able to follow such a vegetation-manipulating foraging strategy. Those species, however, could benefit from the vegetation manipulation strategy of large buffalo herds by following them. Hence, a hypothesis is that a positive association between buffalo herds and individuals of one or more of the other three major herbivore species (that is zebra, wildebeest and elephant) should occur, especially at high consumption levels by buffalo. In that case, a high grazing pressure by buffalo should go hand in hand with a high grazing pressure by the profiting species. The data gathering and analysis necessary for testing this hypothesis were presented in Chapter 6. This hypothesis is supported by the data for zebra and wildebeest, as these two species graze with buffalo in *Cynodon* patches more frequently than expected (Tables 6.3 and 6.5). It appears that buffalo have a despotic foraging strategy: their foraging behaviour dominates, or perhaps even dictates, the foraging strategy of the other species. This is demonstrated by the fact that elephant react negatively to the grazing pattern of buffalo (but not vice versa; Table 6.6) and by the fact that zebra and wildebeest graze together with buffalo (Table 6.5).

7.5 CHANGING PATCH UTILIZATION

During fieldwork in Manyara I observed that some foraging areas with *C. dactylon* were utilized by buffalo, while other areas, apparently quite the same to the human eye, were not accepted as grazing grounds. There was no indication that these areas were inherently more dangerous; at least I never found exceptionally high numbers of kills in those areas nor did I observe lions here more frequently then elsewhere. Moreover, a confusing pattern of buffalo visitation emerged, as during some months certain areas would be utilized by the Manyara buffalo, to be rejected in other months. This resulted in a constantly changing array of accepted and rejected patches (Table 7.7).

The pattern of switching between sites could not be explained solely by a 'win-switch foraging strategy' of the buffalo to allow some regrowth of the grass (Bailey *et al.* 1990), as there was no regrowth of *C. dactylon*

Table 7.7 *Cynodon dactylon* patch utilization by buffalo during the dry season and the beginning of the short rains season (rains started at the end of November)

	Date and buffalo utilization of patches			
Patch name (from north to south)	*31 Aug*	*5 Oct*	*31 Oct*	*5 Dec*
Marere Glade	Yes	Yes	Yes	Yes
Simba River	No	No	Yes	Yes
Mkindu Delta	No	No	No	No
Yellow Fever Tree	No	No	No	No
Albida Corner	No	No	No	No
Msasa Delta	Yes	No	No	Yes
Ndala Pools	Yes	No	Yes	No
Ndala Delta	No	No	No	Yes
Bagayo River	Yes	Yes	Yes	Yes

during August and September (Chapter 2). The hypothesis of Bailey *et al.* (1990) that cattle may be using memory to avoid previously grazed areas instead of remembering where to graze next did not therefore appear to satisfy the case of wild buffalo.

The behaviour of large grazing ruminants, such as buffalo, is difficult to explain by the set of hypotheses and models that are embodied in foraging theory (see for reviews Pyke *et al.* 1977, Stephens and Krebs 1986). The first difficulty is that food is collected during long bouts of grazing, not item for item but more or less as a continuous flow. The second difficulty is that these large grazers have a highly developed rumen in which the food is fermented, often for over 24 hours, before it enters the digestive tract. It thus takes many hours before the animal can 'measure' the quality of its ingested food. Although it is difficult to know the animals' ultimate goals of food selection, large ruminants certainly have to ingest food containing at least 7% crude protein (which is the maintenance level). If the concentration falls below this level then the amount of food which can be ingested is reduced, which results in a decrease in energy intake (Chapter 2). This provides the link with foraging theory, because in most models the net energy intake per unit time is the factor that has to be maximized (Schoener 1971, Pyke *et al.* 1977). In other models it is the rate at which energy becomes available to the animal that has to be maximized (Verlinden and Wiley 1989). Energy, thus, is the 'currency' in which these models are phrased.

Maximization, however, is a difficult task, as it requires perfect information regarding the environment together with an analysis of risks: many economists have come to the conclusion that finding an optimal decision resulting in maximization of currency is nearly impossible (Pottinger 1983). I have presented evidence that buffalo basically ignored the risk of lion predation only, but I did not assess the risk for buffalo of encountering insufficient food, and I do not know a means of measuring this. However, apart from optimal foraging models, economists have also developed other models to predict the behaviour of consumers. One of these is based on strategies only aimed at satisficing certain minimal requirements; such a model requires a less demanding decision-taking model, and it is likely that such strategies can be developed more easily by the consumer than an optimal foraging strategy (Rapport and Turner 1977, Pottinger 1983, *pace* Stephens and Krebs 1986, p. 180).

Whether or not large grazers optimize their energy intake, it appears warranted to suggest that they should at least strive for a diet satisfying the condition of a minimum concentration of crude protein of 7%, that is satisfying the requirement for maintenance since this is closely linked to the intake of digestible energy (Chapter 2). The proximate factors that determine food choices in large grazers remain, however, unknown, and here I try to suggest a mechanism. According to Bailey *et al.* (1989) there is a need for the development of hypothetical mechanisms or spatial decision rules that herbivores might utilize. Ultimately, this will lead to a better understanding of the behaviour and adaptations of the large grazers such as buffalo. The question that is central to the present discussion then is 'What proximate factors were related to the acceptance or rejection of grazing grounds by the buffalo, and how did this (non-)acceptance bear on the (ultimate) satisficing rule of assembling a diet with at least 7% crude protein?' To solve the question of the proximate factors, I reasoned that buffalo would have needed a 'rule of thumb' with which they could assess the quality of the sward before the food entered the digestive tract. This 'rule of thumb' (Janetos and Cole 1981, Stephens and Krebs 1986, p. 172 ff.) should reflect the ultimate quality of the sward, and thus, could be used by buffalo to base their decisions on whether to accept or reject a patch.

Grazing grounds can be considered patches in the sense of foraging theory models if they can be delimited easily in the field. In Manyara, the different swards of *C. dactylon* formed homogeneous monospecific swards with discrete, mappable, boundaries. Thus, the patch concept could be used. I defined patches so that the boundary of a patch was reached when sward species composition dropped to below 90% *C. dactylon*. In the forest glades (vegetation unit G1; see Loth and Prins 1986), this boundary corresponded to the boundary between grassland and forest. On the riverine deltas (G3), it corresponded to the ecotone of grassland with woodland or bushland, or to the transition to *Cyperus laevigatus* (A0) swamp, or to very short *Sporobolus spicatus* (A2) alkaline grassland. In the northern part of the study area, the *C. dactylon* patch boundaries corresponded to Simba River, or a sudden transition to woodland, or sudden transition to (ungrazed) *Chloris gayana* monospecific stands. The location of the patches is shown in Figure 7.8; they all lie between Simba River and Maji Moto.

All patches shown on this map were located within the home range of

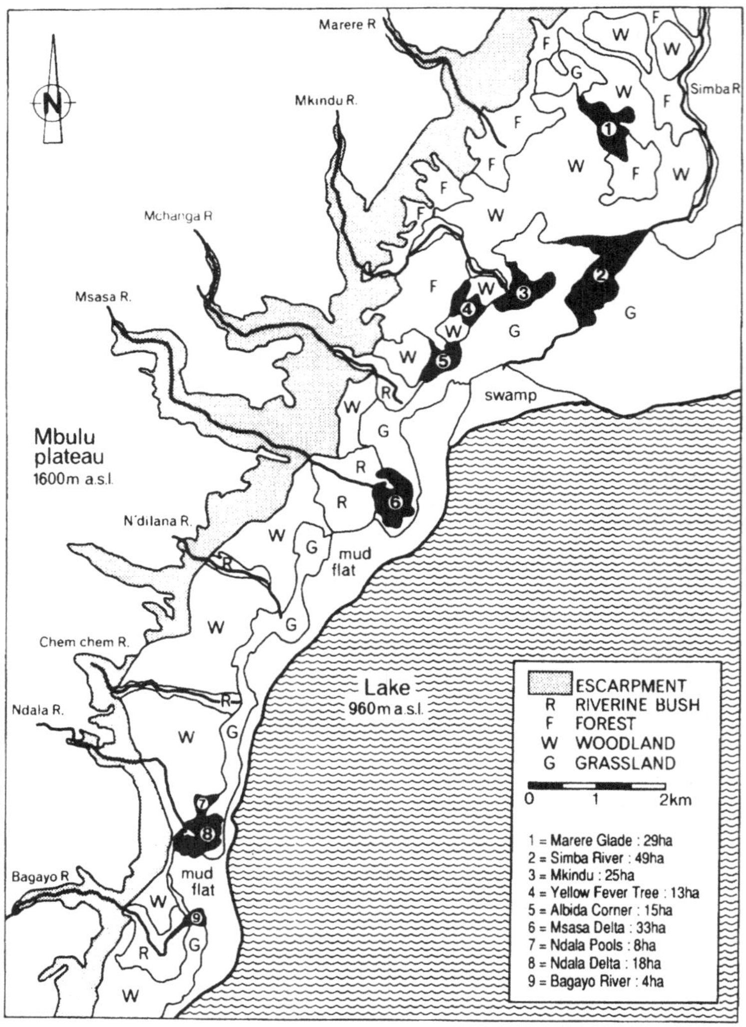

Figure 7.8 Location and size of the *Cynodon dactylon* patches.

the Central Herd, and no other *C. dactylon* patches are found in the area except for two in the extreme southern section of Manyara (approximately 15 km to the south of the southern end of the Central Herd's home range). Another patch, north of the Mkindu and Yellow-fever tree (YFT) patches, was never used by buffalo between 1981 and 1985, because the grass there was always shorter than 2 cm owing to the very heavy grazing by impala. Thus, my analysis was based on nine patches.

I monitored patch utilization by buffalo every day in the 10 days before sampling of the vegetation and in the 10 days following this sampling. Sampling was in 1984 on 31 August, 5 October and 31 October and 5 December. I monitored through direct observation of grazing buffalo, and by searching for dung, tracks or bite marks. On the sampling day I measured sward height 10 times in representative plots in each patch; representativeness was visually determined. I harvested samples of the vegetation at ground level while walking through each plot until I had collected 100 g of fresh weight of grass. I then hand-sorted the samples into four fractions, namely live leaf, live stalk, dead leaf and dead stalk. In the dried subsamples crude protein concentrations were measured (Chapter 2).

7.6 PREVENTING WRONG DECISIONS

The data in Table 7.8 underscore the point made in Chapter 2 that live leaves formed the component of the sward highest in quality. Dead stalk constituted the poorest component. It thus made good sense that buffalo preferred green leaves and avoided dead stalk (Chapter 1). These data also show that sward quality increased again at the end of the long dry season (cf. Chapter 1); in 1984 the rainy season started at the end of November. As I pointed out, I expected that buffalo had a 'rule of thumb' to assess sward quality instantaneously. But how does one find such a rule of thumb? I calculated for each patch and for the four sampling dates a number of potentially usable parameters. These parameters were based on the dry weight ratios of the different components of the sward. The parameters that I tested were 'live leaf divided by live stalk', 'total live biomass divided by total dead biomass', 'live leaf divided by dead leaf', 'live stalk divided by dead stalk', 'total leaf divided by total stalk' and 'live leaf divided by dead stalk'. This was partly an intuitive approach but partly a reasoned one, since (1) any rule of thumb in ecology has to be expressed in quantitative units and (2) ratios can be calculated but also estimated. In a way, this approach was analogous to the initial period in remote sensing techniques when biomass estimates were calibrated by using information from the red and far-red wavebands. It is now a standard approach to use measurements through satellite observation: biomass estimates are derived from the reflected infra-red value plus the reflected red value, and this sum is divided by the

Table 7.8 Mean crude protein concentration in different parts of *Cynodon dactylon* of all patches in Manyara; 95% confidence limits are given in parentheses

	Sampling date			
Sward component	*31 Aug*	*5 Oct*	*31 Oct*	*5 Dec*
Live leaf	11.3 (±1.9)	11.0 (±1.6)	12.1 (±2.1)	17.3 (±2.2)
Live stalk	4.5 (±0.7)	5.0 (±1.1)	7.5 (±1.1)	10.9 (±1.9)
Dead leaf	6.0 (±1.5)	5.2 (±1.3)	7.2 (±2.5)	8.9 (±1.8)
Dead stalk	4.2 (±0.9)	4.0 (±0.8)	4.6 (±0.6)	5.7 (±1.4)
Live sward	6.8 (±1.2)	6.8 (±1.2)	9.6 (±1.3)	14.0 (±2.1)

infra-red value. This index gives a good correlation with the amount of live green biomass density of an area, but the rationale of using this index is based on its utility and not on its physical correctness. The ratio 'live leaf dry weight divided by dead stalk dry weight' was significantly correlated to all other ratios mentioned (Spearman rank correlation, all $P < 0.01$). There was a perfect positive linear relation between the ratio of live leaf over dead stalk as based on dry weights and as based on fresh weights (for example, 31 October 1984: $r = 0.997$, $n = 10$, $P < 0.001$). The ratio of live leaf over dead stalk of the different patches over the four sampling dates was best associated with patch utilization by the buffalo (see below). This particular index made sense because it is positively related to the fraction of the sward for which buffalo had the highest preference (that is green leaf) and is negatively related to the fraction with the lowest utility (that is dead stalk). The advantage, at least for the human observer, is that it can be estimated with only little training (personal observation).

A rather simple (optimal) foraging rule could be 'accept patches that have a ratio of live leaf over dead stalk above the mean of all (potential) patches, and ignore those patches that fall under the mean' (compare Charnov 1976, Verlinden and Wiley 1989). Such a rule has to be based on the current mean, as mean patch quality changed over the season (Figure 7.9).

Because the patch quality parameter was represented by a ratio figure, the current mean had to be calculated as the geometric mean of the ratios (Zar 1984; section 7.7). There was perfect congruity between buffalo accepting or rejecting patches and the *post-hoc* prediction based on the proposed rule of thumb, with the proviso that sward height in a patch was not so low as to prevent grazing by buffalo (Figure 7.9). On 5 October 1984 and on 31 October 1984, the patch named Msasa Delta was too short for buffalo grazing (namely 3 cm) because of intensive wildebeest and zebra grazing. On the same dates, the patch named Bagayo River had a

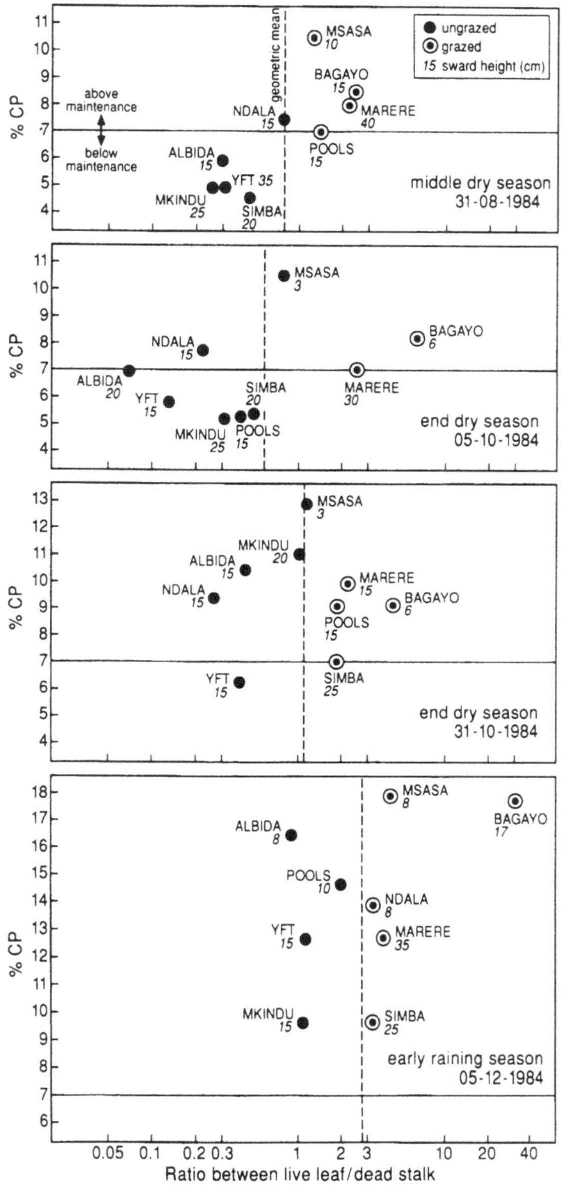

Figure 7.9 Relation between the crude protein concentration in live leaves of *Cynodon dactylon* in different patches (Figure 7.8) and the ratio (on a logarithmic scale) between live leaf and dead stalk in the sward in those patches at different times of the year. The horizontal lines give the buffalo's maintenance requirements, while the vertical broken lines show the geometric means of the ratios between live leaf and dead stalk for the different times. The numbers give sward height in cm. Black dots indicate that buffalo did not graze the sward, while the circled dots indicate that buffalo did graze in the patch. Buffalo only took 'good' or 'neutral' decisions and never took 'bad' decisions' about where to graze (see text).

sward height of 6 cm, which, apparently, was just long enough for buffalo to graze. If the outlined rule of thumb ('accept patches that have a ratio of live leaf over dead stalk above the mean of all patches and ignore those patches that fall under the mean') can be accepted as the proximate rule on which the buffalo based their decision whether to accept or reject patches, then it remains to be determined how this rule would have resulted in the ultimate goal of patch selection. This ultimate goal was only to graze, or to have grazed, in patches with crude protein concentrations higher than the buffalo's maintenance level of 7%. Figure 7.9 shows that buffalo, possibly by applying the proposed rule of thumb, did not utilize patches where the live sward component would provide them with an insufficient concentration of crude protein. Patches with a sward structure (that is a ratio between live leaf and dead stalk) higher than the current mean of all patches always had a crude protein concentration in the live sward component higher than the maintenance requirements of buffalo. This implies that the buffalo's total daily intake in these selected patches was maximal (Chapter 2 and Appendix).

The inferred rule of thumb did not result in the best choices, but it did prevent acceptance of wrong patches. Wrong patches were those patches in which the green sward quality, that is the crude protein concentration, was below the maintenance requirements of buffalo. It thus appears that this rule of thumb minimized the potential loss associated with wrong patch selection (Trueman 1981, p. 126 ff.). Potential loss for a buffalo, in my opinion, occurs when it fills its rumen with food of too low a quality. This would result in a lowered intake and a lowered throughput rate (Chapter 2). The outcome of decisions at the proximate level can be classified as 'patch rejected' and 'patch accepted'. At the ultimate level this outcome can be classified as 'ingested food below the maintenance criterion (of 7% crude protein)' or 'ingested food above the maintenance criterion'. I could qualify the decisions of the buffalo in terms of 'bad', 'good' or 'neutral'. Good decisions were taken when poor-quality patches were rejected and when good-quality patches were accepted, and the quality of the food in that patch would have resulted in ingested food above maintenance. Bad decisions had been made when low-quality patches were accepted. Neutral decisions were taken when good-quality patches were rejected: such decisions would be detrimental to the animal (except in case of food shortage). The data presented in Table 7.9 show that buffalo did not take bad decisions but only neutral and good decisions.

7.7 COGNITION, INTELLIGENCE AND DECISION-MAKING

Making decisions, information processing or using knowledge appears to imply that buffalo have cognitive abilities. Before presenting evidence that the buffalo's behaviour can be explained as if it has cognitive abili-

Table 7.9 Grazing decisions on two levels, namely at the proximate level (rejected or accepted) and at the ultimate level (in relation to the maintenance requirements of buffalo, which is 7% crude protein) and the crude protein content of green leaves of *Cynodon dactylon* growing in the different patches

	Proximate level	*Rejected*				*Accepted*		
	Ultimate level	*(And) below maintenance*		*(But) above maintenance*		*(And) above maintenance*		*(But) below maintenance*
Date	*Qualification of decision*	*Good*		*Neutral*		*Good*		*Bad*
31 Aug 1984		Simba	4.5%	Ndala Delta	7.4%	Marere	8.0%	–
		Mkindu	4.9%			Msasa	10.5%	
		YFT	4.9%			Ndala Pools	7.0%	
		Albida	5.9%			Bagayo	8.4%	
5 Oct 1984		Simba	5.3%	(Msasa	10.5%)	Marare	7.0%	–
		Mkindu	5.1%	Ndala Delta	7.7%	Bagayo	8.4%	
		YFT	5.7%					
		Albida	6.9%					
		Ndala Pools	5.2%					
31 Oct 1984		YFT	6.2%	Mkindu	11.0%	Simba	7.0%	–
				Albida	10.4%	Marere	9.9%	
				(Msasa	12.9%)	Ndala Pools	9.1%	
				Ndala Delta	9.4%	Bagayo	10.2%	
5 Dec 1984		–		Mkindu	9.6%	Simba	9.7%	–
				Albida	16.5%	Marere	12.6%	
				YFT	12.7%	Msasa	17.8%	
				Ndala Pools	14.8%	Ndala Delta	15.3%	
						Bagayo	17.7%	

Decisions are qualified as 'good', 'neutral' and 'bad': buffalo did not take bad decisions. Patches in parentheses: too short sward.

ties, it is, I think, important to resolve the issue of whether it is reasonable to assume that they have cognitive abilities. Moreover, it is important to show that to follow this way of thinking about explaining animals' behaviour is not an outlandish quirk but falls within present-day scientific thinking about animal behaviour.

Let us look at the basic research programme of behavioural ecology so succinctly summarized by Maynard-Smith (1982). The questions are as follows.

1. Why do particular social systems evolve in particular ecological circumstances? (I will come to this question in the final chapter.)
2. If one accepts the rules and customs of a particular society as given, are the actions of different animals in that society those which would be predicted if each individual is behaving, subject to the rules, in the way which would maximize his or her inclusive fitness?

Buffalo cows, like many other animals, have to act so that their fitness is maximized. To achieve this they must exploit their food in such a way that their food intake maximizes their chances of reproduction. Simultaneously, they have to avoid being killed or parasitized, because predation and disease can terminate or reduce their future reproductive output. Their individual food intake depends on the on the food density and food quality on offer, but also the on their ability to maximize their own food intake by preventing competitors removing good food from a patch. Competitors can be herd members or individuals from another species.

The critical assumption in modern behavioural ecology is that variation between individuals, which is the key to understanding natural selection, is due to individual optimization. 'The principle of individual optimization states that each individual should make decisions that are optimal for the particular conditions the individual finds itself in' (Ens 1992, p. 3; see also Drent and Daan 1980). Ens continues by stating that 'clearly, we may only expect animals to profit from an activity when it is their own choice to perform that activity'.

The words 'choice' and 'decisions' are, I think, a reflection of a renewed focus on **causality** of behaviour. In Maynard-Smith's (1982) terminology the word 'strategy' took its meaning from game theory and described a set of actions fully determined by genes. In other words, in an animal-as-phenotype the action determined by a particular genotype came to expression. According to this view 'decision' or 'choice' are not in any way a manifestation of the animal's mind or reflection. However, in the last decade the perception of behavioural ecologists appears to have changed. Now when scientists use words such as 'choice' and 'decision' they are nearly always implying that the animal **has** a mind in which some sort of reflection takes place. Choice and decision are now thought to be cognitive processes. This emancipation of the animal results in the current idea that animals have cognition and perhaps even consciousness – perhaps different from ours but present in some way nevertheless. By cognition psychologists mean that internal representation takes place in the brain, and this requires symbol manipulation. Because of this notion many scientists have concentrated their efforts on finding out whether, for example, primates can use language (Cheney and Seyfarth 1992) or whether they have a concept of number (Davis and Pérusse 1988). Yet, the definition of cognition as used above does not rest easily within the traditional thinking about animal behaviour or its evolution. The definition as used by Cheney and Seyfarth (1992), namely 'cognition [is] the ability to relate different unconnected pieces of information in new ways and to apply the results in an adaptive manner', fits much better. Accepting animal cognition implies that behavioural ecologists should not deny the possibility that animals **do** make decisions or that they **can** make judgements regarding their food.

Whether or not consciousness is involved is the next issue that has to be addressed. Griffin (1992, p. 10 ff.) came to the conclusion that a number of behavioural scientists feel that it is

> likely that animals may sometimes experience perceptual consciousness, but that reflective consciousness is a unique human capability... expressed in the assertion that 'animals may know certain things, but they do not know that they know'.

This agrees with the conclusions reached by Cheney and Seyfarth (1992) in their study of communication in vervet monkeys. This reasoning implies that a trait such as a 'number of offspring' should be viewed not as a trait or a behaviour that has evolved, but as a variable over which individuals may have some control so as to increase its fitness. This control is to be found in choices and decisions made by the animal, choices and decisions that in a restricted group of animal species are under cognitive control.

The key question is then: 'How do we **know** whether these animals exert cognitive control?' Plotkin (1994, p. 150 ff.) came to the conclusion that 'intelligence is an adaptation that allows animals, including ourselves, to track and accommodate to change that occurs at a certain frequency' and 'in general, the kinds of animals that would evolve intelligent adaptations are relatively long-lived and produce relatively few offspring in their lifetime'. He also concluded that

> learning and memory are forms of intelligence common to human beings and quite a wide range of other animals. The ability to manipulate stored knowledge – that is reasoning and thought – is a

> form of intelligence most probably restricted to a small number of species.
>
> (Plotkin 1994, p. 154)

The small group of animals in which this special ability to manipulate stored knowledge developed are social animals; Plotkin (1994, p. 183) claimed that 'the social group is at once both a crucial feature in the survival of social animals, and also the source of that social life which presents the highest intellectual challenge that social animals have to face'. Buffalo meet Plotkin's criteria: they are large, rather long-lived social animals with only few offspring. Hence, it can be assumed that they are quite intelligent animals with the ability to manipulate stored knowledge concerning their environment, both in terms of food and in terms of conspecifics. I thus assume that buffalo have cognitive abilities in the sense of the definition given by Cheney and Seyfarth (1992) above. Even if animal cognition is a thorny subject, the concept of cognition has cropped up time and again in the history of the science of animal behaviour (see also the peer reactions in Davis and Pérusse 1988 and Cheney and Seyfarth 1992). Here I would like to state that when I use terms such as 'decision-making' and 'knowledge' when analysing the Manyara buffalo's behaviour, I use them, with due caution, consciously.

7.8 COLLECTING INFORMATION

If foragers want to reach a judgement about the current mean of sward conditions for the whole home range, they have to collect information on all patches (for example, MacArthur and Pianka 1966, Charnov 1976). The behaviour of the Manyara buffalo appears to indicate that they did this. They visited all patches on a regular basis whether or not they ingested food in a patch (Table 7.6). It should be noted that the northernmost patch (Marere Glade) was in use during all months of the study period, just as the southernmost patch (Bagayo River) was. By shuttling up and down between these two patches the other ones were visited. When the buffalo went through a patch without grazing, they kept their heads low just above the sward; the distance between nose and sward leaf table was approximately 30 cm. This suggests that they were (visually?) inspecting the sward, and thus that they sampled the patch without consuming food. Since the revisitation interval of patches is so short, the permanent herd members (that is excluding adult bulls) are thought to be able to update constantly the information on the status of the different patches relative to each other. Little is known about how animals store this type of information, but it has been demonstrated that in humans a great deal of spatial information is available for retrieval without attention having been directed to it, and it appears that active processing is not required for spatial information to be encoded into long-term storage (Mandler *et al.* 1977). This would imply that it is not essential for buffalo to actively memorize different aspects of patch quality. By analogy with humans it can be assumed that observing the sward structure of patches is sufficient for the necessary information to be stored in their long-term memory. However, it is more important to determine how sward quality is estimated by 'calculating' a ratio of green leaf over dead stalk.

The rule of thumb formulated for grazing buffalo depends on their ability to determine the current mean ratio between live leaf and dead stalk in the swards of all patches in their home range and, perhaps less demanding, on their ability to decide whether a particular patch at a given time has a ratio above that current mean or below that mean. Are buffalo, or other large herbivores, able to make such estimations, and, if so, how do they do this? I can only speculate on possible mechanisms. The easiest assumption is that the ratio between green leaf and dead stalk is related to an index of greenness of the sward. Greenness can then be assumed to be a sensory stimulus. It has been known for a long time that the increment in a stimulus is proportional to the size of the stimulus. Alf and Grossberg (1979) observed that

> although just perceptible increments change as a function of stimulus size, the difference between the logarithms of the stimulus and the just larger stimulus will be constant. Thus... the use of logarithms equalizes the just perceptible differences throughout the usual stimulus range... It can be argued that in combining several independent estimates of a judged quantity [in the present context the greenness of the different patches], one should find the mean of the logarithms of the estimates, and then take the antilogarithm of the result, which is the geometric mean of the observation.

This is what I did when I devised the optimal foraging rule for buffalo in the previous sections.

Foraging theory often demands that the animals know a mean for the foraging area without the authors suggesting how the animals can obtain this knowledge. For example, the marginal value theorem as formulated by Charnov (1976) predicts that a forager will leave a patch if the intake rate (that is food intake per unit invested travel time plus foraging time) drops below the mean intake rate realized for the whole of the foraging area. Another example is given by the model of MacArthur and Pianka (1966), in which patches are ranked according to decreasing expectation of yield; it is assumed that the forager visits the best patch first, than the next best, and so on. At some point a further increase in the number of patches visited will cause an increase in foraging time per unit food larger than the associated decrease in travel time per unit food, and the optimal forager should stop adding more patches to the itinerary at that moment.

It is doubtful that 'mental computations' take place in these cases. First, mental computations in the proper sense of the word need numerical competence. It is not likely that animals have this competence (Davis and Pérusse 1988). Secondly, in humans the outcome of mental computation, for example in judging area or volume, are often erroneous; they frequently lead to underestimation even though the correlation between the true outcome and the judged outcome in these experiments is generally high (Butler and Overshiner 1983). One could, however, claim that the exact numerical correctness of the outcome is not important as long as the judged mean is correctly placed in relation to the judged quantitative estimates of the different patches. Fagot (1978) has addressed the problem of ratio estimation; in the case of the buffalo it then has to be assumed that 'greenness' is not a single stimulus but that 'green leaf' and 'dry stalk' are two stimuli from which ratio estimation has be made. This is possible, but in humans at least this approach is definitely not error-free.

Whatever the exact underlying mechanisms of the behaviour of these 'optimal' foragers, the behaviour has to be based on estimation and on memory. Estimation can be used to make judgements of either numerical or continuous quantities. It is generally assumed that quantitative estimations are based on perceptual features such as length, density or area (so most likely fitting the 'stimulus idea'), but even in humans this process has received scant attention (Chase 1978), and it is far from clear in animals (Houston *et al.* 1982, Green 1984, Kacelnik and Krebs 1985, Davis and Pérusse 1988). Cattle appear to order their choices from larger to smaller rewards and, apparently, can remember not only where they have foraged, but also the amount of food there (Bailey *et al.* 1989). By analogy, it seems to me that buffalo should be able to make similar quantitative or semiquantitative estimates. In a study on optimal foraging in cattle in West Africa (De Boer and Prins 1989) we tried to solve this problem by looking at decisions of cattle herdsmen when they were grazing their cattle. In that study we could at least ask the herdsmen why they took a certain decision, which is not possible when studying free-ranging animals. But even in that study it was not easy to formulate decision criteria within the rules of optimal foraging theory.

Rapport and Turner (1977) made a distinction between satisficer models and threshold models. Since the cost of acquiring information and determining the responses to changes in the environment may be considerable, organisms may be responsive only to stimuli which exceed thresholds rather than to continuous variation. The finding that buffalo appear to base their foraging decisions on the current mean ratio of live leaf to dead stalk, which changes over the season, suggests that a simple threshold model does not apply in their case. The alternative is a satisficing model in which the satisfier examines alternative courses of action and then chooses the first one that satisfies a set of minimal requirements. This alternative seems to fit the buffalo's behaviour better. The ultimate satisfaction requirement had to be a live sward component with a crude protein concentration higher than the buffalo's maintenance level, while the proximate rule of thumb would have been 'accept patches that have a ratio of live leaf over dead stalk higher than the mean ratio of all patches if the sward is longer than approximately 5 cm'. This resulted in grazing buffalo taking good or neutral decisions only, and avoiding bad ones. Since the adaptation of overt behaviour is what counts in the study of animal behaviour, the proposed grazing rule is a workable hypothesis.

On the basis of the formulation of the rule of thumb that buffalo could have applied and on the result that buffalo avoid bad decisions about where to graze, two further questions can be raised. First, are all individuals equally good in assessing sward quality and remembering the quality of other patches? Secondly, can the accuracy of the estimate of both the current mean of the sward quality of all patches and the index of sward quality (that is the ratio of live leaves over dead stalk) be improved through sociality? If so, there could be an answer to the question of why buffalo have been selected for to live in large herd and why animals suffer social inequality. I will try to shed some light on this in the final two chapters.

8

Selecting grazing grounds: a case of voting

8.1 KEY FINDINGS TO UNDERSTAND COMMUNAL DECISION-TAKING

In the previous chapters I have discussed a number of essential details of the life of the Manyara buffalo. These details are of importance to understand their sociality in foraging and the associated decision-taking process. The key findings that relevant are to this can be summarized as follows.

- The Manyara buffalo live in a fine-grained landscape mosaic of good grazing patches, relatively small in comparison with their grazing needs, interspersed with areas bare of vegetation or where shrubs and trees exclude grasses (Chapter 1).
- Food selection by buffalo is aimed at satisfying their requirements for a balanced diet, that is the crude protein content of the diet should be between about 9 and 12% but at least 7% (Chapter 2 and Appendix)
- Out of the vast array of potential food plants only a few species are selected by the buffalo. These food plants offer a crude protein content high enough to satisfy their dietary needs and/or these occur in dense enough swards to satisfy their intake rate requirements (Chapters 1 and 2).
- During much of the year many of these patches, even if they are covered with the right food plants, cannot satisfy the buffalo's requirements, but normally at least some patches are good enough (Chapters 1 and 7).
- Selection of feeding areas is not constrained by the simultaneous need to search for drinking water because watering points are regularly spaced at short intervals of each other (Chapter 1).
- Selection of feeding areas is also not constrained by the need for

predator avoidance since for group-living buffalo the risk of predation is low and, apparently, within an acceptable range (Chapters 5 and 7).

- Selection of areas is not constrained by disease avoidance since the occurrence of diseases (for example rinderpest, anthrax) cannot be avoided. Parasite avoidance is to some extent possible for high-ranking animals but not for others (Chapters 3 and 5).
- Patches that are used by buffalo are also grazed by elephant (and zebra and wildebeest). Grazing by these competitors can severely reduce the amount of food on offer for the buffalo (Chapter 6).
- Timing of patch depletion by competitors is apparently unpredictable and can only be avoided through pre-emptive feeding (Chapters 6 and 7).
- Buffalo cows (but not adult bulls) have an extreme fidelity to their home range. This home range is maintained over generations and is only temporarily vacated during very wet or very dry periods, perhaps only once in a lifetime (Chapter 3, 4 and 7).
- Buffalo cows (but not adult bulls) have a strong fidelity to the herd in which they were born (Chapters 3 and 4).
- Buffalo cows in the rear of a large herd can only rarely attain good enough condition for breeding. They suffer from inequality through a reduced intake rate, most likely a lower food quality and an increased chance of parasite infestation (Chapters 3 and 5).
- Yet these rear buffalo do not readily split off from the herd. When they temporarily leave the main herd, they do so only in groups, and permanent fission takes many years. I have suggested that this splitting off is prevented by the (rear) buffalo's need for information from the other buffalo cows (Chapter 3).
- Decisions about grazing patches to be visited are rarely wrong but nearly always either neutral or correct when it comes to feeding on grass that satisfies the buffalo's needs (Chapter 7).

This then is essentially the herd buffalo's life in Manyara. Most of these are adult cows, but subadult cows and subadult bulls are members too; adult bulls are temporary members only. Since these cows permanently stick to their home range, it must be assumed that fidelity has a high adaptive value. It can also safely be assumed that buffalo cows, in common with many other animals, have a mental map of their home range: indeed they move around their home range, which varies in size between about 5 and 100 km^2 depending on herd size, for on average 12 years, at which age they die.

8.2 A DIFFICULT TASK: INDIVIDUAL JUDGEMENT

Within this home range some areas are good for grazing and others are not. Whether a patch is good for grazing depends on whether or not the patch will meet the intake requirements (in terms of crude protein content and sward density). This will depend on (1) previous grazing history, when and if it was grazed by themselves or by herd members before, (2) patch depletion by competitors, (3) regrowth speed, and (4) soil fertility. The choice of where to feed next also might depend on the spatial relation between the place where the herd currently is in relation to potentially acceptable patches. When this choice has been made, it can be said that a decision has been taken about where to feed. In Chapter 7 the cognitive abilities of animals were discussed. I adopted the definition of Cheney and Seyfarth (1992) that cognition is the ability to relate different unconnected pieces of information in new ways and to apply the results in an adaptive manner.

The task of the buffalo cows is now clear: they have to relate grazing history of a patch, regrowth speed and quality to their own needs in such a way that they ensure maximal body condition. For assessing the grazing history of a patch they can use their memory since I assume that their abilities are at least equal to that of cattle. Indeed, cattle have been observed to order their choices from larger to smaller rewards. From observations and experiments it was also deduced that cattle can remember where they have foraged and also the amount of food they consumed in different places (Bailey *et al.* 1989). They learn regrowth speed and quality through observing regrowth and food quality in the different patches. Both regrowth speed and quality are interdependently related to specific patches, since each patch has different soil fertility characteristics and water availability differs too. Because each patch is slightly different they have to be aware of the spatiotemporal relationship of the patches. This leads to the inference that they have need of a cognitive map and, as I discussed before, it can safely be assumed that they possess such a cognitive map. Assessing their own body condition is apparently possible for buffalo because bulls that have a different condition but are otherwise matched for body size do not engage in dominance interactions, while buffalo bulls that have equal condition can start aggressive interactions (Chapter 4). This I can only explain by assuming some awareness about the ego's individual condition.

The key feature in understanding decision-taking *vis-à-vis* patches to be grazed or not is that only animals in good condition can breed; the adaptive value of reaching good decisions is thus clear. Selecting grazing grounds on the basis of the judgements that I outlined is a task from which, I am sure, many people would refrain. However, for pastoralists such a task is a daily fact of life (De Boer and Prins 1989). I dare to state that I believe that buffalo are at least on a par with these pastoralists when it comes to judging pastures.

What proof is there that buffalo take these decisions and how do they do it? It should be realized that in optimal foraging theory it is assumed that animals can take optimal decisions regarding their food intake (Chapter 7). Most researchers in this field have no difficulty in accepting that 'experimenters have demonstrated that predators often optimise their food intake by regulating the amount of time spent within each discrete foraging patch [or that] many predators will change parameters of their foraging strategies according to the exact distribution, density, and quality of food items' (Johnson 1988). Many, however, would frown when reading that 'animals are natural mathematicians' (Johnson 1988), and few have speculated about the mechanism of this observed behaviour. Yet this is what concerns me here in this and the previous chapter. In Chapter 7 I came to the conclusion that if greenness can be thought of as representing one stimulus, and if this reflects the ratio between green leaf and dead stalk, then this leads to the buffalo avoiding bad decisions regarding patch utilization. The result is that they only graze patches with minimally 7% crude protein.

A major problem remains: when it is concluded that an animal takes a decision based on its judgement of a situation, then the researcher still has two problems. First, how good is the judgement and, secondly, how good is the decision that is based on the judgement? These two questions are not easily separated. However, as an alternative it can be surmised that the decision is based not on judgement but on a random event (like tossing up a coin if one does not know whether to turn left or right) or is related to a proximate environmental variable (such as moving against the wind or with the sun at the back). Randomness is not difficult to discover in a series of observations, but to disprove a relation with a proximate variable is extremely difficult, since one often does not know which particular environmental variable was used by the decision-taker: the number of potential proximate factors is seemingly endless. If a series of decisions (1) are not random, (2) cannot be related to particular proximate environmental variables and (3) are good in realizing the prerequisites of a satisficing diet as related to all possible and alternative choices, I assume that (a) a decision has been taken and (b) that it was based on judgement. In the previous chapter I have shown already that condition (3) is met by the buffalo. Later I will come to the quality of judgement itself and whether there are means of improving its quality, but now I first want to discuss the decision-taking process.

8.3 FIRST EVIDENCE FOR COMMUNAL DECISION-TAKING

Let me first describe what takes place when a buffalo herd in Manyara starts grazing in the evening. Before they start, they are resting in the afternoon. This nearly always means that the herd is lying down somewhere on the mudflats along the lake shore. Typically it is about 28 to 30°C, and there is a slight breeze from the lake towards the escarpment (that is from the east to southeast). Most animals have stopped chewing the cud (about 70%), a few buffalo are standing (1%), and some are lying on their sides and using their horns as a neck-rest (5%). Adult bulls are not searching for adult cows in oestrus and calves are not pestering their mother for milk. It is normally completely quiet and it seems that nothing is happening. However, this is not the case.

Around 17.00 to 17.30 hours a subtle change takes place. It also occurs in less pronounced form in the morning before the midday grazing bout. It took me 2 years of observation to notice this change of behaviour: some buffalo cows arise, shuffle around a bit and bed down again. At first I interpreted this as 'stretching the legs', but one day I noticed that the cows adopt a particular stance after the shuffling and before lying down again. They seem to gaze in one direction and keep their head higher than the normal resting position but lower than the alert. This they do for about 1 minute only. That day I also noticed that most of these cows 'gazed', sequentially, in the same direction: sequentially, because only very few cows are standing at the same time – in the large Central Herd (about 950 animals) one will see at most 15 animals standing but normally only about five. This standing up, gazing and lying down behaviour continues for about an hour, but the overall impression remains that of a herd totally at rest.

Then at about 18.00 hours there is a sudden energizing of the herd. This nearly always occurs when the sun hits the escarpment to the west of them; when the shadow of the escarpment moves over the herd, everywhere in the herd animals become active. Animals that sleep on their side (mostly subadult and adult bulls) suddenly lift their head, calves rise and start nudging their mothers to rise, many buffalo literally stretch their legs when standing and many animals defecate or urinate. The sun hitting the escarpment appears to work as a *Zeitgeber* for the buffalo, indicating that they have to go 'back to work'. Within a few minutes, here and there in the herd some subadult bulls start sparring, adult bulls look around for oestrus cows and adult cows allow their calves to suckle, but grazing rarely takes place. A few moments later, everywhere in the herd buffalo start trekking. The exciting thing is that they start trekking, at the beginning independently of each other, **in the same direction**. Within seconds, the animals that initiate these movements are followed by other individuals, clusters of movement arise, and within about 3–5 minutes the whole herd of hundreds of individuals moves as if conducted by one master. They totally give the impression that they know where they are going to: apparently, some decision has been taken in the group.

A few additional remarks are necessary. First, after many years of observation of individually known buffalo, it is absolutely clear to me that the Manyara buffalo do not have leaders (Chapter 3). Some individual cows are often observed in a forward position within the herd, but they will never be in the leading position for longer than a few minutes if the herd moves in Indian file. Buffalo bulls can keep this position much longer (the longest I observed was for about 1 hour over a distance of 6 km) but they also are individually not leaders: I have observed the same bull in the lead position only three times in a period of 5 years – all other times that he was in the herd he was not in the lead. So 'leadership' changes approximately every minute if the first animal is a cow or every quarter of an hour if the leader is a bull. I never observed subadults or juveniles as 'leader'. Secondly, at the first moments of this mass movement different individuals set off in the same direction while hundreds of often standing animals may separate these initial movers. I think that these initial movers can hardly (if at all) see each other because of buffalo bodies around them obscuring their view: yet they move in the same direction. Thirdly, although this is the general picture of the start of this movement, there are exceptions. I will come to the exceptional situations later.

Once the herd starts moving it is nearly dark. Since Manyara is situated only 3° south of the equator, twilight is short and darkness falls at about 18.45 hours. Night grazing is very important for buffalo (Chapter 2), and generally the next day they graze close to the area where they have been grazing at night. Since the buffalo are always very silent, they will have difficulty maintaining contact when they move through or graze in the woodlands and bushlands of Manyara (Chapter 7), where visibility is frequently less than a few metres. If the buffalo knew from the outset of their leaving the mudflats where they were headed, this would be advantageous if they were to lose contact at night. The concerted movement of the animals appears to suggest that they know where they are going.

When the buffalo start moving at dusk adult bulls normally form the van of the herd. These vanguard bulls are those that are in a good condition; bulls in a poor condition form the rearguard. Sometimes bulls in a poor condition do not leave the grazing ground at all, and become bachelor bulls (Chapter 4). Not all bulls in good condition will always move to the front. Sometimes a number of them will form a sort of 'screen' through which the cows have to pass; the bulls are checking all these females for signs of oestrus by sniffing to the left and right as the cows pass. But most frequently the bulls will start walking at the front of the herd. To a casual observer these bulls seem to lead the herd in a particular direction. This, however, is not the case. I have often observed that when these bulls enter a seemingly adequate patch of grass and start grazing the adult cows just pass these bulls to continue their trek without even paying attention to the grass that the buffalo bulls had deemed good enough for grazing. Within a few minutes, the 'leading' bulls then discovered that their 'followers' have left them. This suggests that the adult females have somehow come to a different decision from the bulls, again suggesting that the cows know where they are going.

From this description it appears that, when buffalo rest in the afternoon, a communal decision about where to graze at night has been taken **before** the movement towards the grazing ground starts. I have several times tried to block the route of the buffalo by placing my car between a cliff and the lake, where the level ground is only about 50 m wide. The buffalo just waited half an hour and ultimately passed by, to continue their trek. I also have chased them back, only to observe that about 1 hour later they came back to continue their trek in the chosen direction. Finally, I have forced the herd to split by letting about 50% of the animals pass. Again, the animals that were forced back by me joined the others later. All these observations enforce the idea that the buffalo have made up their mind about where to go to. I will thus present a formal analysis of my observations of what I have called the buffalo's voting behaviour since this is so closely related to how the buffalo make up their mind.

8.4 ORIENTATIONS OF BUFFALO WHEN GRAZING OR RESTING

I have analysed the orientations of buffalo from approximately 1 hour before their departure until the departure itself. Orientations were scored as compass bearings of the body axis (and head) of lying, standing, grazing and walking animals. In 1983 and 1984 sectors of 45° were used (that is north, northeast, east, etc.) and in 1991 sectors of 22.5° (N, NNE, NE, ENE, E, etc.). These observations were made from a long distance, at least 600 m, because otherwise the buffalo will start looking in the direction of the observer. In addition to scoring compass bearings, a distinction was made between the sexes and ages (subadult, adult), and the animals were classified on the condition scale as explained in Chapter 3. The direction of departure was estimated in the field. The direction of the grazing ground to be visited relative to the area where the animals rested was determined from the map later. When there were relatively few animals in the group it was possible to determine the duration of standing; this was impossible in groups larger than approximately 150 buffalo. Circular statistics was used (Zar 1984) to calculate a number of relevant parameters.

1. The mean angle. In nearly all cases the angular distributions were unimodal or nearly so.
2. The length of the resulting vector. This ranges between 0 when equal numbers of animals are facing outwards from each sector and 1 when

all animals are facing in the same direction. A correction factor to compensate for underestimation of the vector length r was applied, since the data were collected for each sector.

3. The angular deviation as measure of angular dispersion. The angular deviation is similar to the standard deviation but it has a maximum (of 81.03°; its minimum is 0).
4. The 95% confidence limits of the mean angle (with a maximum of 90°).
5. Rayleigh's z was used for testing the null hypothesis of no population mean direction.
6. Finally, the one-sample test for the mean angle was used to determine whether the direction of the vector deviated from direction of movement or the direction of the grazing ground; this test simply checks whether a direction to be tested falls within the 95% confidence region or not. It should be noted that N = 0°, E = 90°, S = 180°, W = 270°.

In the period before departure there are sometimes, but not always, a few buffalo already grazing. In a large herd like the Central Herd (approximately 1000 animals) this ranges between 3–4 animals to 90. These grazing buffalo walk between the lying and sleeping buffalo but do not leave the resting area. As a result, the vector is very short, which coincides with a wide angular dispersion. A long vector would mean that all grazing animals were drifting in the same direction. Unless they were walking in one cohort up and down the resting herd or in one short cohort in a circle, such a relatively long vector would then result in the grazing animals leaving the resting herd within a few minutes, because even if buffalo graze their walking speed is still some 3 km h^{-1}, while the diameter of a large resting herd is at most only 200 m. Because these grazing animals do not leave the resting area, they graze in a criss-cross pattern between their recumbent herd companions.

As an example, on 9 February 1984 in the late afternoon between 17.00 and 18.15 hours the parameters were as follows: mean angle = 73°, the angular deviation 72°, z = 9.3, which shows that the animals were not totally randomly oriented during grazing as $P < 0.001$, and vector length $r = 0.221$, $n = 190$. This drifting of a few grazing animals at a resting site is also frequently seen during the midday resting period. For example, on the same day between 11.15 and 12.15 hours: mean angle = 65°, angular deviation = 77°, $z = 3.93$ with $P < 0.02$; and vector length $r = 0.090$, $n =$ 491. The very short vector shows that the point of gravity of the grazing animals moved only very little. On other days the movements of the grazing buffalo were truly random, as on 8 February 1984 in the late afternoon: mean angle = 116°, angular deviation = 80°, $z = 0.025$, which is not significant, and vector length $r = 0.016$, $n = 96$.

Resting animals are sometimes randomly orientated. For example, on 27 June 1983 between 17.00 and 18.00 hours: mean angle = 351°, angular deviation = 76°, z = 1.928, which indicates randomness, and vector length r = 0.119, n = 137. On other days they are not lying randomly but show some orientation, as on 14 October 1984: mean angle = 155°, angular deviation = 69°, z = 9.58 with $P < 0.0001$, and vector length r = 0.275, n = 129. The wide angular deviations and the short vector lengths indicated that these animals showed a wide distribution in compass bearings.

8.5 WHICH OF THE SEXES OR AGE CLASSES SHOW 'VOTING' BEHAVIOUR?

However, when the animals rise, shuffle around and adopt their 'voting posture', the resulting compass bearings are very different. A clear unidirectionality of these animals is the rule in these voting buffalo. A good example occurred on 8 June 1983, when I was able to score the bearing of the same individual lying animals immediately before rising and their subsequent direction when voting. The parameters of the lying animals before the voting were as follows: mean angle = 172°, angular deviation = 66°, z = 6.63 with $P < 0.002$, and vector length r = 0.344, n = 56. However, during standing the parameters were as follows: mean angle = 314°, angular deviation = 41°, z = 31.73 with $P < 0.00001$, and vector length r = 0.746, n = 56. The 95% confidence limits decreased from 30 to 12°, and the buffalo nearly fully turned around by 140°. The result is that the two angular distributions are significantly different ($\chi^2 = 54.193$, $P < 0.0001$). The most interesting observation, however, is that the vector length increased drastically: the buffalo showed a much greater consensus in their voting than in their resting orientations. This is a general observation. Lengths of vectors of voting animals are quite long (Table 8.1). I will come to this again later.

Table 8.1 Relationship between the compass bearing of the voting vector, the direction of departure from the resting area and the direction of the grazing area in relation to the resting area

Date	*Vote*	*Departure direction*	*Grazing ground direction*	*Remarks*
8/6/83	314°±12° r=0.746 z=31.73; n=57 P<0.00001	315°	320° to 310°	
2/12/83	None	270°	0° to 180°	Herd split
6/12/83 (morning)	253°±5° r=0.953 z=38.12; n=42 P<0.00001	230°	250° to 225°	Voting extremely short: about 5 minutes
6/12/83 (evening)	None	180°	225°	Herd split

8/12/83	241°±9° r=0.658 z=48.87; n=113 P<0.00001	205°	210° to 250°	
12/12/83	350°±8° r=0.971 z=16.01; n=17 P<0.00001	350°	345°	
8/2/84	266°±8° r=0.484 z=72.72; n=310 P<0.00001	260°	260° to 225°	Herd split
9/2/84	231°±10° r=0.735 z=41.71; n=76 P<0.00001	241°	230° to 225°	
10/2/84	229°±14° r=0.671 z=25, 24; n=56 P<0.00001	340°	220°	
11/2/84	13°±8° r=0.596 z=69.11; n=185 P<0.00001	348°	10°	Herd probably split
19/5/91	316°±13° r=0.855 z=28.54; n=39 P<0.00001	290°	322°	
3/6/91	294°±13° r=0.685 z=38.02; n= 81 P<0.00001	240°	280° to 285°	Herd split
19/6/91	63°±18° r=0.571 z=16.32; n=50 P<0.00001	89°	55°	Herd was split already
21/6/91	267°±11° r=0.613 z=42.45; n=113 P<0.00001	247°	270°	Herd split
11/7/91	307°±10° r=0.613 z=42.45; n=113 P<0.00001	0°	310°	Herd split and returned to the same location later

Directions are expressed in degrees; north is 0°, which is identical to 360°, east is 90°, south is 180°, west is 270°. The angular distance between the compass bearing of the voting vector and that of the grazing grounds is on average 3° only. The parameter r gives the length of the vector, which can vary between 0 (every animal points in a different direction) and 1 (all animals vote for the same direction). z is a probability parameter (see text).

Up until now I have reported data on adult cows only. This I have done on purpose. During the observations it became clear that it is the adult cows that are of interest when it comes to this 'voting behaviour'. Subadult cows or subadult bulls hardly play a role in it. For seven dates I have good censuses of all age–sex categories that could potentially have played a role in voting (8 June 1983, 8 and 10 February 1984, 3, 19 and 21 June 1991 and 11 July 1991). Of all permanent herd members (thus excluding adult bulls), the voters comprised 3.2% subadult cows, 1.5% subadult males and 95.3% adult cows. This distribution of categories deviates significantly of the distribution within the herd (7.8, 11.7 and 80.5%, respectively; $\chi^2 = 106.6$, d.f. = 2, $P < 0.0001$). Standing juveniles or calves were extremely rare. This makes it impossible to determine whether voters were cows with dependent offspring or cows without such offspring.

Adult bulls, being transient members of the herd, quite often become active at the end of a resting period. Sometimes they start grazing and at other times they wander around in the herd, apparently searching for females in oestrus. If they are standing, their orientation is quite often a random one. For example, on 19 May 1991: mean angle = 132°, angular deviation = 76°, 95% confidence limits > 90°, $z = 0.60$, which is not significant, and vector length $r = 0.123$, $n = 40$. Simultaneously the adult cows showed a very clear orientation and a high degree of consensus: mean angle = 316°, angular deviation = 31°, 95% confidence limits = 13°, $z = 28.54$ with $P < 0.00001$, and vector length $r = 0.855$, $n = 39$. A similar situation could be measured on 10 February 1984. The bulls showed a random orientation, but the adult cows had a high degree of consensus (adult bulls, mean angle = 252°, angular deviation = 74°, 95% confidence limits > 90°, $z = 0.41$, which is not significant, and vector length $r = 0.177$, $n = 13$; adult cows: mean angle = 229°, angular deviation = 46°, 95% confidence limits = 14°, $z = 25.20$ with $P < 0.00001$, and vector length $r = 0.678$, $n = 56$). Sometimes adult bulls show a significant mean orientation but then the degree of consensus is still very small, as on 8 February 1984: mean angle = 180°, angular deviation = 67°, 95% confidence limits = 33°, $z = 6.47$ with $P < 0.002$, and vector length $r = 0.328$, $n = 60$. As a result of these and other observations, I have excluded adult buffalo bulls from the voting session analyses, just as the subadult bulls and the subadult cows. It is particularly the adult cows, lifelong members of the herds, that show the voting behaviour.

As outlined above, I have not been able to determine whether or not these voters had calves. I do have, however, some good data on body condition of these voting cows. On the five-point scale that was used in this study (Chapter 3) the adult cows had the following average condition: 3 June 1991, 3.29 (95% confidence limits 0.11, $n = 11$), 20 June, 3.30

(95% confidence limits 0.09, n = 148) and 11 July, 3.18 (95% confidence limits 0.06, n = 98). These average conditions are equal to the average condition for all cows at these times of year (Chapter 3). The associated standard deviations of the body conditions are quite large (0.50, 0.59 and 0.60 respectively), which indicates that adult cows from all condition classes join in voting.

So the typical voters are adult cows of all condition classes. Whether they have dependent offspring or not I do not know, but calves and juveniles do not vote. Subadult bulls and subadult cows also do not vote, and adult bulls, although they may be standing during the time of voting, apparently are more interested in finding oestrus females than in expressing an opinion about where to graze.

8.6 VOTING

Why do I call this behaviour 'voting'? When adult cows have been lying around in the resting herd they have a more or less random orientation (above sections). After they have risen, they shuffle around and then adopt a specific posture, in which their head is raised higher than in the ordinary 'at ease' position but lower than in the 'alert' position. They give the strong impression that they are gazing in a particular direction. The duration of this standing bout is quite short, namely about 1 minute only (8 June 1983: 44 seconds, SD = 40 seconds, n = 51; 3 February 1985: 105 seconds, SD = 86 seconds, n = 23). Once, in the very small Msasa Herd I saw that the individual cows do this repeatedly: bouts of lying down alternated with standing about every 8 minutes. This and other observations plead against the interpretation that the short standing bouts are to be interpreted as comfort behaviour or that they are necessary for defecation or urination. Indeed, if this were the case one would expect at least subadult individuals to do it equally frequent as adult cows.

The orientation of the standing cows shows two remarkable features. First, the consensus in directionality is often high (with long resulting vectors) and, secondly, the resulting confidence region of the mean compass bearing nearly always overlaps with the compass bearing of the area where they are going to graze that night (Table 8.1). For a long time I was reluctant to believe this interpretation. To remove any possible bias I asked Drs Jan Beekman and Henk van der Jeugd to collect data completely independent of me on the same Manyara buffalo. Their observations are in agreement with my own (and have been included in the presented results).

A number of alternative interpretations can be rejected on the basis of the findings. First, that the behaviour could be interpreted as 'comfort behaviour' has been rejected already. Secondly, suppose that the front animals strive to be in the front again and again (Chapter 3). They could do this by ensuring that when the herd moves they are the ones that initiate the movement, in this case they should rise earlier than the other ones. This interpretation is not borne out by (1) standing animals lie down again and (2) the average condition of the standing animals is equal to that of the herd as whole; in the moving herd the condition of the front animals is higher the average (Chapter 3). The third possible interpretation of the described behaviour is that it is not voting behaviour but a so-called 'intention movement'. If that were the case, then I would not expect the following.

- I would expect they would not lie down again.
- An even stronger expression of this intention would be the observed grazing behaviour during the period that the rest of the herd is still resting. However, the orientation of the grazing vector was not related to the direction of departure of the herd.
- If 'voting' were to be interpreted as an intention 'movement' stance, then the direction of its vector would always be equal to the direction of departure; this is not the case (Table 8.1).

If the adoption of this 'voting' stance were to be correlated with a proximate environmental factor, then a number of specific directions of the resulting vector could be postulated. These are (1) facing away from the sun, that is due east in the late afternoon (that is 90°); (2) facing the setting sun, that is due west (270°); (3) facing the wind, for example to be able to smell lions (the wind near-invariably blows from the east to southeast, that is 90 to 135°); and perhaps (4) standing with their rump against the wind (that is 270 to 315°). Inspection of the calculated vector directions of the voting buffalo on the different days (Table 8.1) shows that these vectors do not agree with the interpretation that a proximate environmental factor plays a role in the voting behaviour of buffalo cows.

From the resting location of the herd the relative direction of the grazing ground was established from the map. The mean angular distance between this direction and the direction of the voting vector is only 3°. The angular distance is defined as the shortest distance, in angles, between two points on a circle. In this calculation I have used the absolute angular distances, that is an angular distance of, say, 350° was taken to be equal to an angular distance of 10°. The difference between the voting direction and the direction of the grazing ground is not significantly different (one-sided Hotelling test: $F_{2,11} = 2.371$, $P > 0.1$; on original angular distances). Out of 13 cases, the direction of the grazing ground falls within the 95% confidence limits of the voting vector in 11 cases (Table 8.1). Given the crudeness of our estimates of the direction of the individual voting buffalo cows, this small angular distance is remarkable. On the other hand, the mean angular distance (again, on the basis of absolute distances) between the direction of the initial movement of the herd vacating its resting location and the direction of the voting vector is 25°. The difference between the voting direction and the direction of the initial movement is almost significant (one-sided Hotelling test: $F_{2,11} = 3.151$, $P < 0.1$; on original angular distances). Although the hypothesis that the direction of the voting vector indicates an intention movement, stance cannot be rejected totally on the basis of this test: the finding that in nine out of the 13 cases the direction of the initial movement falls **outside** the 95% confidence limits of the voting vector (Table 8.1) pleads against this interpretation. The much closer agreement between the direction of the voting vector and the direction of the grazing ground leads me to conclude that the hypothesis that voting indicates the direction of the grazing ground is to be preferred.

Thus the behaviour described by me as voting indicates that buffalo made a decision about where to graze before leaving their resting area. This decision was based on judgement by a number of adult cows, because I showed that the outcome of the decision (1) was not random, (2) cannot be related to particular proximate environmental variables and (3) was good in realizing the prerequisites of an satisficing diet as related to all possible and alternative choices. A corollary is that voting is a form of **communal** decision-making.

8.7 ELECTION OR CONSENSUS?

It is now time to look at the precise terminology of the processes that I have described. I have called the behaviour of the buffalo cows at the resting ground 'voting'. The term 'vote' is defined (*Oxford Dictionary*) as (1) a formal expression of will or opinion in regard to election and (2) opinion expressed. The term 'election' is derived from 'to elect', which is defined as (1) to choose (things, to do) and (2) to choose by vote. Thus 'voting' does not necessarily result in 'electing'. The observed behaviour of buffalo when they adopt the described posture of apparently gazing in a particular direction can be inferred to be, at least, the expression of an opinion. **Expression** comes about by the behaviour of the buffalo: a particular stance that is maintained for about 1 minute in a particular direction which is characterized by the posture of the raised head. **Opinion** comes about by the directionality of the stance towards the area to be grazed. This implies that at least some individual adult buffalo cows have an opinion about the area where grazing should take place (see also Chapter 7).

However, is it possible that these buffalo also have an **election** about where to graze? The evidence for this is too meagre. If an election were to take place I would have expected that the direction of the grazing ground would be in the compass bearing of a majority of the votes. This is not the case, since the direction of the grazing ground does not lie in the direction of the majority of the votes but in the direction of the weighed **mean** of the votes (that is the vector). Secondly, I would expect that at least in some instances a bimodal distribution of the directionalities of votes would be apparent in the data, but this is not the case in the present data set.

Thus, selection of grazing grounds appears to take place by means of finding a **consensus**. 'Consensus' is defined as (1) agreement and (2) majority view. It is the first definition that particularly appears to fit the buffalo's behaviour. In a consensus model one tries to reach a communal opinion in which all opinions are weighed equally. The calculation of the vector of all votes is a formal approach to this weighing of votes.

It is interesting to note that I have observed cases in which this reaching of consensus appears to have been constrained. First, on some days a voting procedure was not observed at all (Table 8.1). On these days the herd just started grazing on the resting location itself and subsequently moved off slowly during the night. Secondly, it should be recalled that vector length can vary between 0 and 1. If the vector length is 0, then an equal number of votes have been cast in each direction. If the vector length is 1, then all votes were in the same direction. On average (Table 8.1) the vector length is approximately 0.7 (a second-order mean vector length cannot be calculated because sample sizes are so different). Inspection of the data in Table 8.1 shows that when the vector is longer than 0.65 the herd did not split that night in seven cases, but on one night it did. When the vector was shorter than 0.65, the herd split in three cases and on one night it probably did. There thus appears to be a significant relation between the vector length of the communal votes and herd-splitting propensity ($\chi^2 = 4.215$, d.f. $= 1$, $P < 0.05$). This suggests that if the consensus of the buffalo's opinions is too small the herd will split. However, the selected grazing ground is still in the direction of the vector of all votes, implying that the votes that were cast still did not result in an election procedure based on a majority rule. Further speculation about a possible election procedure is, I think, not possible with the present data however tempting it might be.

8.8 POOLED EXPERT OPINION

Animals and people are at all times faced with the task of figuring out where they are going and how they are going to get there. There is ample evidence that some internal representation of the environment plays a major role in guiding locomotive behaviour for both animals and people. This is called the cognitive map (Loftus 1978). A system of compass bearings forms a fundamental tool by which location information is transmitted from person to person, and possibly also – which is of relevance to buffalo – a fundamental structure within which one constructs an internal representation of one's geographical surroundings (Loftus 1978). With their vote the Manyara buffalo bring to expression their opinion about the exact location of the grazing ground to be visited relative to their present geographical location. How exact the direction of their vote is depends on the precision of their estimates about where they presently are and where the grazing ground is located according to their cognitive map. In humans it has been demonstrated that a great deal of spatial information is available for retrieval without attention having been directed to it, and it appears that active processing is not required for spatial information to be encoded into long-term storage (Mandler *et al.* 1977). It seems likely that the same applies to other large vertebrates such as buffalo. The resulting judgement regarding the direction of the grazing ground is subjective, and, thus, quite likely not free of error. In humans inconsistency can occur in judging information both by cognitive maps acquired from real maps and by direct experience from the area (Moar and Bower 1983).

During the period that an individual buffalo cow expresses her opinion, other adult cows do the same. The resulting vector is, as was shown in the previous section, nearly error-free. Indeed, the mean angular distance between the voting direction and the compass bearing of the grazing ground, which was 3°, is even better than shown by test groups of humans that had to estimate angles from a map (compare Moar and Bower 1983). Apparently, the amalgamation of the different voting directions of the individual buffalo into one consensus direction, as approached by my calculating the mathematic vector, is an adequate approximation of what happens in a buffalo herd. Vector algebra is mathematically akin to regression analysis, in which the sum of squared distances is minimized. The latter was found to match closely the visually perceived line of orientation of a two-dimensional array of dots by humans (Yakimoff and Mitrani 1979). This implies, I think, that individual cows also make an assessment of the vote directions of herd members because they appear to know in which direction the vector of all votes is pointing. I do not know how good they are at this, since humans can both overestimate and underestimate angles, and there are many instances in human perceptual research in which Euclidean relations do not obtain perceptual judgements (Wenderoth and Johnson 1982a). Humans frequently reproduce acute angles larger than they should be and obtuse angles too small (MacRae 1983), but on average the mean error associated with angle estimation does not depend on angle size (Wenderoth and Johnson 1982b). I assume that buffalo **on average** are able to judge correctly the difference in angular distance between the compass bearing of their own body and that of other buffalo.

There are few parallels to be found in ethological studies which show that other animal species employ similar communal decision processes through voting. Of course, the present results in buffalo are reminiscent of the studies of Karl van Frisch on honey bees. Honey bees were shown to be able to communicate both direction and distance of a food source to nest mates. These nest mates were able to 'read' the angle between food source, beehive and the sun. Likewise ideas have been postulated for birds under the so-called 'information centre' hypothesis: the idea that congregations of birds can function as such was originally proposed by Ward (1965) as a result of his observations on quelea birds. Later it was formulated as a more general hypothesis by Ward and Zahavi (1973). Basically the idea is that relatively non-successful individuals are able to collect information from successful individuals and follow them to the resource that will be exploited. It is important to realize that birds 'requesting' information about resources cannot learn anything **directly** by watching their conspecifics' feeding activities as is the case when individuals start foraging where other individuals forage already (so-called local enhancement).

The idea of information centres has attracted quite some attention in ornithology, but to prove the idea has been difficult (Bayer 1982, Ydenberg *et al.* 1983, Ydenberg and Prins 1984). Recent work by Brown (1986) on cliff swallows, Greene (1987) on ospreys and Waltz (1987) on terns has validated the idea that relatively unsuccessful birds follow successful individuals. However, neither the studies on bees nor those on birds showed a communal decision-taking process but can be interpreted as revealing the occurrence of information centres in a number of animal species. The experiments with quelea birds in captivity by De Groot (1980), however, are very revealing. He showed that birds in need of a resource (food or water) of which they do not know the location can find it in the **presence** of knowledgeable flock companions. At least part of this information is probably acquired by following knowledgeable birds (which is the 'normal' component of the information centre hypothesis), but naive birds might even be the **first** to go to a resource. This last observation might perhaps be explained by the reaction of experienced birds, for example through hopping, to the explorative behaviour of the naive birds. So, in De Groot's experiments there is a clear suggestion of information transfer before feeding that could have influenced decision-taking by naive individuals.

The study of Stolba (1979) on the decision-making process of hamadryas baboons in Ethiopia describes a situation that is most akin to the situation in buffalo. Stolba reported that male hamadryas baboons, after sleeping on a kopje, descend to ridges where they persistently gaze in a particular direction. Male clan leaders can gaze either in a different direction or in the same one. When the clans disperse in the early morning, each with its own leader, they reconvene at noon at a waterhole that lies on the line of gazing of the majority of the males some 6 hours before. Apparently, the decision to meet at this particular waterhole was made during the early morning gazing session of the adult males.

I think that in reasonably predictable environments, such as Manyara, where discrete patches have to be grazed at particular times to best meet the nutritional demands of buffalo, information has to be pooled to reach the best conclusion about where to feed next (Chapter 7). Uncertainty may exist about patch depletion by competitors such as elephant (Chapter 6). In human societies, the pooling of expert opinions is a proven method of reaching a good solution when perfect information is not available. This method, for example, has been used in war rooms in different countries to estimate the risk of attack by an enemy: individual members of a panel are asked to make individual guesstimates of the risk of such potential events, and the pooled (that is averaged) guesstimate is used as an estimate of the potential risk. The same method is used in decision processes in companies (for example Dalrymple 1987). The specific pooling of judgement that the buffalo employ can even be given a special name: it closely resembles a 'jury of executive opinion' system (Makridakis and Wheelwright 1989), because there is feedback from voting buffalo to other buffalo and because they can cast their vote again and again during a voting session.

8.9 A REMARKABLE ADAPTATION: SOCIALITY IN OPTIMAL FORM

My conclusion is that the buffalo herd functions both as an information centre, in which different opinions about feeding grounds are expressed, and a communal decision centre where, through consensus, a common opinion about where to feed is reached. Throughout this book I have ventured the opinion that rear buffalo stay in a large herd, perhaps even through their whole life, because they gain something. This 'something' is information: information about where to feed, and especially where to feed best.

It appears likely that buffalo bulls do not share in this pooling of expert opinions because first they are not experts and secondly they are not necessarily interested in locating the best feeding grounds. Bulls are not 'experts' because they wander from home range to home range (Chapter 4) and, thus, have much less information about the recent grazing history of the herd. They are not necessarily interested in locating the best feeding ground because they search primarily for oestrus females and because they can leave the herd to graze on their own (Chapter 4). Nor do subadults cast a vote, perhaps because they are not yet experienced enough. It would be useful to know whether especially successfully breeding cows were voting. One particular individual buffalo cow, named 'the Judge' because her drooping horns resembled the wig of an English judge, is a case in point. She always walked in the front section of the herd and was normally in good condition, but for one reason or another she lost her calves within 2 years of birth. Nor were juveniles attached to her, so apparently she had previously been unable to raise calves. Thus, in fitness terms she was not successful. I never have seen her vote. On the other hand, a particular cow with a broken horn had raised a calf at least twice, and was seen to vote regularly. These cases are too few to draw firm conclusions, but these observations might have revealed some glimpses of the finer details of the decision-taking process in buffalo. The emerging science of cognitive ethology, when conducted along these lines, might in the future give more insights in the proximate rules of behavioural ecology. I am convinced that if this branch of science comes to full development, there could be other startling discoveries of animal adaptation. Perhaps the results that I have presented will be interpreted differently then, but I hope that at least more work along these lines will be done.

African buffalo exhibit cognition as defined in Chapter 7, since their voting and pooling of opinions then can be viewed as proof of their ability to relate different unconnected pieces of information in new ways and to apply the results in an adaptive manner. They do this socially through the consensus of adult female herd members. This has the result that the effects of social inequality are partly mitigated through information sharing. The general image of the buffalo as a savage brute of an animal (Preface) is, I am convinced, wrong. I am not suggesting that they should be seen as docile, peaceful animals because they definitely know how to defend themselves. However, in their social organization and especially their communal decision-making through voting they show a remarkable adaptation to their environment.

9

The effects of ecology on social organization

9.1 PHYLOGENY AND SOCIAL ORGANIZATION OF BUFFALO

One of the main themes in recent ecological studies has been the search for a model to explain sociality in animals. My study on the African buffalo has been propelled by this drive too. A few years ago, researchers interested in the behaviour of animals stressed its genetic basis. It was a time when the behavioural repertoire of a species was considered more or less programmed by the genome. For example, Mayr (1970) thought that behaviour was programmed by relatively few genes as compared with morphological traits since closely related species that (almost by definition) morphologically resembled each other could show a different behavioural repertoire, indicating the relative ease of change; the alternative, as he saw it, was that the same behaviour could be produced by different combinations of genes. This way of thinking resulted in the idea that, if the genetic relatedness of species was known, phylogeny could yield insight in the behaviour and even social organization of a given species on the basis of knowledge of a related species (Tinbergen 1962). This approach was merged with the increasing knowledge of the behaviour and social organization of free-living animals gained through field studies.

In East Africa the 1960s were a very active period of research into the ecology and social organization of antelope but hyena, lion, zebra, elephant and giraffe were also studied. All this research resulted in a treasure house filled with information, especially on ruminants (see also Hofmann 1973). It enabled Jarman (1974) to make his classification of different types of social organization in relation to the ecology of African bovids and a number of generalizations regarding the relation between social organization and feeding ecology could be made. Many years before, Crook (1962) had already laid the basis for this way of thinking through his study on the relation between ecology, habitat and social organization in weaver birds. Also in 1974 Estes published his classification of social organization of African bovids. He drew attention to the phylogenetic pattern that underlaid the ecological basis of the social organization of these bovids, by tabulating degree of gregariousness for the species within the eight major bovid tribes of Africa. Estes' publication enabled thinking about the **evolution** of social behaviour, because the different tribes were phylogenetically related. This development led to the idea that social behaviour was, as it were, a trait that could be inherited and was, thus, to a large extent the result of evolution and was the logical outcome of ideas put forward about the genetic basis of behaviour. This idea is still explored, although perhaps implicitly, as demonstrated by palaeo-ecologists who use specific information on living large mammals to reconstruct the behaviour of extinct species such as mammoth and mastodonts (Haynes 1991) or *Bison priscus* (Guthrie 1990).

Yet, more or less simultaneously, another development in scientific thinking took place. This had as its starting point the idea that behaviour could have an adaptive value, that is that individuals with a particular behaviour or set of behaviours would have a higher fitness than other individuals within the population. In this way of thinking the manifestation of past natural selection, that is evolution, is taken more or less for granted but the study of current natural selection is the main field of interest. This new focus on Darwinian fitness started with Nico Tinbergen and David Lack in the 1950s. Tinbergen's statement that the 'student of survival value tries to find out whether any effect of the observed [behaviour] contributes to the survival [and] if so how survival is promoted and whether it is promoted better by the observed process than by slightly different processes' (Tinbergen 1963, p. 418) is indicative of this way of thinking. Tinbergen not only meant 'survival' as continuance of life of a particular individual but also its successful rearing of offspring. Lack conducted experiments to show that clutch size in birds was adaptive, and could be manipulated by the researcher to show this. So, a trait such as 'clutch-size' was to be viewed as not just a trait or a behaviour that had evolved, but as a variable, with a genetic basis, that was differentially distributed over individuals in a population. In different individuals this could come differentially to expression and could thus, through natural selection, contribute to fitness differences. Experiments by Tinbergen and his students on herring gulls and kittiwakes resulted in a clear understanding of the adaptive value of behaviour. This finally resulted in a new branch of science called behavioural ecology, but now Tinbergen's question would be formulated somewhat differently: would a particular behaviour have adaptive value and can it be shown to have importance for the fitness of the individuals involved? (Dunbar 1982).

So, two main questions have been guiding current thoughts and theories in behavioural ecology, namely (1) why do particular social systems evolve in particular ecological circumstances and (2) if one accepts the rules and customs of a particular society as given, are the actions of different animals in that society those which would be predicted if each individual is behaving, subject to the rules, in the way which would maximize his or her inclusive fitness? (Maynard-Smith 1982).

The classic approach to answering the first question is by comparing different related species with regard to their ecological circumstances and their social organization (Tinbergen 1963, Clutton-Brock and Harvey 1978, Struhsaker and Leland 1979, Jarman 1982). For African members of the family Bovidae (antelopes and buffalo), interspecific comparisons have been carried out by Jarman (1974), Estes (1974, 1991) and Leuthold (1977). These comparisons point out that, within this group, buffalo deviate from the 'normal' pattern. Most bovids are territorial, and form open social groups, that is individuals move from a social unit to another. With regards to inter-individual distance most bovids are of the 'distance type' (the individuals infrequently touch one another). Buffalo are not territorial, the females form closed groups and both sexes are of the 'contact type'. Not withstanding the wealth of information, the interspecific comparisons, however, do not give much more information than that buffalo, in common with other typical consumers of grass, form large aggregations of individuals, although this is not a universal rule (for example Burchell's zebra): they do not give answers to questions about causality of behaviour or about its function.

Information from the social organization of phylogenetic sibling species outside Africa is also not particularly helpful to answering the question of why particular social systems evolve in particular ecological circumstances. None of the Bovini or the related Boselaphini (nilgai and four-horned antelope) and Tragelaphini (bushbuck, nyala, greater and lesser kudu, sitatunga, bongo, eland and Lord Derby's eland) are territorial. So non-territoriality might be a phylogenetically determined behavioural trait. Whether the three species of anoa – dwarf buffalo from Sulawesi in Indonesia and the Philippines – are territorial is not known. However, by stating that some trait is phylogenetically determined is not answering the question about **why** territoriality did or did not evolve. What more information can we glean from information on the buffalo's sibling species? Anoa most likely live in pairs in the rain forests. Anoa could represent the lifestyle and social organization of the ancestral Bovini. The other typical grazing sibling species, such as yak, banteng and bison form family groups, while water buffalo form harems. This pattern is also shown by the gaur and perhaps kouprey, wisent and aurochs/cattle (McHugh 1958, Schaller 1967, Heinemann and Thenius 1972, Tulloch 1979). All sibling species of the African buffalo are non-territorial, live off grass and (except anoa) live in extended family groups which can increase in size to large herds. Jarman's (1974) or Estes' (1974) correlative classifications already yielded this insight, but it does not solve the question about functionality of the social organization that is shown by a given species.

9.2 INTRASPECIFIC VARIATION IN SOCIAL ORGANIZATION

A more powerful way to appreciate the value of the social organization of a given species or of its behaviour could be through an intraspecific comparison of populations. This was appreciated by Crook (1962). At that time the first systematic reports of different social systems in different populations of the same species came to the fore (Leyhausen 1965). The African buffalo is an extremely variable species that ranged over nearly all Africa south of the Sahara (Figure 9.1): it appears to be an ideal subject for intraspecific comparison.

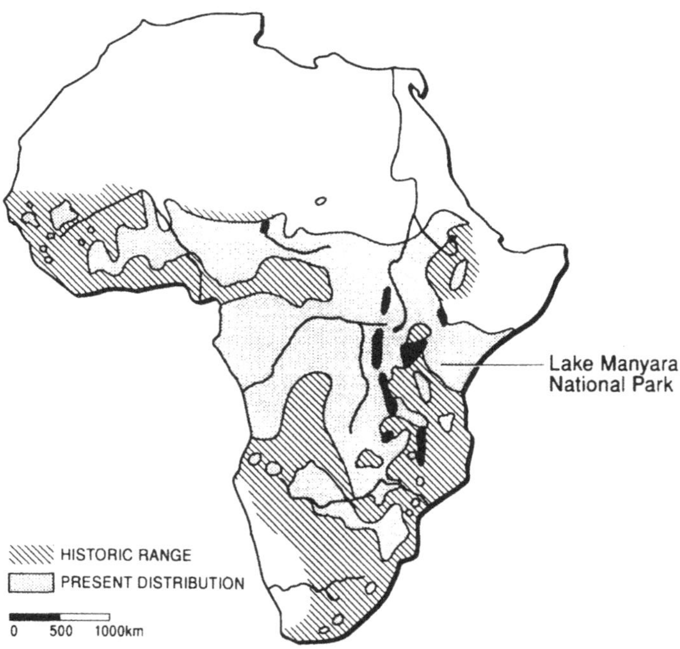

Figure 9.1 Map of Africa with the present generalized distribution of buffalo (stippled) and the generalized distribution before the Modern period (hatched plus stippled). After Sinclair (1977), Haltenorth and Diller (1979) and Mloszewski (1983). Major lakes in black.

The morphological variability of the species has led to views that 52 subspecies occurred divided over two species (Christy 1929). An even more extreme subdivision into 12 different species took place, but later these were considered as so many subspecies (Blancou 1935; Figure 9.2) because extensive intergrading between the forms was observed.

These forms could be grouped into three groups of subspecies (Figure 9.3): the *nanus* group living in forests of West and Central Africa, the *brachyceros* group from the Sahel-Soudan savannah, and the *caffer* group from the East African and southern African savannah.

Blancou's (1935) grouping already created a great deal of simplicity. Dalimier (1955) distinguished five species. For this he fused some subspecies, namely *aequinoctialis* of the Sudan with *azrakenis* of Ethiopia into *aequinoctialis*, and *cottoni* of southern Zaire with *mathewsi* of eastern Zaire into *mathewsi*. For this group *mathewsi–aequinoctialis* he introduced a new species name: *Syncerus aequinoctialis*. New confusion arose, since *cottoni*

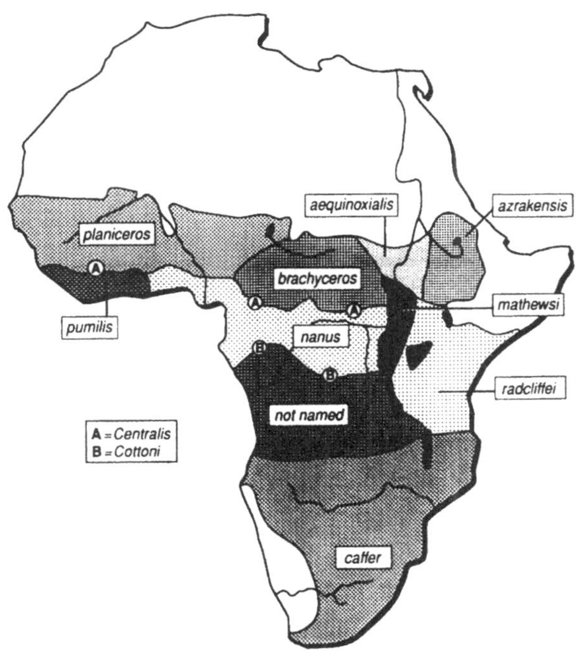

Figure 9.2 Map of Africa with the distribution of the 12 (sub)species of buffalo as recognized by Blancou (1935). The range of the (sub)species *Syncerus (caffer) centralis* and *S.(c.) cottoni* was not exactly given by him and hence can only be indicated imprecisely. Range boundary based on the pre-Modern distribution of buffalo of Figure 9.1.

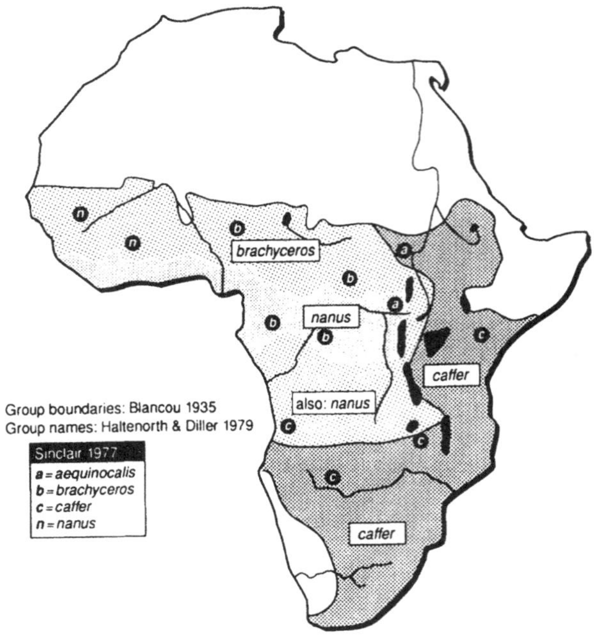

Figure 9.3 Map of Africa with the distribution of the presently recognized (Haltenorth and Diller 1979) three subspecies of *Syncerus caffer*, the African buffalo, based on Figure 9.2. The taxonomic position of buffalo of north Angola and south Zaire is unclear: they are often considered to belong to the subspecies *nanus*. Sinclair (1977) discerns a fourth subspecies, namely *S. c. aequinoctialis*.

was also made equivalent to *simpsoni*, for which he introduced a new species name: *Syncerus sylvestris*. The third species distinguished by him was *Syncerus savanensis*, which consisted of the subspecies *brachyceros* plus, I think, *planiceros*. The fourth species, *Syncerus caffer*, was a straightforward fusion of the subspecies *caffer* and *radcliffei*. Finally, a fifth species was distinguished consisting of the subspecies *nanus* and a new form from the Ituri rain forest of eastern Zaire, namely *adolfi-friederici*. Apparently, the idea of five different African buffalo species did not appeal to taxonomists, because at a later stage the three groups of Blancou (1935) were classified as subspecies with the original group names (*nanus*, *brachyceros* and *caffer*) now as subspecies names (Haltenorth and Diller 1979). Sinclair (1977) still discerned a fourth subspecies, *aequinoxalis* (or *aequinoctialis*) (Figure 9.3).

Why this excursion into the taxonomy of the African buffalo? It mainly serves to prove the point of the extreme morphological variability of this species, which is reflected in the difficulty that taxonomists have in coming to grips with. The present-day view is that the African buffalo is a single species in a separate genus, distinct from the water buffalo, anoa, *Bison* and the many *Bos* species. This is borne out by cross-breeding trials reported by Dalimier (1955): apparently *Bos* and *Bison* form a supergenus because within this group nearly all crosses are possible, but neither the African buffalo nor the water buffalo can be crossed with any other species. I have not found reports on attempts to cross-breed the different forms of African buffalo, but the occurrence of all types of transitions between the different 'subspecies' is accepted as proof of the occurrence of a single species: *Syncerus caffer*.

As indicated before, comparison of populations within a species might indicate the survival value of the social organization of a given species or of its behaviour. The buffalo seems, thus, to be an ideal candidate for such a comparison. Dwarf buffalo of the Zairese rain forest are small (shoulder height 100–110 cm), light animals of maximally some 200 kg, while buffalo of the East and South African savannahs have a shoulder height of 150–165 cm and bulls weigh on average 700 kg (Dalimier 1955). To make meaningful intraspecific comparisons there is, however, a problem: only little is known about both the rain forest buffalo and its conspecifics from West Africa. Dalimier (1955) gives a tantalizing glimpse of the social organization of these animals, and mentions that groups can split and reunite, that groups are led by a bull and that average group size is about 10–20 individuals (which can increase to 70; Jobaert 1952). However, this information is not detailed enough. Intensive field studies to answer the most pressing question have not been conducted on West or Central African buffalo: only information within the subspecies *S. c. caffer* from East Africa has been published in enough detail to make a comparison. The excellent study of Sinclair (1977) on the buffalo of the Serengeti and my own work on the buffalo of Lake Manyara National Park (about 100 km to the southeast) offers this possibility. It is a pity that Mloszewski (1983) did not report on a single population in southern Africa, because then the comparison could have been even more meaningful.

Apart from the remark on the use of intraspecific variation to study the adaptiveness of a social organization, it is important to realize that the survival value of the social organization adopted by a given species can only be fully appreciated if its relationship with food exploitation has been studied (Clutton-Brock and Harvey 1978). In my thinking on the function of social organization I have kept this statement as a guiding principle because Jarman's (1974) classification of social organization also showed this to be strongly correlated with the food exploitation strategy. The emphasis of Sinclair's study was on food utilization and resource limitation on the population level. My own work was aimed more at understanding individuals too. Nevertheless, the two studies give ample scope for comparison.

9.3 BUFFALO OF MANYARA AS COMPARED WITH THOSE OF THE SERENGETI

In Table 9.1 the differences and similarities between the Manyara and the Serengeti buffalo are given.

In the first two chapters information on rainfall patterns, plant phenology, and a description of the vegetation provide a background to the measurements of the vegetation on offer (biomass, species composition, quality) against which the diet actually assembled by the Manyara buffalo was assessed. The salient features are embraced by seasonality, spatial heterogeneity and how these impinge on food quality as reflected in crude protein content. As the research progressed, it became clear that predation, diseases and parasites (Chapter 5), and competition with other herbivores (Chapter 6), could not be ignored as they also have a shaping role to play in the evolution of the social organization of herd-living animals such as buffalo; however, for an understanding of the similarities and differences between the Manyara and Serengeti buffalo these data are of less relevance. Central to understanding how the social organization of the Manyara buffalo functions is the finding that the cows (with their calves and subadult offspring) are locked in a herd life with a discrete home range and even a predictable position within it (Chapter 3), whereas the adult bulls are only temporary intruders into this cow-structured herd, living otherwise in loose bands of bachelors (Chapter 4). Seasonal trends were shown in the growth pattern of the food resources, the quality and the quantity of the food on offer and of the assembled diet, the physical condition of especially the herd-living buffalo and even parturition and conception. The cycle of body condition in buffalo, namely a clearly annual cycle in the cows and a much more short-term (*c.* 3 months) in the adult bulls, with the additional striking difference that the cows are synchronized and the bulls are not, brings the marked differentiation between the sexes into sharp focus and will serve as a convenient point of departure for discussing the adaptive features of the social organization of the Manyara buffalo.

It is simplest to consider the cows first (Figure 9.4). Because the condition of the adult cows is seasonal, and as this condition determines whether a cow comes into oestrus, most cows will come into oestrus during the rainy season. This results in most conceptions occurring during the rains and, as gestation in buffalo is nearly one full year, most calves will be born when food conditions are optimal for the mothers to produce milk (Figure 9.4B and C). As female buffalo live in herds, for the

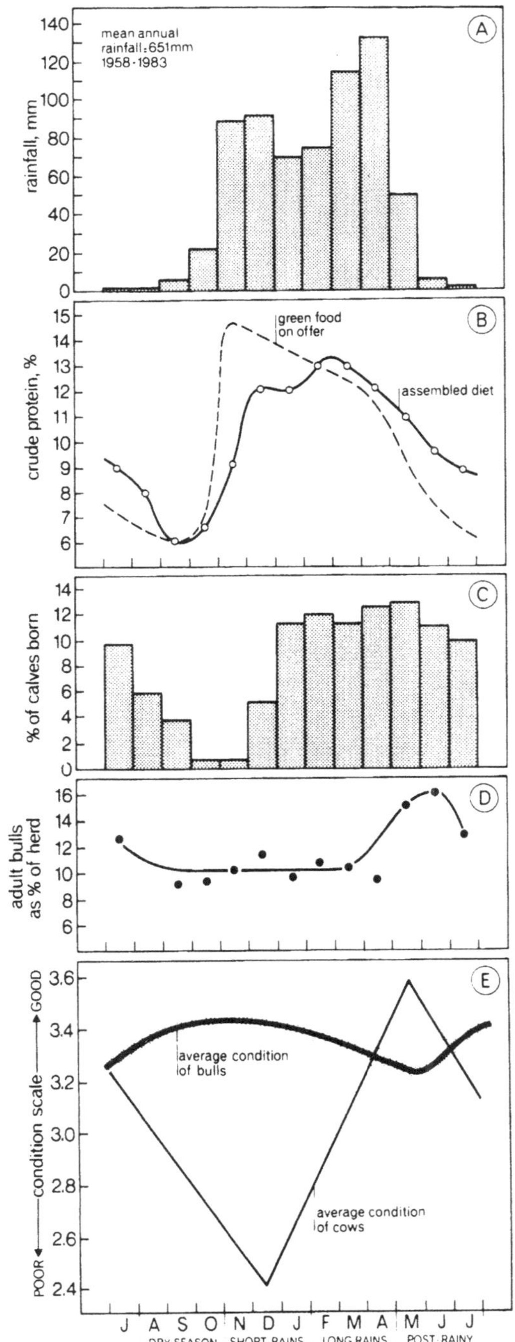

Figure 9.4 Summary of the seasonal patterns of relevance for Manyara and its buffalo.

Table 9.1 Comparison of the ecology and social organization of buffalo of Manyara with those of the Serengeti

Manyara	*Serengeti* [a]
Environment	
Rainfall seasonal, with dry season lasting approximately 5 months	Rainfall seasonal with dry season lasting approximately 5 months
Total annual rainfall is approximately 650 mm	Total annual rainfall is approximately 800 mm
Owing to areas with upwelling groundwater or high water tables, quite some land units remain green in dry season	Owing to fire green flush can occur in patches in dry season but infrequently so
Water is not limiting and is widely available in home range	Water distribution is rather restricted in dry season
Many vegetation types with different phenology patterns in close proximity The patches are relatively small	Not studied on this fine scale in Serengeti but my impression is that diversity is less pronounced here and that the patches are relatively large
Quality of assembled diet approximates a balanced diet; in dry season quality is high enough to cover maintenance	Quality of assembled diet approximates a balanced diet only in February to April (long rains); for 6 months it is below maintenance
Grazing competitor: especially elephant	Grazing competitor: especially wildebeest
Average age at death 11.5 years, mainly due to predation by lions	Average age at death 11.5 years, mainly due to starvation
Reproduction	
Calving and conceptions do not show a clear peak as 50% of births occur in approximately 17 weeks	Clear peak in calving and conceptions as 50% of births occur in approximately 7 weeks
Very few births in dry season and births take place when condition of mothers is improving after the dry season	Very few births in dry season and births take place when mothers are in peak condition
Calving interval close to 36 months	Calving interval is 15 months
Mothers continue to produce milk for offspring until calf is 18 months old, sometimes even longer	Mothers stop producing milk for offspring when calf is 10 months old; cessation normally coincides with end phase of renewed pregnancy

Behaviour of adult bulls

There is no bull rutting season although most conceptions take place during 17 weeks from February to July (long rains and post-rainy season)	There appears to be a bull rutting season in April and May (long rains), although some conceptions take place even in the dry season
Fighting between males is extremely rare: threatening sequences are very rare; sparring by adult bulls not infrequent	Fighting between males is infrequent; threatening sequences are often seen; sparring by adult bulls never observed
Males direct aggression towards other males, sometimes to subadult males, never to females	Males direct aggression towards other males, sometimes to subadult males, never to females
Yearlings remain close to their mother; male calves become gradually independent but do not form clear subadult male groups	Yearlings remain close to their mother but male calves become gradually independent and join males of similar age to form subadult male groups
Young adult males (as is the case for all other adult bulls) leave a herd at any time (dependent on their condition) and never as a group	Young adult males leave the herd in the dry season as a group
Adult bulls stay a few weeks with a herd then become bachelor for some weeks, to return to a herd, etc.	Sinclair (p.124) suggests that 'the males move into the herds only during the rutting season, moving to their own groups when reproductive activity is at a low ebb'
Hierarchy of adult bulls in a herd is rotating: bulls enter in prime condition at the top of the hierarchy and lose both condition and rank during herd life, after which they become bachelor again	Hierarchy of adult bulls (especially in herds but also in bachelor groups) is essentially linear, and based on size and age
Submissive behaviour is rarely observed in bachelor groups; in herds only evasion is noticed	Bulls with a low rank stay in the herd: 'the maintenance of the hierarchy depends on the lower ranking bulls remaining with the herd and adopting submissive behaviour towards the dominants; this suggests that there must be some advantage in staying with the herd rather than leaving to avoid the aggression of larger, dominant bulls' (p.119)

	Switching between herd life and bachelor state, however (as in Manyara), is explicitly mentioned for one bull in the Serengeti (p.122)
As the hierarchy is rotating, it is conceivable that many adult bulls obtain matings dependent on the number of other bulls and on females in oestrus	A few adult bulls achieve most matings, but subordinate bulls may achieve an occasional mating
Old males continue to play a role in the rotating hierarchy in the herd	Old males never return to the herd
Bachelor bulls do not form stable associations; they graze at other times	Old bulls form bachelor groups in which they stay together continuously, especially in areas where cows do not graze
Bachelor bulls do not have particular home ranges	A group of bachelors has a home range of its own and there are few relations between individuals of different bachelor groups
Predation risk is high for bachelors	Predation risk is high for bachelors
Social organization of adult bulls can be characterized as a re-entrant consecutive polygyny	Social organization of adult bulls can be characterized as a polygyny
Adult bulls frequently switch between herds so there is a fair chance of outbreeding	This is unclear but I get the impression that Sinclair's idea (p.152) is that there is a fair amount of inbreeding
Behaviour of adult cows	
Calves are sometimes born outside the herd but are never left hidden; mother moves to the herd during daytime and after catching up mother and calf stay with the herd	Calves always born in the herd but mother and calf stay behind if calf is very young and herd moves to another place but calves are never left hidden; mother moves during daytime to catch up with the herd
Aggression of cows towards mothers with calves less than towards other females, and mothers with calves more aggressive towards other females	Aggression of cows towards mothers with calves less than towards other females, but females with calves do not make distinction
Herd members rarely observed to pay any attention when a calf (or another herd member) is attacked by predators	All herd members come to defend a calf against predators; Sinclair's observations partly based on experiments in Manyara!
Yearlings remain close to their mother and my impression is that subadult cows still stay close to their mother	Yearlings remain close to their mother and Sinclair favours the interpretation that once females

(and quite often subadult bulls as well)	reach 'maturity' they lose contact with their mother (subadult bulls go to their age groups)
Aggression of mothers to their offspring or that of others is very rare but more frequent to subadult bulls than to subadult cows	Aggression towards 18- to 24-month-old buffalo is very frequent by females
Perhaps a hierarchy between adult cows: there is a strong fidelity to location in the herd and some locations are better than others	No hierarchy (at least not linear) between adult cows (except perhaps within family groups but the existence of such groups is unclear). Location fidelity was not studied
Competition within the herd is severe when the herd is large	Competition within the herd was not studied
Location in the herd is probably inherited from the mother. If so, this might explain long investment of mothers in calves and equal investment in sons and daughters as competition in herd appears severe	Investment is much shorter (one would expect more investment in sons than in daughters)
Density (buffalo per km^2 of home range) 17.9 ± 8.5	Density (buffalo per km^2 of home range) 6.8 ± 3.8
Home range of mixed herd not defended	Home range of mixed herd not defended
Small overlap between home ranges of herds, although incipient herds in the process of splitting off still show considerable overlap with 'original' herd	Very small overlap between home ranges of herds, although incipient herds in the process of splitting off still show considerable overlap with 'original herd'
Home range in the dry season is the same size as in the rainy seasons (because the distribution of green grass is not limited to riverine grasslands as forest glades and areas with upwelling groundwater are widely distributed)	Home range in the dry season is much smaller than in the rainy seasons (because in the dry season the herds are restricted to riverine grasslands)
It is unclear why there is sometimes overlap in home ranges of 'unrelated' herds	Overlap in home ranges caused by 'rare accidents of local rainfall' causing transgressions (p. 136)
Females of a herd form a closed group (individually known cows were never observed in a herd other than their own)	Females of a herd form a closed group (strange cows are harassed if they enter a herd by accident)
No sign of aggression between herds; herds occasionally mingle for a few hours	No sign of aggression between herds

Herds can be identified over many years as discrete entities	Herd history has not been studied
Herds are structured entities with distinct concentration areas of the age–sex categories; adult cows have fixed locations (and perhaps subadults too?) but adult bulls have no fixed locations	In a herd subgroups can be identified; location fidelity was not studied. The nature of the subgroups is not clear: do cows form subgroups too?
Herds show a fusion–fission pattern which is related to herd size: large herds split more frequently than smaller ones (which has no relation with adult bull behaviour – there is no rutting season – but competition appears to be more severe in large herds than in small herds)	Herds show a fusion–fission pattern which is related to season: herds split more frequently in dry season (which coincides with the departure of young adult bulls, the shrinking of the available habitat of green riverine grassland and with the end of the rutting season)
In a large herd the rear animals suffer most from inequality because they have to graze in depleted areas, are in a worse condition and are more infected with intestinal parasites. These animals perhaps form subgroups of 100–200 animals which frequently graze on their own in other parts of the herd's home range but fuse with the main herd frequently too	Possibly a large herd is made up of subgroups of about 100–200 animals, each of which prefers certain parts of the complete range. Occasionally they may join other subgroups and move into other parts of their range
Among adult cows inequality is pronounced and coupled to location within the herd; it is likely that this causes inequality in lifetime reproductive success between females. It is also probable that mothers can improve survival of their offspring by long maternal investment to ensure a good location in the herd. This would explain the long calving interval and low recruitment (lifetime reproductive success is thought to be on average two calves only). The buffalo population has been stable between 1959 and 1985 and is perhaps kept at the present level by competition with elephant, but since 1991 poaching is important	Inequality in the herd was not studied but, in view of the relatively short investment in calves and short calving intervals, I would suggest that inequality is less pronounced in the Serengeti, and also that intra-herd competition is less pronounced (because the herds split?). Total lifetime reproductive success of females is likely to be five calves. The Serengeti buffalo population has shown a threefold increase between 1961 and 1976, and the population can continue growing, especially if poaching is checked

[a] The data on the Serengeti are from Sinclair (1977) or are inferred from that study. The distance between Manyara and Serengeti is only 100 km as the crow flies, and both populations belong to the same subspecies of buffalo.

bulls the oestrus females are a clumped resource, especially in the rainy season. However, not all females come into oestrus during the rainy season since some calves are even born during the long dry season, which means that for an adult bull it can be advantageous to be with a herd even during the dry season. An important difference with the Serengeti buffalo emerges here, as 50% of the Manyara buffalo calves are born over an extended period of approximately 17 weeks while in the Serengeti 50% of the calves are born in approximately 7 weeks (Table 9.1).

Several factors play a role in determining why some cows come into oestrus during the dry season, that is why some cows achieve a condition enabling them to come into oestrus then (Table 9.1). At the level of resource availability, the extended period of births in Manyara may well be explained by the occurrence of patches of reasonably high quality (Figure 9.4B) during the dry season. The near absence of areas with a high groundwater table in the Serengeti can thus be seen as the cause of a sharply delimited conception period there. During the study it was concluded that the chance of coming into oestrus will be dependent on the location in the herd, as living in the best location is related to having the highest condition (Chapter 4). Possibly regardless of position some individual females are perhaps better adapted in a physiological sense to make use of these resources during the dry season, or possess physiological mechanisms to reduce the cost of maintaining a good condition during that period. It would be worth investigating if there are benefits to producing offspring during the dry season instead of during the rainy season: one could imagine that having a calf in the dry season could enable weaning of the calf in the rainy season when food quality is high (Figure 9.4B), so that a mother could have a shorter calving interval (which would then have to result in a higher lifetime reproductive success to be selected for). It is noteworthy that even in the Serengeti some females calve in the dry season although in that area the rigours of the dry season are more pronounced than in Manyara (Table 9.1).

Regardless of the reasons, bulls in Manyara can **always** 'expect' some cows to be in oestrus; in the Serengeti cows in oestrus will be much rarer outside the peak conception time. This is a first step in understanding why the Manyara bulls do not show a 'rutting' season while the Serengeti bulls appear to have such a season (Table 9.1). Territoriality did not evolve in buffalo (or in other Bovini), and bulls can expect many competing bulls when entering a large mixed herd. It has been shown that adult bulls in a mixed herd are more vigilant than other age–sex categories (Chapter 7), and the conclusion was reached that this is mainly because these bulls are searching for mates and at the same time are watching for competitors. This competition between bulls interferes with feeding opportunities and is seen as the cause of loss of physical condition of the adult bulls in a mixed herd. The bulls are able to regain condition when outside the herd during their bachelor phase (Chapter 4). Since an individual bull in Manyara can recoup condition on average in 6 weeks of bachelor living, he still can 'expect' oestrus females in a Manyara herd. The best tactic for such a bull would be to rejoin a herd, and a 'rotating dominance hierarchy' is the result. As the Serengeti bulls have to compete for a limited period only (in that area there is a clustering of oestrus females in time as there is a real calving peak; Table 9.1), nearly all adult bulls are in the herds at the same time and nearly all adult bulls are bachelor at the same time again, which explains Sinclair's (1977) observation that a more or less linear hierarchy between adult bulls can be observed in the Serengeti. To explain these forms of hierarchy fully, one has to assume that oestrus females exert female choice for the bull (in Manyara at least this means the bull in the best condition). From the male's point of view the result is that the Manyara bulls have a special form of polygyny, namely re-entrant consecutive polygyny, while the Serengeti bulls appear to have a simple polygyny; from the female's standpoint their social organization can be thought of as serial monogamy in both areas. It now can be envisaged that in Manyara there will be a strong selection for adult males to maintain their position in the herd as long as possible and to regain condition rapidly thereafter when in the bachelor phase (Chapter 4). The pressure to minimize the duration of the bachelor phase is accentuated by the finding that males run a three times higher risk of predation by lions during their stay outside the herd than when they are herd bulls (Chapter 7). Reproduction in the male is at the cost of bachelor existence. So, seasonality impinges on the bulls' social organization via the females' reaction to resource seasonality.

As is clear from the present study, the Manyara area comprises a variable landscape with much spatial heterogeneity (Chapters 1 and 8); in other words, Manyara comprises an ecologically fine-grained environment, which appears to contrast to with the more coarse-grained Serengeti environment. Although even within Manyara itself there are differences with regard to rainfall (the southern section of the park receives on average more rain than particularly the Msasa area; a difference of approximately 200 mm per year), the phenological reaction of different plant species and vegetation types is mainly determined by the depth of the groundwater table and infiltration rate of precipitation (Chapters 1 and 2). These factors have several far-reaching consequences for the buffalo, especially for the females. First, they can assemble a diet approximating a balanced diet over most of the year (Chapter 2). Secondly, suitable patches appear to be small relative to the number of buffalo in a herd (Chapters 3 and 8). Thirdly, buffalo live in a complicated environment with patches differing in quality at any one time (Chapter 8). Because the Manyara buffalo can assemble a more or less balanced diet, and because their diet, on average, does not drop to below maintenance requirements during the dry season (in contrast to the nearby Serengeti; see Table 9.1), the main message of the present work is not in terms of population regulation through resource limitation (Sinclair 1977) but about how the individual functions in the social system. Intraherd competition between individual buffalo even explains density-dependent growth rates of herds although competitor species, such as elephant, also play a role too (Chapter 5).

I started with two questions, of which the first was 'Why do particular social systems evolve in particular ecological circumstances?'. I think that the comparison of Sinclair's (1977) study on the Serengeti buffalo with the results from my study on the buffalo of Manyara answers the question quite satisfactorily. It shows that a particular social organization can be explained to a large extent by differences in local ecology. Food exploitation occupies a central place in this way of explaining a social organization, but the distribution of resources in space and time, in which accessible females are seen as a resource for the males, is the key factor in the explanation. This, I think, is a breakthrough in the thinking about social organization: it needs causal and functional explanations at the level of scarce resources. However, with this the present social organization can be understood at a certain level of understanding only. It does not give much insight into the **evolution** of social organization. The missing information is about natural selection, and the relation between individual behaviours and reproductive output.

9.4 SOCIAL INEQUALITY

It now is time to look at Maynard-Smith's (1982) second question: 'If one accepts the rules and customs of a particular society as given, are the actions of different animals in that society those which would be predicted if each individual is behaving, subject to the rules, in the way which would maximize his or her inclusive fitness?'. Because of the emphasis on individuals, I put much effort into recognizing as many individual buffalo as possible during my research. Nevertheless, the question cannot really be answered for the Manyara buffalo. Even though I collected many data on differences between individuals, I do not have data on fitness (a problem I share with most behavioural ecologists) and too few on reproductive success. I mainly have data on condition of individuals, which could be indicative of probabilities of reproduction. This information on condition, and also on the social environment of individuals, I have for hundreds of individually known adult buffalo cows and bulls (Chapters 3 and 4).

Without speculation at the level of the individual, which is dangerous in view of the paucity of observations, a satisfactory frame of reference for interpretation of the social organization cannot be achieved. Here lies the main difference between present-day behavioural ecology and the across-species correlative studies of the 1960s and 1970s: since the individual is considered the unit of selection, interpretation and formulation of rules regarding the function of behaviour and especially social organization have to be carried out within the framework of individual adaptation and selection (for example, Drent and Van Eerden 1980, Clutton-Brock *et al.* 1982, Ens and Goss-Custard 1984, Goss-Custard 1985, Teunissen *et al.* 1985). Indeed, my work on the Manyara buffalo shows that individual buffalo bulls do not form a stable company with females, and as they are not attached to a fixed home range (Chapter 4) the movements of adult bulls have to be studied on a scale equivalent to the size of the Masai ecosystem (35 000 km^2; Figure 9.5).

Current data allow just a glimpse of how decision rules for the bulls might operate to predict herd presence or absence on this scale; perhaps we will never know because buffalo bull movement appears to have stopped owing to excessive hunting and poaching in northern Tanzania. The buffalo cows, in contrast to the adult bulls, show a strong spatial fixation; the mixed herds have home ranges which can be more or less permanent for generations (Chapter 3). Although defence of such a spatial area did not arise in Bovini, indeed they are non-territorial, this permanence suggests that a concrete land unit and its permanent use is a valuable asset to these females. A review of the history of the Manyara herds shows that a herd appears to cling to its home range in spite of pressure exerted by, for example, humans or rising lake levels. This suggests two things: first the mixed herd is not only a collection of individuals but has some characteristics of a supraindividual organization and, second, occupying an area (home range) over an extended period offers some benefit to the individuals involved. Solving the puzzle of possible benefits critically depends on understanding the food exploitation strategy of these buffalo. The key to solving this problem is, I think, understanding what prevents large buffalo herds splitting permanently, since in Chapter 3 it was shown that these herds typically show a fusion–fission pattern.

Buffalo are non-territorial, rather long-lived animals, and because resources are of good quality, closed groups of overlapping generations can arise. Because of the risk of predation by lions there will be a strong selection against individual cows splitting off from the herd (Chapter 7). On the other hand, because of the relatively small patches of suitable vegetation, and because of the many mouths to be fed in a large buffalo herd, there will be a rapid depletion of resources during a herd's visit to a particular patch (Chapters 3 and 6). Apart from competition with other herbivores, mainly elephant, patch depletion will lead to competition within the mixed herd (Chapters 3, 5 and 7). As individual cows in

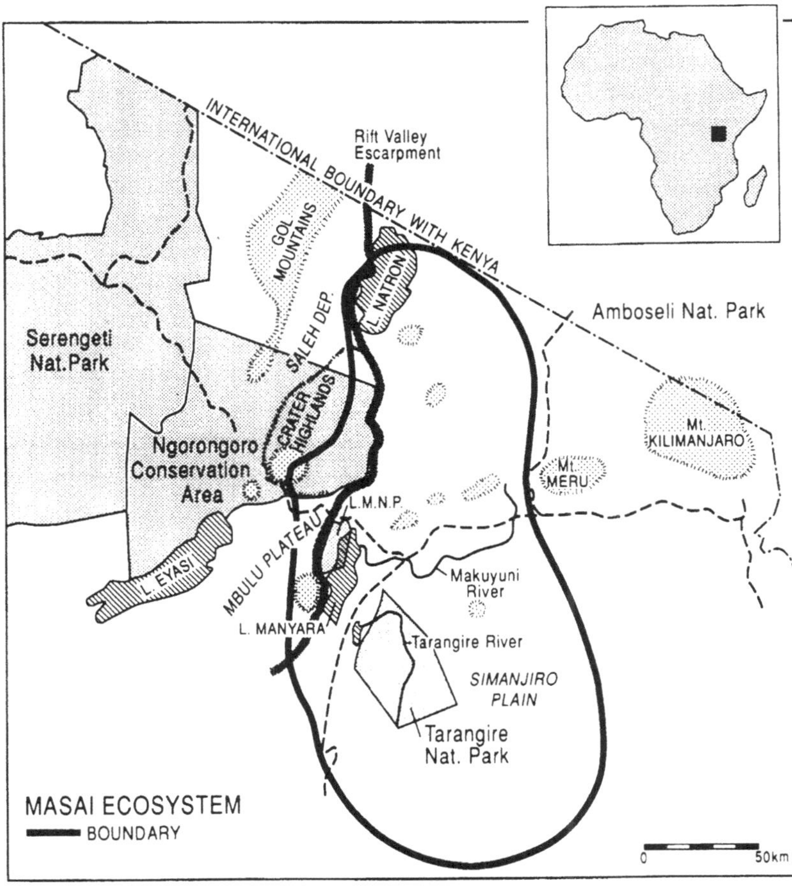

Figure 9.5 The approximate boundaries of the Masai ecosystem (Prins 1987; 35 000 km^2) of northern Tanzania. Broken lines = major roads, stippled = mountains. LMNP, Lake Manyara National Park.

Manyara have a more or less fixed position within such a herd, this competition results in some positions in the herd being more attractive than others, as revealed in differences in physical condition of individual cows within the herd. Coupled with an increased chance of infestation by intestinal (and other) parasites, this inequality within a large herd can become very pronounced (Chapters 3 and 5). Because of this competition, it can be envisaged that there will be a strong selection against accepting strange females in such a herd, and this could give (and has given) rise to the feature of closed groups in female buffalo.

Two features of herd life deserve further comment. Because of inequality in a large herd one would expect to see the rear animals of a large herd split off to form a new herd. Secondly, females in a herd having a fixed advantageous position should 'try' to monopolize that position for their offspring, and one would expect that there will be selection pressure for mothers in a good position to ensure that their offspring will inherit that advantageous position. Mothers with offspring show an improvement in position within the herd although (most of them?) cannot maintain that position. Cows in the large Central Herd of Manyara show long calving intervals and high weaning ages, and it is thought that in this competitive environment the mothers have to invest heavily in their current offspring including influencing their future position in the herd. At least during lactation, investment in sons and daughters appears to be equal. This apparent contradiction with current theory (Clutton-Brock *et al.* 1982; Chapter 5) may be partly explained by the observation that Manyara buffalo bulls are not strictly polygynous in the sense that some males can monopolize females against age mates (compare Clutton-Brock *et al.* 1982).

A particularly thorny problem to tackle is posed by the long-term history of the Central Herd in Manyara. Despite the vicissitudes of changing lake levels, this herd has existed as a recognizable entity for at least 35 years, during most of this period being composed of roughly 900 individuals, and maintaining the same boundaries. What has prevented the animals in the rear of the herd from separating themselves and forming a new but smaller herd on their own, similar for instance to the Escarpment Herd (*c.* 200 individuals)? Permanent fission and formation of new herds does occur occasionally, but this splitting off takes a long time and is a relatively rare event (Chapter 3). Relative to a large herd, the patches of suitable vegetation are small and these herds deplete patches more rapidly than smaller herds. For example, the Msasa Herd (12 head) never moves outside its home range of 100 ha, but the Central Herd (980 head) moves through its home range (4000 ha) in approximately 5 days, during which time it visits all of the available feeding grounds (Chapter 7). In large herds buffalo graze closer together than in small herds and, hence, in large herds competition is more severe and there is a strong negative correlation between herd size and herd rate of increase (Chapter 5). Despite the fact that large herds live in a fission state more frequently than small herds, there appears to be 'something' which prevents individuals easily vacating their home range and their herd.

9.5 THE EVOLUTION OF SOCIALITY AND INFORMATION SHARING

Sinclair (1977, p. 137) was struck by the fact that individuals appear to be 'conservative' when he studied the Serengeti buffalo; indeed, he wondered why a home range boundary is so rarely transgressed and he concluded 'that there is little evidence that the limits of a range are the maximum that the animals can physically cover or learn'. He ventured that

> as the buffalo prefers isolated pockets or thin strips of riverine grass, which are less influenced by accidents of rainfall [in contrast to the vegetation units preferred by wildebeest], ...wildebeest can search for their food source by moving towards rainstorms [but] buffalo cannot. If the latter were to search for small pockets of food and water in the dry season by random movements and without prior knowledge of their location, the food supply would not be used efficiently; by chance some areas would be used by too many herds (to the detriment of those herds), while other areas would be missed. ...In other words, the animals do better to stay in the same place, the home range.

Sinclair (1977, p. 138) remarked that 'answers to the question of why there is little or no overlap in home range must lie in the disadvantage of meeting other herds or using their range' but, I am afraid, these possible answers are not very satisfactory. The Manyara findings and the ideas that I have developed over the years regarding information sharing are probably more relevant to this question.

It appears likely that predation pressure sets some lower limit to herd size, as buffalo living in herds have a lower risk of predation than buffalo living apart (Chapter 5; see also Rowell 1979). If there were no predation, and if food were evenly distributed (both in terms of quantity and quality), the individuals should be better off foraging alone, but if food is in large clumps, group foraging is not deleterious. An upper limit is then set by the number of individuals that can move around together and still find enough food; large groups would then be merely a byproduct of a period of good survival and recruitment coupled with a tendency for families to stay together (Rowell 1979). If this were true, the upper limit should be reached when animals in the rear of the herd produce fewer offspring than if they were in a smaller herd, and at this break-even point they will in fact split off. The inescapable conclusion is that, even in very large herds, individuals in the rear still gain something. One possibility might be that they gain through kin selection (for example Wrangham 1977) and

> a case can be made for the... group as an educational institution in which learned information about feeding in an exceedingly complex environment can be shared to general advantage. This would provide a food-related mechanism for extending the family groups into adulthood, but is difficult to stretch to the maintenance for larger groups.
>
> (Rowell 1975)

More recent work, however, points to information sharing as the benefit cementing the larger groups as cohesive force (Fossey and Harcourt 1977, Wrangham 1977, Stolba 1979, Van Roosmalen 1980, Ydenberg *et al.* 1983, Ydenberg and Prins 1984). As perfect information is hardly ever available to one individual, information sharing can lead to better decisions. Why there should be an upper limit to the number of individuals bringing in their 'expert opinion' in the case of information sharing is a problem in itself. However, the time cost of sharing information may set an upper boundary, as might conceivably be reached by a declining accuracy of prediction if the herd reaches such proportions that the potential feeding sites are no longer visited frequently enough to ensure up-to-date information.

As observed repeatedly during the study, neither individual cows nor individual bulls are leaders of a herd (Chapter 8). I also observed that being at the forefront is more dangerous than being further down the herd because of the increased risk of lion predation. As bulls are only temporary members of the herd, and because many of them have no long-term acquaintance with the home range of the herd (Chapter 4), it appears unlikely that they have much knowledge about where the herd has been grazing previously or about the location of the most suitable patch to graze next; patches have different phenological reactions relative to each other (Chapter 8), and elephant can have caused depletion (Chapter 6). Moreover, I frequently observed that if a bull is 'leading' he is not followed by the cows when he starts grazing in a particular sward – the cows just continue walking to another patch, leaving the 'leader' behind. In my view, bulls are not leading a herd in the sense that they make decisions about where to graze. On the other hand, because individual cows also do not lead a herd in Manyara, it should be concluded that buffalo herds do not have individual leaders. Nevertheless, a number of movement rules are necessary to move around in a home range preferably to avoid predators (Chapter 7), risk of infection (Chapter 5), the effects of depletion by competitors (especially elephant) and by the herd itself on its previous foraging circuit (Chapter 8), and at the same time to assemble a balanced diet (Chapters 1 and 2). Having a fixed long-term home range appears advantageous now: individual buffalo, through intimacy with older buffalo cows or with more successful ones, can profit from a vast stock of knowledge about **their** home range. Decisions about where to graze appear to be taken collectively by female buffalo, through a process which can be described as 'voting' and which appears to be based on communal judgement and for which an intimate knowledge of the home range is a prerequisite (Chapter 8): because of the joining of information a better judgement about patch quality in relation to that of other patches can be reached.

It has often been stated that herbivores eating grass have access to an easily obtainable, abundant food source. However, quality appears partially genetically determined in grass, and seemingly equal grass swards often consist of different clones, which would suggest that different patches of even the same food species at the same time can be different from the herbivores' point of view. Whatever the causes, patches of *Cynodon dactylon* are different regarding quality in Manyara, and a number of them would not even meet the quality requirements of buffalo. This, undoubtedly, is also the case in other ecosystems and for other herbivores. The problem for the herbivore is then to discriminate between these patches, which to the human observer look alike.

The fact that there is a strong correlation between eating grass and sociality both in mammals and in birds (e.g. geese) suggests that sociality facilitates information sharing on these often only marginally different food patches, some of which, but not others, will fulfil their needs. I venture the opinion that one of the ultimate benefits of sociality in these graminivorous species is that the individuals involved can utilize grass as a food resource, both through information sharing and also through keeping swards in an optimal grazing condition (Vesey-FitzGerald 1969, 1974, Ydenberg and Prins 1981). Undoubtedly, sociality in these species could have evolved without the risk of predation, but in most situations has been strengthened and accentuated by it. In Manyara seemingly equal grazing grounds are not equal in terms of protein or sward structure (Chapter 7) and an important conclusion is that in this complex environment information is gained at the cost of food competition. Even the rear animals, by staying with the main herd, are led to areas where they can sustain themselves better. The repeated fusion–fission pattern then could be understood as a means of staying in contact with the information from the successful members of the (main) herd as a means of accumulating information on their 'future' home range in the event that a fission group becomes the nucleus of a new herd. These speculations are at the core of the conclusion reached: that a herd of buffalo can be viewed as a supraindividual organization in which individuals gain from being social. The statement of Schaller (1967) that 'there is no existing and generally accepted theory regarding the purpose of the herding habit in mammals, and it may simply be the alternative to a solitary life' can then be rejected for herbivores.

If it is correct to see food competition as a potential cost of sociality and information sharing as a benefit, then it is clear that sociality can be shaped through natural selection. Through the buffalo's communal judgement by 'voting' for the best feeding grounds and through staying in touch with the information from the successful members of the herd, the buffalo apparently also show that they have cognition, since this is defined as the ability to relate different unconnected pieces of information in new ways and to apply the results in an adaptive manner. Thus, natural selection brought about cognition in buffalo, which enabled information sharing, communal judgement and communal decision-making because this increased the fitness of females that otherwise suffered from social inequality. The evolution of sociality and cognition are apparently intertwined with inequality and information sharing, perhaps not only in bovids.

Appendix: Protein and energy requirements

Protein requirements at maintenance for ruminants, such as buffalo, can be calculated as:

$$\text{Digestible protein (g day}^{-1}) = 3.150 \times W^{0.75} \qquad (1a)$$

where $W^{0.75}$ stands for metabolic weight in kg (Lloyd *et al.* 1978, p. 425 ff.; this formula is almost the same as that derived by Crampton and Harris 1969, p. 174 ff.). For non-ruminant herbivores, such as zebra and elephant, this formula is different, namely:

$$\text{Digestible protein (g day}^{-1}) = 2.800 \times W^{0.75} \qquad (1b)$$

This difference between ruminants and non-ruminant herbivores is due to the loss resulting from the use of protein in the rumen by mainly bacteria. For the weight range of 400–600 kg (adult cattle and adult buffalo), the empirical relationship as derived for dairy cattle in The Netherlands (Van Es and Boekholt 1976) predicts the same requirements as equation 1a, namely:

$$\text{Digestible protein (g day}^{-1}) = 0.13 \times (3.33 \times W + 1000) + (L \times 63) \qquad (2)$$

in which W is body weight (kg) and L is kg of 4% fat-corrected milk. Both equations 1a and 2 lead to a dietary requirement of 0.5 g digestible N per day per kg metabolic weight in ruminants such as buffalo, which compares well with the value of 0.4 g as derived by Ketelaars (1986) (however, he calls this ingested N when he probably means ingested digestible N). The crude protein can be measured easily in plants through Kjeldahl analyses in which total N is measured. The total N concentration multiplied by 6.25 is by definition the crude protein concentration (Crampton and Harris 1969, p. 48 ff.). Digestible protein is difficult to determine without trials of many stall-fed animals. However, the concentration of digestible protein (DP) in the food is found to depend on the concentration of crude protein (CP) in that food. This relation is, according Crampton and Harris (1969, Tables 7.5 and Table 7.6), on average:

$DP\% = 0.54 \times CP\%$ (for roughage) (3a)

$DP\% = 0.79 \times CP\%$ (for concentrate feed) (3b)

For (sub)tropical grasses, most of which are C_4 grasses and which are classified as roughage (just as all grass), equation 3a can be refined. I incorporated a number of published formulas, all with the same form (Elliott and Fokkema 1966 cited in Rees 1974, Milford and Minson 1966, one calculated from data in Crampton and Harris 1969, Table A3-1, Boudet 1975, Barrett and Larkin 1974), into:

$DP\% = (0.91 \times CP\%) - 3.22$ [for (sub)tropical grasses] (4)

A first approximation of the crude protein demand for tropical ruminants to maintain body weight can be given by combining equations 1a and 3a or 3b as respectively:

Crude protein intake (g day^{-1}) $= 5.83 \times W^{0.75}$ (for ruminants on roughage) (5a)

Crude protein intake (g day^{-1}) $= 3.99 \times W^{0.75}$ (for ruminants on concentrate feed) (5b)

Hence, an average buffalo with a live weight of 500 kg needs 333 g of digestible protein per day (equation 1a), which represents approximately 617 g of crude protein from roughage (equation 5a). However, to determine the crude protein concentration in food necessary to maintain body weight, one has to know this buffalo's daily food intake. This can be calculated by determining the necessary energy intake at maintenance. The energy requirements of an adult animal can be calculated according to the formula derived by Kleiber for basal metabolic rate (BMR) as:

Basal metabolic rate (kcal day^{-1}) $= 70 \times W^{0.75}$ (6)

(Lloyd *et al.* 1978, p. 396 ff., for derivation). There is some indication that the factor 70 is too low for ruminants (Moen 1973, p. 116 ff., for a review on wild ruminants), but because this formula is so widely used I adhere to it here. Owing to 'unavoidable activities', the energy requirements will be higher for maintenance, which can be calculated as:

Digestible $energy_m$ (kcal day^{-1}) $= 2.0 \times 70 \times W^{0.75}$ (for ruminants) (7a)

Digestible $energy_m$ (kcal day^{-1}) $= 1.4 \times 70 \times W^{0.75}$ (for omnivores) (7b)

In these formulas, the subscript m stands for 'at maintenance' (further Crampton and Harris 1969, p. 147, Lloyd *et al.* 1978, p. 408 ff.). Note that the relation between digestible energy (DE) and metabolizable energy

(ME) is ME = 0.82 × DE (Crampton and Harris 1969, p. 476). For a number of wild ruminants, the ME requirements have been measured. The mean of 23 values quoted by Moen (1973, Tables 16.12–16.16) for males and non-lactating females is 1.57 × BMR, so DE can be calculated as (1/0.82) × 1.57 × BMR = 1.91 × BMR, which is very close to equation 7a.

In his calculations regarding Tarangire Game Reserve, Lamprey (1964a) used an overall figure of 3 × BMR for the energy requirements of wild large herbivores, which was followed by Eltringham (1974), and in the free-living state values of 5 × BMR have been measured (Hume 1982, p. 16). As lactating females need approximately 2.10 × BMR for maintenance (Moen 1973, p. 352 ff.) a value of 3 × BMR seems too high. The activity metabolic rate in deer is 1.23–1.98 × BMR, increasing to 2.3 × BMR during lactation and for thermoregulation in summer even to 3 BMR (Wallmo *et al.* 1977), but it can be assumed that ruminants in Africa have to expend less energy on staying warm; it is thus within this expectation that Coppock *et al.* (1986) derived a value of 2.1 × BMR for nomadic zebu, sheep, goats and camels in Kenya.

I assume that an adult non-lactating, non-pregnant buffalo of 500 kg needs 14 803 kcal DE per day (equation 7a) for maintenance purposes. This equals 2 × BMR. As the energy contents of roughage can be approximated by DE (kcal kg^{-1} dry matter) = percentage TDN/100 × 4409, where TDN is total digestible nutrients (Crampton and Harris 1969, p. 213), and as the percentage total digestible nutrients of roughage is on average 50% (*op. cit.*, p. 213):

$$\text{Digestible energy (kcal kg}^{-1}\text{)} = 2204 \text{ (for roughage)} \qquad (8)$$

This is very close to the average value for East African grasses of 2282 kcal kg^{-1} (95% confidence limits 191, $n = 26$) (calculated from Table 6.2 in Barrett and Larkin 1974). The combination of equations 7a and 8 gives an estimate of dry matter intake for a ruminant that maintains body weight while feeding on roughage, as:

$$\text{Dry matter intake (kg day}^{-1}\text{)} = 0.0635 \times W^{0.75} \text{ (for ruminants on roughage)} \qquad (9a)$$

This means that a 500-kg non-lactating buffalo needs a dry matter intake of 6.71 kg day^{-1} (that is 1.3% of its body weight) to cover its energy needs for maintenance. As this food has to contain approximately 617 g of crude protein (equation 5a), this means that the percentage crude protein of the food has to be 9.2%. Note that for all roughage grazers the percentage crude protein for maintenance can be calculated by equation 5a divided by equation 9a × 0.1, which is constant at 9.2% crude protein (or 1.5% N). A refinement to the calculation of this ratio is that, according to equation 1, the buffalo specified will need 333 g digestible protein in its dry matter intake of 6.71 kg (equation 9a) and, according to equation

4, this means that the crude protein concentration is 9.0% to maintain body weight if dry matter intake is sufficient to meet energy demands. Note that the combination of these formulas again leads to a constant (9.0% crude protein) which does not depend on body size.

Under most circumstances cattle ingest more food than 1.3% of their body weight (Crampton and Harris 1969, Table 14.5). Sinclair (1975) proposes:

$$\text{Dry matter intake (kg day}^{-1}) = 0.09 \times W^{0.75} \qquad (9b)$$

This formula leads to a crude protein concentration for maintenance of 7.4% (which is higher than the value of 5% which Sinclair 1977 supposes to be sufficient). This value is a constant, independent of body weight, as it can be calculated by substituting into equation 4 the outcome of the division of equation 1a by equation 9b × 0.1, which is constant at 3.5% digestible protein. Van Wijngaarden (1985) and Van Soest (1982) propose:

$$\text{Dry matter intake (kg day}^{-1}) = 0.025 \times W \qquad (9c)$$

which compares reasonably with the published data for food intake of East African herbivores (Delany and Happold 1979, Table 11.12). Equation 9c leads to a crude protein concentration of 6.5% for a 500-kg buffalo to maintain body weight. Note that this is not a constant, as the outcome of the digestible protein percentage needed for maintenance as derived from equation 1a divided by equation 9c × 0.1 is not constant, and so the outcome after substitution in equation 4 is not constant. Digestible protein percentage is then decreasing with increasing body weight according to the formula DP% = 12.6% × $W^{-0.25}$, that is equation 1a divided by equation 9c × 0.1.

Evidence in favour of crude protein concentrations of 8–9% for maintenance purposes is the required concentration of protein in the rumen needed by the microbes for adequate fermentation. The Agricultural Research Council (1965) considers a level of 9% adequate for normal digestion in ruminants fed roughage diets, and more recent work shows the same level of crude protein needed in the rumen for fermentation (7.5–11%) (McMeniman *et al.* 1976, Orskov 1976). As a consequence, voluntary feed intake falls when the crude protein concentration is below this level. For temperate feeds this critical level is 8.5% crude protein (Blaxter and Wilson 1963 cited in Whiteman 1980) and for tropical feeds 7% (Milford and Minson 1966) or 8% crude protein (Van Soest 1982, p. 280). The NRC (1975 to 1981 cited in Van Wijngaarden 1985) feeding standards for a variety of domestic herbivores prescribe 7–8% crude protein.

As several authors pointed out, there is a relationship between energy content and (crude) protein concentration in the feed. Bredon *et al.* (1963 cited in Pratt and Gwynne 1977) give:

Digestible organic matter % (DOM) = 46.05 + 1.4152 CP% (10)

while Van der Honing *et al.* (1977) provide:

Metabolizable energy (kJ) = 15.06 × g DOM (11)

Combining these two formulas gives the metabolizable energy of grass as:

Metabolizable energy (kcal kg^{-1}) = 1657 + (50.9208 × CP%) (12)

The voluntary food intake (kg dry matter day^{-1}) of cattle varies between 1.5% of their body weight to maximally slightly over 3.5%, depending on the digestibility and on the energy content of the food: when the metabolizable energy content is high, the fibre concentration is low and the voluntary intake will be high, and vice versa (Crampton and Harris 1969, Table 14.5; compare Van Soest 1982, p. 280).

References

Agricultural Research Council (1965) *The Nutrient Requirements of Farm Live Stock*, No. 2, *Ruminants*, ARC, London.

Alados, C. (1985) An analysis of vigilance in the Spanish ibex (*Capra pyrenaica*). *Z. Tierpsychol.* **68**, 58–64.

Alf, E.F. and Grossberg, J.M. (1979) The geometric mean: confidence limits and significance tests. *Percept. Psychophys.* **26**, 419–421.

Allden, W.G. and Whittaker, I.A.McD. (1970) The determinants of herbage intake by grazing sheep: the interrelationship of factors influencing herbage intake and availability. *Aust. J. Agric. Res.* **21**, 755–766.

Anderson, R.M. (1981) Population ecology of infectious disease agents, in *Theoretical Ecology: Principles and Applications* (ed. R.M. May), Blackwell Scientific Publications, Oxford, pp. 318–355.

Anderson, R.M and May, R.M. (1979) Population biology of infectious diseases: Part I. *Nature* **280**, 361–367.

Anderson, R.M. and May, R.M. (1985) Vaccination and herd immunity to infectious diseases. *Nature* **318**, 323–329.

Anderson, S.S. and Fedak, M.A. (1985) Grey seal males: energetic and behavioural links between size and sexual success. *Anim. Behav.* **33**, 829–838.

Anonymous (1960) Annual report of the biologist, Kruger National Park, 1958/1959. *Koedoe* **3**, 1–260.

Anonymous (1966) Preliminary helminthon survey on Manyara buffalo. Unpublished report of the Veterinary Investigation Centre, Arusha, to the Director of Tanzania National Parks, Arusha (Tanzania).

Anonymous (1989) Selous Census. Unpublished report of the Tanzania Wildlife Conservation Monitoring Programme, Dar-es-Salam, Tanzania.

Appleby, M.C. (1980) Social rank and food access in red deer stags. *Behaviour* **74**, 294–309.

Arbuthnot, T.S. (1954) *Grand Safari*, Wiliam Kimber, London.

Arens, W. (1979) *On the Frontiers of Change: Mto-wa-Mbu, Tanzania*, University of Michigan Press, Ann Arbor.

Armitage, P. (1971) *Statistical Methods in Medical Research*, Blackwell Scientific Publications, Oxford.

Ash, C., Crompton, D. and Keymer, A. (1984) Nature's unfair food tax. *New Sci.* 8 April, 17–20.

Bailey, D.W., Rittenhouse, L.R., Hart, R.J., Swift, D.M. and Richards, R.W. (1989) Association of relative food availabilities and locations by cattle. *J. Range Mgt.* **42**, 480–482.

Bailey, D.W., Walker, J.W. and Rittenhouse, L.R. (1990) Sequential analysis of cattle location: day-to-day movement patterns. *Appl. Anim. Behav. Sci.* **25**, 137–148.

Baker, M.C., Belcher, C.S., Deutch, L.C., Sherman, G.L. and Thompson, D.B. (1981) Foraging success in Junco flocks and the effect of social hierarchy. *Anim. Behav.* **29**, 137–142.

Banks, E.M., Wood-Gush, D.G.M., Hughes, B.O. and Mankovich, N.J. (1979) Social rank and priority of access to resources in domestic fowl. *Behav. Processes* **4**, 197–209.

Barash, D.P. (1980) The influence of reproductive status on foraging by hoary marmots (*Marmota caligata*). *Behav. Ecol. Sociobiol.* **7**, 201–205.

Barnes, R.F.W. (1983) The elephant problem in Ruaha National Park, Tanzania. *Biol. Cons.* **26**, 127–148.

Barrett, M.A. and Larkin, P.J. (1974) *Milk and Beef Production in the Tropics*, Oxford University Press, Oxford.

Basson, P.A., McCully, R.M., Kruger, S.P., Van Niekerk, J.W., Young, E. and De Vos, V. (1970) Parasitic and other diseases of the African buffalo in the Kruger National Park. *Onderstepoort J. Vet. Res.* **37**, 11–28.

Baumann, O. (1894) *Durch Masailand zur Nilquelle: Reisen und Forschungen der Masai Expedition des Deutschen Antisklaverei Komite in den Jahren 1891–1893*, Reimer, Berlin.

Bayer, R.D. (1982) How important are bird colonies as information centres? *Auk* **99**, 31–40.

Beard, P.H. (1988) *The End of the Game, the Lost World of Paradise*, Chronicle Books, San Francisco.

Beekman, J.H. and Prins, H.H.T. (1989) Feeding strategies of sedentary large herbivores in East Africa, with emphasis on the African buffalo. *Afr. J. Ecol.* **27**, 129–147.

Beesley, J.S.S. (1965) *Report on a Brief Survey of Aquatic Snails in the Lake Manyara National Park*, Miscellaneous Report 521, Tropical Pesticides Research Institute, Arusha.

Beilharz, R.G. and Mylrea, P.J. (1963) Social position and movement orders of dairy heifers. *Anim. Behav.* 11, 529–533.

Bell, R.H.V. (1970) The use of the herb layer by grazing ungulates in the Serengeti, in *Animal Populations in Relation to their Food Resource* (ed. A. Watson), Blackwell Scientific Publications, Oxford, pp. 111–113.

Bell, R.H.V. (1971) A grazing ecosystem in the Serengeti. *Sci. Am.* **224**, 86–93.

Belsky, A.J. (1986) Does herbivory benefit plants? A review of the evidence. *Am. Nat.* **127**, 870–892.

Belsky, A.J. (1987) The effects of grazing: confounding of ecosystem, community and organism scales. *Am. Nat.* **129**, 777–783.

Belsky, A.J., Carson, W.P., Jensen, C.L. and Fox, G.A. (1993) Overcompensation by plants: herbivore optimization or red herring. *Evol. Ecol.* **7**, 109–121.

Bergerud, A.T., Nolan, M.J., Curnew, K. and Mercer, W.E. (1983a) Growth of the Avalon Peninsula, Newfoundland caribou herd. *J. Wildl. Mgt.* **47**, 989–998.

Bergerud, A.T., Wyett, W. and Snider, B. (1983b) The role of wolf predation in limiting a moose population. *J. Wildl. Mgt.* **47**, 977–988.

Berry, H.H. and Louw, G.N. (1982) Nutritional balance between grassland productivity and large herbivore demand in the Etosha National Park. *Madoqua* **13**, 141–150.

Bertram, B.C.R. (1973) Lion population regulation. *E. Afr. Wildl. J.* **11**, 215–225.

Bertram, B.C.R. (1978) Living in groups: predators and prey, in *Behavioural Ecology: An Evolutionary Approach* (eds J.R. Krebs and N.B. Davies), Blackwell Scientific Publications, Oxford, pp. 64–96.

Bertram, B.C.R. (1980) Vigilance and group size in ostriches. *Anim. Behav.* **28**, 278–286.

Bhattacharyya, N.K. and Mullick, D.N. (1965) Comparative study of mechanical factors in ruminant digestion: Part II. Pattern of rumen movements in ox and buffalo under similar conditions. *Ind. J. Exp. Biol.* **3**, 255–259.

Blancou, L. (1935) Buffles de L'Oubangui-Chari-Tschad. *Terre et Vie* **5**, 202–221.

Borgia, G. (1986) Satin bowerbird parasites: a test of the bright male hypothesis. *Behav. Ecol. Sociobiol.* **19**, 355–358.

Borner, M., FitzGibbon, C.D., Borner, Mo., Caro, T.M., Lindsay, W.K., Collins, D.A. and Holt, M.E. (1987) The decline of the Serengeti Thomson's gazelle population. *Oecologia* **73**, 32–40.

Boudet, G. (1975) *Manuel sur les Pâturages Tropicaux et les Cultures Fouragères, Série Manuels et Précis d'Elevage*, No. 4, IEMVT, Ministère de la Coopération, Paris.

Branagan, D. and Hammond, J.A. (1965) Rinderpest in Tanganyika: a review. *Bull. Epizoot. Dis. Afr.* **13**, 225–246.

Brantas, G.C. (1968) On the dominance order in Frisian–Dutch dairy cows. *Z. Tierzüchtung Züchtungsbiol.* **84**, 127–151.

Breman, H. (1982) La productivité des pâturages, in *La Productivité des Pâturages Sahéliens: une étude des sols, des végétations et de l'exploitation de cette resource naturelle* (eds F.W.T. Penning de Vries and M.A.Djiteye), Centre for Agricultural Publishing and Documentation PUDOC, Wageningen, pp. 13–17.

Breman, H. and Krul, J.M. (1982) La pluviosité et la production de fourage sur les pâturages naturels, in *La Productivité des Pâturages Sahéliens: une étude des sols, des végétations et de l'exploitation de cette resource naturelle* (eds F.W.T. Penning de Vries and M.A. Djiteye), Centre for Agricultural Publishing and Documentation PUDOC, Wageningen, pp. 304–322.

Brown, C.R. (1986) Cliff swallow colonies as information centres. *Science* **234**, 83–85.

Bryant, J.P., Kuropat, P.J., Cooper, S.M., Frisby, K. and Owen-Smith, N. (1989) Resource availability hypotheses of plant antiherbivore defence tested in a South-African ecosystem. *Nature* **340**, 227–229.

Buss, I.O. (1961) Some observations on food habits and behavior of the African elephant. *J. Wildl. Mgt.* **25**, 131–148.

Butler, D.L. and Overshiner, C. (1983) The role of mental computations in judgements of area and volume. *Percept. Pscychophys.* **34**, 593–598.

Capstick, P.H. (1983) *Death in the Dark Continent*, St Martin's Press, London.

Cargill, S.M. and Jefferies, R.L. (1984) The effects of grazing by lesser snow geese on the vegetation of a sub-arctic salt marsh. *J. Appl. Ecol.* 21, 669–686.

Caughley, G. (1976) The elephant problem: an alternative hypothesis. *E. Afr. Wildl. J.* **14**, 265–283.

Caughley, G. (1980) *Analysis of Vertebrate Populations*, John Wiley, New York.

Caughley, G. and Krebs, C.J. (1983) Are big mammals simply little mammals writ large? *Oecologia* **59**, 7–17.

Chacon, E.A. (1976) The effects of sward characteristics upon grazing behaviour, intake and animal production from tropical pastures, PhD Thesis, University of Queensland.

Chacon, E. and Stobbs, T.H. (1976) Influence of progressive defoliation of a grass sward on the eating behaviour of cattle. *Aust. J. Agric. Res.* **27**, 709–727.

Chandra, R.K. (1976) Nutrition as a critical determinant in susceptibility to infection. *Wld. Rev. Nutr. Diet.* **25**, 166–188.

Chandra, R.K. (1977) Immunoglobulins and antibody response in protein–caloric malnutrition – a review, in *Malnutrition and the Immune Response* (ed. R.M. Suskind), Kroc Foundation Series, Vol. 7, Raven Press, New York, pp. 155–168.

Chandra, R.K. (1979) Interaction of nutrition, infection and immune response. *Acta Paediatr. Scand.* **68**, 137–144.

Chapin, F.S. and McNaughton, S.J. (1989) Lack of compensatory growth under phosphorus deficiency in grazing-adapted grasses from the Serengeti. *Oecologia* **79**, 551–557.

Chaplin, R.E. and White, R.W.G. (1972) The influence of age and season on the activity of the testes and epididymides of the Fallow deer, *Dama dama*. *J. Reprod. Fertil.* **30**, 361–369.

Charnov, E.L. (1976) Optimal foraging, the marginal value theorem. *Theor. Pop. Biol.* **9**, 129–136.

Chase, W.G. (1978) Elementary information processes, in *Handbook of Learning and Cognitive Processes: Human Information Processing*, Vol. 5 (ed. W.K. Estes), Lawrence Erlbaum, Hillsdale, NJ, pp. 19–90.

Chaturvedi. M.L., Singh, U.B. and Ranjhan, S.K. (1973) Effect of feeding water-soaked and dry wheat straw on feed intake, digestibility of nutrients and VFA production in growing zebu and buffalo calves. *J. Agric. Sci.* **80**, 393–397.

Cheney, D.L. and Seyfarth, R.M. (1992) Précis of how monkeys see the world. *Behav. Brain Sci.* **15**, 135–182.

Chivers, D.J. (1969) On the daily behaviour and spacing of howling monkey groups. *Folia Primat.* **10**, 48–102.

Christy, C. (1929) The African buffaloes. *Proc. Zool. Soc. Lond.* **30**, 445–462.

Clutton-Brock, T.H., Greenwood, P.J. and Rowell, R.P. (1976) Rank and relationships in highland ponies and highland cows. *Z. Tierpsychol.* **41**, 202–216.

Clutton-Brock, T.H., Albon, S.D., Gibson, R.M. and Guiness, F.E. (1979) The logical stag: adaptive aspects of fighting in red deer (*Cervus elaphus* L.). *Anim. Behav.* **27**, 211–225.

Clutton-Brock, T.H., Guiness, F.E. and Albon, S.D. (1982) *Red Deer: Behaviour and Ecology of Two Sexes*, Chicago University Press, Chicago.

Clutton-Brock, T.H. and Harvey, P.H. (1978) Primate ecology and social organization, in *Readings in Sociobiology* (eds T.H. Clutton-Brock and P.H. Harvey), Freeman, San Francisco, pp. 342–383.

Coe, M.J., Cumming, D.H. and Phillipson, J. (1976) Biomass and production of large African herbivores in relation to rainfall and primary production. *Oecologia* **22**, 341–354.

Coppock, D.L., Swift, D.M., Ellis, J.E. and Calvin, K. (1986) Seasonal patterns of energy allocation to basal metabolism activity and production for livestock in a nomadic pastoral system. *J. Agric. Sci.* **107**, 357–365.

Cornet, A. (1984) Utilisation de modaèles simples de bilan hydrique et de production de biomasse pour déterminer les potentialités de production de parcours en zone sahélienne Senegalaise, in *Proceedings of the Workshop on Land Evaluation for Extensive Grazing* (ed. W. Siderius), International Institute for Land Reclamation and Improvement ILRI, Wageningen, pp. 207–227.

Crampton, E.W. and Harris, L.E. (1969) *Applied Animal Nutrition – The Use of Feedstuffs in the Formulation of Livestock Rations*, Freeman, San Francisco.

Crook, J.H. (1962) The adaptive significance of pair formation types in weaver birds. *Symp. Zool. Soc. Lond.* **8**, 57–70.

Croze, H., Hillman, A.K.K. and Lang, E.M. (1981) Elephants and their habitats: how do they tolerate each other, in *Dynamics of Large Mammal Populations* (eds C.W. Fowler and T.D. Smith), John Wiley, New York, pp. 297–316.

Dalimier, P. (1955) *Les Buffles du Congo Belge*, Institut des Parc Nationaux du Congo Belge, Bruxelles.

Dalrymple, D.J. (1987) Sales forecasting practices: results from a United States survey. *Int. J. Forecasting* **3**, 379–392.

Davis, H. and Pérusse, R. (1988) Numerical competence in animals: definitional issues, current evidence, and a new research agenda. *Behav. Brain Sci.* **11**, 561–615.

De Boer, W.F. and Prins, H.H.T. (1989) Decisions of cattle herdsmen in Burkina Faso and optimal foraging models. *Hum. Ecol.* **17**, 445-464.

De Boer, W.F. and Prins, H.H.T. (1990) Large herbivores that strive mightily but eat and drink as friends.*Oecologia* **82**, 264–274.

De Groot, P. (1980) Information transfer in a socially roosting weaver (*Quelea quelea*; Ploceinae): an experimental study. *Anim. Behav.* **28**, 1249–1254.

Delany, M.J. and Happold, D.C.D. (1979) *Ecology of African Mammals*, Longman, London.

Demment, M.W. and Van Soest, P.J. (1985) A nutritional explanation for body-size patterns of ruminant and nonruminant herbivores. *Am. Nat.* 125, 641–672.

DeVaccaro, L., Garcia, M., Bazan, O. and Bardales, E. (1977) Fertility and body weight at first mating of zebu cattle grazing cleared jungle land in the Amazon basin. *Trop. Agric.* **54**, 223–227.

Dhillon, J.S., Acharya, R.M., Tiwana, M.S. and Aggarwal, S.C. (1970) Factors affecting the interval between calving and conception in Hariana cattle. *Anim. Prod.* **12**, 81–97.

Dirksen, G. (1970) Acidiosis, in *Physiology of Digestion and Metabolism in Ruminants* (ed. A.T. Phillipson), Oriel Press, Newcastle-upon-Tyne, pp. 612–625.

Doreau, M., Martin-Rosset, W. and Petit, D. (1980) Activités alimentaires nocturnes du cheval au pâturage. *Ann. Zootech.* **19**, 299–304.

Douglas-Hamilton, I. (1972) On the ecology and behaviour of the African elephant: the elephant of Manyara, PhD Thesis, Oxford University.

Dradu, E.A.A. and Harrington, G.N. (1972) Seasonal crude protein content of samples obtained from a tropical range pasture using oesophageal fistulated steers. *Trop. Agric.* **49**, 15–21.

Drent, R.H. and Daan, S. (1980) The prudent parent: energetic adjustments in avian breeding. *Ardea* **68**, 225–252.

Drent, R.H. and Prins, H.H.T. (1987) The herbivore as prisoner of its food supply, in *Disturbance in Grasslands: Species and Population Responses* (eds J. van Andel, J.P. Bakker and R.W. Snaydon), Dr W. Junk Publishing, Dordrecht, pp. 133–149.

Drent, R.H. and Van Eerden, M.R. (1980) Goose flocks and food exploitation: how to have your cake and eat it. *Acta XVII Congress. Int. Ornithol.*, pp. 800–806.

Dunbar, R.I.M. (1982) Adaptation, fitness and the evolutionary tautology, in *Current Problems in Sociobiology* (eds King's College Sociobiology Group), Cambridge University Press, Cambridge, pp. 9–28.

Dunbar, R.I.M. (1985) Population consequences of social structure, in *Behavioural Ecology: Ecological Consequences of Adaptive Behaviour* (eds R.M. Sibley and R.H. Smith), Blackwell Scientific Publications, Oxford, pp. 507–519.

Duncan, P. (1992) *Horses and Grasses: The Nutritional Ecology of Equids and their Impact on the Camargue*, Springer, New York.

Elgar, M.A. and Catterall, C.P. (1981) Flocking and predator surveillance in house sparrows: a test of an hypothesis. *Anim. Behav.* **29**, 868–872.

Elliott, J.P., McTaggart-Cowan, I. and Holling, C.S. (1977) Prey capture by the African lion. *Can. J. Zool.* **55**, 1811–1828.

Eloff, F.C. (1984) Food ecology of the Kalahari lion *Panthera leo vernayi*. *Koedoe* (Suppl. 1984), 249–258.

Eltringham, S.K. (1974) Changes in the large mammal community of Mweya Peninsula, Rwenzori National Park, Uganda, following removal of hippopotamus. *J. Appl. Ecol.* 11, 855–865.

Ens, B.J. (1992) The social prisoner: causes of natural variation in reproductive success of the oystercatcher, PhD Thesis, University of Groningen.

Ens, B.J. and Goss-Custard, J.D. (1984) Interference among oystercatchers, *Haematopus ostralegus* L., feeding on mussels, *Mytilus edulis* L., on the Exe estuary. *J. Anim. Ecol.* **53**, 217–232.

Esselink, P. and Van Gils, H.A.M.J. (1985) Ground-based reflectance measurement for standing crop estimates. *ITC J.* **1985**(1), 47–52.

Estes, R.D. (1974) Social organisation of the African Bovidae, in *The Behaviour of Ungulates and its Relation to Management* (eds V. Geist and F.R. Walther), IUCN Publications, New Series 24, IUCN, Morges, pp.166–205.

Estes, R.D. (1991) *The Behavior Guide to African Mammals, including Hoofed Mammals, Carnivores, Primates*, University California Press, Berkeley.

Everitt, B.S. (1977) *The Analysis of Contingency Tables*, Chapman & Hall, London.

Fagot, R.F. (1978) A theory of relative judgement. *Percept. Psychophys.* **24**, 243–252.

Field, C.R. (1968) The food habits of some wild ungulates in Uganda, PhD Thesis, Cambridge University.

Field, C.R. (1971) Elephant ecology in the Queen Elizabeth National Park, Uganda. *E. Afr. Wildl. J.* **9**, 99–123.

Field, C.R. (1972) The food habits of wild ungulates in Uganda by analysis of stomach contents. *E. Afr. Wildl. J.* **10**, 17–42.

Field, C.R. and Blankenship, L.H. (1973) Nutrition and reproduction of Grant's and Thomson's gazelles, Cokes's hartebeest and giraffe in Kenya. *J. Reprod. Fertil.* **19** (Suppl.), 287–301.

Ford, J. (1971) *The Role of the Trypanosomiases in African Ecology: a Study of the Tsetse Fly Problem*, Oxford University Press, Oxford.

Fossey, D. and Harcourt, A.H. (1977) Feeding ecology of free-ranging mountain gorilla (*Gorilla gorilla beringei*), in *Primate Ecology Studies of Feeding and Ranging Behaviour in Lemurs, Monkeys and Apes* (ed. T.H. Clutton-Brock), Academic Press, New York, pp. 415–447.

Fryxell, J.M. (1987) Food limitation and demography of a migratory antelope, the white-eared kob. *Oecologia* **72**, 83–91.

Fuller, W. A. (1960) Behaviour and social organization of the wild bison of Wood Buffalo National Park, Canada. *Arctic* **13**, 3–19.

Gakahu, C. (1984) Feeding ecology of plains zebra in the Serengeti, PhD Thesis, University of Nairobi.

Galaty, J. (1989) Cattle and cognition: aspects of Maasai practical reasoning, in *The Walking Larder: Patterns of Domestication, Pastoralism and Predation* (ed. J. Clutton-Brock), Unwin Hyman, London, pp. 215–230.

Garrett, W.N. (1970) Energy utilization of high protein diets. *J. Anim. Sci.* **31**, 242–249.

Garland, T. (1983) Scaling the ecological costs of transport to body mass in terrestrial mammals. *Am. Nat.* **121**, 571–587.

Gauthier, D. and Coulard, G. (1986) Effect of underfeeding on testosterone–LH feedback in the bull. *J. Endocrinol.* **110**, 233–238.

Geist, V. (1971) *Mountain Sheep: A Study in Behavior and Evolution*, University Chicago Press, Chicago.

Georgiadis, N.J., Ruess, R.W., McNaughton, S.J. and Western, D. (1989) Ecological conditions that determine when grazing stimulates grass production. *Oecologia* **81**, 316–322.

Giblin, J. (1990) Trypanosomiasis control in African history: an evaded issue? *J. Afr. Hist.* **31**, 59–80.

Gittleman, J.L. (1989) Carnivore group living: comparative trends, in *Carnivore Behavior, Ecology and Evolution* (ed. J.L. Gittleman), Chapman & Hall, London, pp. 183–207.

Goodrich, S.G. (1849) *A Pictoral Geography of the World*, 19th edn, C.D. Strong, Boston.

Goodrich, S.G. (1875) *Johnson's Natural History, Comprehensive, Scientific, and Popular: Illustrating and Describing the Animal Kingdom with its Wonders and Curiosities, etc.*, Vol. I, A.J. Johnson, NewYork.

Goss-Custard, J.D. (1985) Foraging behaviour of wading birds and the carrying capacity of estuaries, in *Behavioural Ecology: Ecological Consequences of Adaptive Behaviour* (eds R.M. Sibley and R.H. Smith), Blackwell Scientific Publications, Oxford, pp. 169–188.

Green, R.F. (1984) Stopping rules for optimal foragers. *Am. Nat.* **123**, 30–44.

Greene, E. (1987) Individuals in an osprey colony discriminate between high and low quality information. *Nature* **329**, 239–241.

Greenway, P.J. and Vesey-FitzGerald, D.F. (1969) The vegetation of Lake Manyara National Park. *J. Ecol.* **57**, 127–149.

Greenway, P.J. and Vesey-FitzGerald, D.F. (1972) Annotated checklist of plants occurring in Lake Manyara National Park. *J. East Afr. Nat. Hist. Soc. Nat. Mus.* **28**, 1–29.

Griffin, D.R. (1992) *Animal Minds*, University Chicago Press, Chicago.

Grimsdell, J. (1969) The ecology of the buffalo, *Syncerus caffer*, in Western Uganda, PhD Thesis, Cambridge University.

Grogan, P. (1978) Feeding ecology of the plains zebra in the Serengeti, MSc Thesis, Texas A & M University.

Grubb, P. (1974) Population dynamics of the Soay sheep, in *Island Survivors: the Ecology of Soay Sheep on St. Kilda* (eds P.A. Jewell, C. Milner and J.M. Boyd), Athlone Press, London, pp. 242–272.

Guthrie, R.D. (1990) *Frozen Fauna of the Mammoth Steppe: The Story of Blue Babe*, Chicago University Press, Chicago.

Guy, P.R. (1975) The daily food intake of the African elephant, *Loxodonta africana* Blumenbach, in Rhodesia. *Arnoldia* **7**, 1–8.

Gwynne, M.D. and Bell, R.H.V. (1968) Selection of vegetation components by grazing ungulates in the Serengeti National Park. *Nature* **220**, 390–393.

Haltenorth, T. and Diller, H. (1979) *Elseviers Gids van de Afrikaanse Zoogdieren* (revised and enlarged by C. Smeenk), Elsevier, Amsterdam.

Halvorsen, O. (1986) On the relationship between social status of host and risk of parasitic infection. *Oikos* **47**, 71–74.

Hamilton, W.D. (1971) Geometry of the selfish herd. *J. Theor. Biol.* **31**, 295–311.

Hanby, J.P. and Bygott, J.D. (1979) Population changes in lions and other predators, in *Serengeti: Dynamics of an Ecosystem* (eds A.R.E. Sinclair and M. Norton-Griffiths), Chicago University Press, Chicago, pp. 249–262.

Hanley, G. (1971) *Warriors and Strangers*, Hamish Hamilton, London.

Hansen, R.M., Mugambi, M.M. and Bauni, S.M. (1985) Diets and trophic ranking of ungulates in the northern Serengeti. *J. Wildl. Mgt.* **49**, 823–829.

Harris, J.H. (1951) Lake Manyara, the phenomenon of a salt crust which disappears as it is approached. *Tanganyika Notes Rec.* January–June 1951.

Hausfater, G. (1975) Dominance and reproduction in baboons *(Papio cynocephalus)*, a quantitative analysis. *Contrib. Primatol.* **7**, 1–150.

Hausfater, G. and Watson, D.F. (1976) Social and reproductive correlates of parasite ova emissions by baboons. *Nature* **262**, 688–689.

Hawley, A.W.L., Peden, D.G., Reynolds, H.W. and Stricklin, W. (1981) Bison and cattle digestion of forage from the Slave River lowlands, Northwest Territories, Canada. *J. Range Mgt.* **34**, 126–130.

Haynes, G. (1991) *Mammoths, Mastodonts, and Elephants*, Cambridge University Press, Cambridge.

Heinemann, D. and Thenius, E. (1972) The even-toed ungulates, in *Grzimek's Animal Life Encyclopedia* (ed. H.C.B. Grzimek), *Mammals* IV, Vol. 13, Van Nostrand Reinhold, New York, pp. 71–75.

Herd, D.B. and Sprott, L.R. (1986) *Body Condition, Nutrition and Reproduction of Beef Cows*, Texas Agricultural Extension Service B-1526, Texas A & M University, College Station, TX.

Hilborn, R. and Sinclair, A.R.E. (1979) A simulation of the wildebeest population, other ungulates, and their predators, in *Serengeti: Dynamics of an Ecosystem* (eds A.R.E. Sinclair and M. Norton-Griffiths), Chicago University Press, Chicago, pp. 287–309.

Hirst, S.M. (1969) Populations in a Transvaal lowland nature reserve. *Zool. Afr.* **4**, 199–230.

Hirth, D.H. and McCullough, D.R. (1977) Evolution of alarm signals in ungulates with special reference to white-tailed deer. *Am. Nat.* **111**, 31–42.

Hofmann, R.R. (1973) *The Ruminant Stomach: Stomach Structure and Feeding Habits of East African Gameruminants*, East African Literature Bureau, Nairobi.

Hofmann, R.R. and Stewart, D.R.M. (1972) Grazer or browser: a classification based on the stomach-structure and feeding habits of East African ruminants. *Mammalia* **36**, 226–240.

Holmes, P.H. (1986) Pathophysiology of nematode infections, in Parasitology–quo vadit?, in *Proceedings of the 6th International Congress of Parasitology* (ed. M.J. Howell), Australian Academy of Science, Canberra, pp. 443–451.

Holness, D.H. and Hopley, J.D.H. (1978) The effects of plane of nutrition, live weight, temporary weaning and breed on occurrence of oestrus in beef cows during the post-partum period. *Anim. Prod.* **18**, 43–48.

Homewood, K.M. and Rodgers, W.A. (1991) *Maasailand Ecology: Pastoralist Development and Wildlife Conservation in Ngorongoro, Tanzania*, Cambridge University Press, Cambridge.

Hoppe, P.P. (1977) Rumen fermentation and body weight in African ruminants, in *Game Biologists, Proceedings of the XIIIth International Congress* (ed. T.J. Peterle), Wildlife Society, Washington DC.

Hornocker, M.G. (1970) An analysis of mountain lion predation upon mule deer and elk in the Idaho Primitive Area. *Wildl. Monogr.* **21**.

Houston, D.C. (1979) The adaptations of scavengers, in *Serengeti: Dynamics of an Ecosystem* (eds A.R.E. Sinclair and M. Norton-Griffiths), Chicago University Press, Chicago, pp. 263–286.

Houston, A.I., Kacelnik, A. and McNamara, J. (1982) Some learning rules for acquiring information, in *Functional Ontogeny* (ed. D. McFarland), Pitman Books, London, pp. 140–191.

Hume, I.D. (1982) *Digestive Physiology and Nutrition of Marsupials*, Cambridge University Press, Cambridge.

Humphreys, K.R. (1981) *Environmental Adaptations of Tropical Pasture Plants*, Macmillan Publishers, London.

Janetos, A.C. and Cole, B.J. (1981) Imperfectly optimal animals. *Behav. Ecol. Sociobiol.* **9**, 203–209.

Janis, C. (1976) The evolutionary strategy of the Equidae and the origins of rumen and cecal digestion. *Evolution* **30**, 757–774.

Jarman, P.J. (1971) Diets of large mammals in the woodlands around Lake Kariba, Rhodesia. *Oecologia* **8**, 157–187.

Jarman, P.J. (1974) The social organisation of antelope in relation to their ecology. *Behaviour* **48**, 215–267.

Jarman, P.J. (1982) Prospects for interspecific comparison in sociobiology, in *Current Problems in Sociobiology* (ed. King's College Sociobiology Group), Cambridge University Press, Cambridge, pp. 323–342.

Jarman, P.J. and Sinclair, A.R.E. (1979) Feeding strategy and the pattern of resource partitioning in ungulates, in *Serengeti: Dynamics of an Ecosystem* (eds A.R.E. Sinclair and M. Norton-Griffiths), University of Chicago Press, Chigago, pp. 130-163.

Jennings, T. and Evans, S.M. (1980) Influence of position in the flock and flock size in the starling (*Sturnus vulgaris*). *Anim. Behav.* **28**, 634–635.

Jobaert, A.J. (1952) Le Buffle de Cafrerie ou Buffle noir du Sud du Congo Belge. *Zooleo* **17**, 399–406.

Johnson, D.H. (1989) Political ecology in the Upper Nile: the twentieth century expansion of the pastoral 'common economy'. *J. Afr. Hist.* **30**, 463–486.

Johnson, M. (1928) *Safari: a Saga of the African Blue*, G.P. Putnam's Sons, New York.

Johnson, M. (1988) Out for the count. *Behav. Brain Sci.* **11**, 589.

Johnson, W.L., Ordoveza, A.L., Hardison, W.A. and Castillo, L.S. (1967) The nutritive value of *Panicum maximum* (Guinea grass). II. Digestibility by cattle and water buffaloes, related to season and herbage growth stage. *J. Agric. Sci.* **69**, 161 -170.

Joubert, S.C.T. (1974) The social organization of the Roan antelope, *Hippotragus equinus*, and its influence on the special distribution of herds in the Kruger National Park, in *The Behaviour of Ungulates and its Relation to Management*, Vol. II (eds V. Geist and F. Walter), IUCN, Morges, pp. 661–675.

Kacelnik, A. and Krebs, J.R. (1985) Learning to exploit patchily distributed food, in *Behavioural Ecology: Ecological Consequences of Adaptive Behaviour* (eds R.M. Sibly and R.H. Smith), Blackwell Scientific Publications, Oxford, pp. 189–205.

Kalmbacher, R.S., Long, K.R., Johnson, M.K. and Martin, F.G. (1984) Botanical composition of diets of cattle grazing South Florida rangelands. *J. Range Mgt.* **37**, 334–340.

Ketelaars, J.J.M.H. (1986) Prediction of feeding intake in ruminants, in *Modelling of Extensive Livestock Production Systems* (eds N. de Ridder, H. van Keulen, N.G. Seligman and P.J.H. Neate), International Livestock Centre for Africa, Addis Ababa, pp. 108–161.

Kikkawa, J. (1980a) Weight changes in relation to social hierarchy in captive flocks of Silvereyes, *Zosterops lateralis*. *Behaviour* **74**, 92–99.

Kikkawa, J. (1980b) Winter survival in relation to dominance classes among Silvereyes, *Zosterops lateralis chlorocephala* of Heron Island, Great Barrier Reef. *Ibis* **122**, 437–446.

Kingdon, J. (1982) *East African Mammals: An Atlas of Evolution in Africa*, Vol. III (Part C) *Bovids*, Academic Press, London.

Kjekshus, H. (1977) *Ecology Control and Economic Development in East African History: the case of Tanganyika, 1850–1950*, Heinemann, London.

Klein, G.R. (1988) The archaeological significance of animal bones from Acheulean sites in southern Africa. *Afr. Arch. Rev.* **6**, 3–25.

Klingel, H. (1967) Soziale Organization und Verhalten freilebender Steppen zebras. *Z. Tierpsychol.* **24**, 580–624.

Kruuk, H. (1972) *The Spotted Hyena*, Chicago University Press, Chicago.

Kruuk, H. and Turner, M. (1967) Comparative notes on predation by lion, leopard, cheetah and wild dog in the Serengeti area, East Africa. *Mammalia* **31**, 1–27.

Lamphear, J. (1988) The people of the grey bull: the origin and expansion of the Turkana. *J. Afr. Hist.* **29**, 27–39.

Lamprey, H.F. (1963) Ecological separation of the large mammal species in Tarangire Game Reserve, Tanganyika. *E. Afr. Wildl. J.* **1**, 63–92.

Lamprey, H.F. (1964a) Estimation of the large mammal densities, biomass and energy exchange in the Tarangire Game Reserve and the Masai steppe. *E. Afr. Wildl. J.* **2**, 1–46.

Lamprey, H.F. (1964b) Ecological separation of the large mammal species in the Tarangire Game Reserve, Tanganyika. *E. Afr. Wildl. J.* **1**, 63–92.

Le Boeuf, B.J. (1974) Male–male competition and reproductive success in elephant seals. *Am. Zool.* **14**, 163–176.

Leader-Williams, N. (1979) Age-related changes in the testicular and antler cycles of Reindeer, *Rangifer tarandus*. *J. Reprod. Fertil.* **57**, 117–126.

Leader-Williams, N. (1982) Relationship between a disease, host density and mortality in a free-living deer population. *J. Anim. Ecol.* **51**, 235–240.

Lee, P.C. and Moss, C.J. (1986) Early maternal investment in male and female African elephant calves. *Behav. Ecol. Sociobiol.* **18**, 353–363.

Leslie, D.M. and Starkey, E.E. (1985) Fecal indices to dietary quality of cervids in old-growth forests. *J. Wildl. Mgt.* **49**, 142–151.

Leuthold, W. (1977) African ungulates: a comparative review of their ethology and behavioral ecology. *Zoophysiol. Ecol.* **8**.

Leuthold, W. (1978) Ecological seperation among browsing ungulates in Tsavo East National Park, Kenya. *Oecologia* **35**, 241–252.

Leyhausen, P. (1965) *The Communal Organization of Solitary Mammals*, Symposia of the Zoological Society of London No. 14, pp. 249–263.

Lincoln, G.A., Guiness, F. and Short, R.V. (1972) The way in which testosterone controls the social and sexual behavior of the red deer stag (*Cervus elaphus*). *Hormones Behav.* **3**, 375–396.

Lloyd, L.E., McDonnald, B.E. and Crampton, E.W. (1978) *Fundamentals of Nutrition*, Freeman, San Francisco.

Lochmiller, R.L., Hellgren, E.C. and Grant, W.E. (1986) Reproductive responses to nutritional stress in adult female collared pecaries. *J. Wildl. Mgt.* **50**, 295–300.

Loftus, G.R. (1978) Comprehending compass directions. *Memory Cogn.* **6**, 416–422.

Loth, P.E. and Prins, H.H.T. (1986) Spatial patterns of the landscape and vegetation of Lake Manyara National Park. *ITC J.* **1986**(2), 115–130.

Lott, D.F. (1979) Dominance relations and breeding rate in mature male American bison. *Z. Tierpsychol.* **49**, 418–432.

Lott, D.F. (1981) Sexual behavior and intersexual strategies in American bison. *Z. Tierpsychol.* **56**, 97–114.

MacArthur, R.H. and Pianka, E.R. (1966) An optimal use of a patchy environment. *Am. Nat.* **100**, 603–609.

McBride, C.J. (1984) Age and size categories of lion prey in Chobe National Park, Botswana. *Bots. Notes Rec.* **16**, 139–143.

McDonald, D.R. (1946) Enemy property in Tanganyika: including a catalogue of enemy property (prepared under direction of D.R. McDonald, Custodian of Enemy Property in Tanganyika) Cape Town.

McGregor Ross, W. (1911) Two finds on Mount Kenia. *J. E. Afr. Ugan. Nat. Hist. Soc.* **2**, 60–63.

McHugh, T. (1958) Social behaviour of the American buffalo (*Bison bison bison*). *Zoologica* **43**, 1–40.

Mack, R. (1970) The great African cattle plague epidemic of the 1890s. *Trop. Anim. Health Prod.* **2**, 210–219.

McKay, G.M. (1973) Behavior and ecology of the Asiatic elephant in south eastern Ceylon. *Smithsonian Contrib. Zool.* **125**, 1–113.

MacLatchy, A.R. (1932) Les buffles du Gabon (Régions du Fernan-Vaz et de la N'Gounié). *Terre et Vie* **2**, 584–596.

McMeniman, N.P., Ben-Ghedalia, D. and Armstrong, D.G. (1976) Nitrogen energy interaction in the rumen fermentation, in *Protein Metabolism and Nutrition* (eds D.J.A. Cole, K.N. Boorman, P.J. Buttery, D.Lewis, R.J. Neale and H. Swan), Butterworths, London, pp. 217–229.

McNaughton, S.J. (1976) Serengeti migratory wildebeest: facilitation of energy flow by grazing. *Science* **191**, 92–94.

McNaughton, S.J. (1979) Grazing as an optimization process: grass–ungulate relationships in the Serengeti. *Am. Nat.* **113**, 691–703.

McNaughton, S.J. (1983) Compensatory plant growth as a response to herbivory. *Oikos* **40**, 329–336.

McNaughton, S.J. (1984) Grazing lawns: animals in herds, pland form and coevolution. *Am. Nat.* **124**, 863–886.

MacRae, A.W. (1983) Constant errors in simultaneous matching of angles are not an artifact of the starting point for adjustement. *Percept. Psychophys.* **33**, 96–98.

Maddock, L. (1979) The 'migration' and grazing succession, in *Serengeti: Dynamics of an Ecosystem* (eds A.R.E.Sinclair and M. Norton-Griffiths), University of Chicago Press, Chigago, pp. 104–129.

Maddox, G. (1990) Mtunga: famine in Central Tanzania, 1917–1920. *J. Afr. Hist.* **31**, 181–197.

Makacha, S. and Schaller, G.B. (1969) Observations on lions in the Lake Manyara National Park, Tanzania. *E. Afr. Wild. J.* **7**, 99–103.

Makridakis, S. and Wheelwright, S.C. (1989) *Forecasting Methods for Management*, 5th edn, John Wiley, Chichester.

Malpas, R.C. (1977) Diet and the condition and growth of elephants in Uganda. *J. Appl. Ecol.* 14, 489–504.

Mandler, J.M., Seegmiller, D. and Day, J. (1977) On the coding of spatial information. *Memory Cogn.* **5**, 10–16.

Mannetje, L. 't (1978) Measuring quantity of grassland vegetation, in *Measurement of Grassland Vegetation and Animal Production* (ed. L. 't Mannetje), Commonwealth Bureau of Pastures and Field Crops, Hurley, pp. 63–95.

Manzolillo, D.L. (1986) Factors affecting intertroop transfer by adult male *Papio anubis*, in *Primate Ontogeny, Cognition and Social Behaviour* (eds J.G. Else and P.C. Lee), Cambridge University Press, Cambridge, pp. 371–380.

Marburger, R.G. and Thomas, J.W. (1965) A die-off in white-tailed deer of the Central Mineral Region. *J. Wildl. Mgt.* **29**, 706–716.

Mattson, W.J. (1980) Herbivory in relation to plant nitrogen content. *Ann. Rev. Ecol. Syst.* **11**, 119–161.

May, R.M. and Anderson, R.M. (1979) Population biology of infectious diseases: Part II. *Nature* **280**, 455–461.

Maynard-Smith, J. (1982) Introduction, in *Current Problems in Sociobiology* (ed. King's College Sociobiology Group), Cambridge University Press, Cambridge, pp. 1–3.

Mayr, E. (1970) Evolution und Verhalten. *Verh. Dtsch. Zool. Ges.* **64**, 322–336.

Milford, R. and Minson, D.J. (1966) The feeding value of tropical pastures, in *Tropical Pastures* (eds W. Davies and C.L. Skidmore), Faber & Faber, London, pp. 106–114.

Mills, M.G.L. (1984) Prey selection and feeding habits of the large carnivores in the southern Kalahari. *Koedoe* **1984** (Suppl.), 281–294.

Mills, M.G.L., Wolff, P., LeRiche, E.A.N. and Meyer, I.J. (1978) Some population characteristics of the lion *Panthera leo* in the Kalahari Gemsbok National Park. *Koedoe* **21**, 163–171.

Mitchell, B.L., Shenton, J.B. and Uys, J.C.M. (1965) Predation on large mammals in the Kafue National Park, Zambia. *Zool. Afr.* **1**, 297–318.

Mittermeier, R.A. (1973) Group activity and population dynamics of the Howler monkey of Barro Colorado Island. *Primates* **14**, 1–19.

Mloszewski, M.J. (1983) *The Behavior and Ecology of the African Buffalo*, Cambridge University Press, Cambridge.

Moar, I. and Bower, G.H. (1983) Inconsistency in spatial knowledge. *Memory Cogn.* **11**, 107–113.

Moen, A.N. (1973) *Wildlife Ecology: An Analytical Approach*, Freeman, San Francisco.

Mohan, R.N. and Gotts, M.G. (1970) Diseases and parasites of the African buffalo. *Vet. Bull.* **40**, 157–165.

Monro, R.H. and Skinner, J.D. (1979) A note on condition indices for adult male impala, *Aepyceros melampus*. *S. Afr. J. Anim. Sci.* **9**, 47–51.

Morton-Boyd, J. and Jewell, P.A. (1974) The Soay sheep and their environment: a synthesis, in *Island Survivors: the Ecology of Soay Sheep on St. Kilda* (eds P.A. Jewell, C. Milner and J.M. Boyd), Athlone Press, London, pp. 360–373.

Mueller-Dombois, D. and Ellenberger, H. (1974) *Aims and Methods in Vegetation Ecology*, John Wiley, New York.

Murton, R.K. (1971) Why do some birds feed in flocks? *Ibis* **113**, 534–536.

Mwalyosi, R.B.B. (1977) A count of large mammals in Lake Manyara National Park. *E. Afr. Wildl. J.* **15**, 333–335.

Myers, N. (1973) Tsavo National Park, Kenya, and its elephants: an interim appraisal. *Biol. Cons.* **5**, 123–132.

Naga, M.A. and El-Shazly, K. (1969) Activities of rumen microorganisms in water buffalo (*Bos bubalus* L.) and in zebu cattle. *J. Dairy Res.* **36**, 1–10.

Ndhlovu, D.E. and Balakrishnan, M. (1991) Large herbivores in Upper Lupande Game Management Area, Luangwa Valley, Zambia. *Afr. J. Ecol.* **29**, 93–104.

Neumann, G.C. (1977) Nonspecific host factors and infection in malnutrition – a review, in *Malnutrition and the Immune Response*, Kroc Foundation Series, Vol. 7 (ed. R.M. Suskind), Raven Press, New York, pp. 355–374.

Norton-Griffiths, M. (1979) The influence of grazing, browsing, and fire on the vegetation dynamics of the Serengeti, in *Serengeti: Dynamics of an Ecosystem* (eds A.R.E. Sinclair and M. Norton-Griffiths), Chicago, University Press, Chicago, pp. 310–352.

Orians, G.H. (1975) Diversity, stability and maturity in natural ecosystems, in *Unifying Concepts in Ecology* (eds W.H. van Dobben and R.H. Lowe-McConell), Dr W. Junk Publishers, The Hague, and PUDOC, Wageningen, pp. 139–151.

Orskov, E.R. (1976) Factors influencing protein and nonprotein nitrogen utilization in young ruminants, in *Protein Metabolism and Nutrition* (eds D.J.A. Cole, K.N. Boorman, P.J. Buttery, D. Lewis, R.J. Neale and H. Swan), Butterworths, London, pp. 457–476.

Osmond, C.B., Winter, K. and Ziegler, H. (1982) Functional significance of different pathways of CO_2 fixation in photosynthesis, in *Physiological Plant Ecology*, Vol. II, *Water Relations and Carbon Assimilation* (eds O.L.Lange, P.S. Nobel, C.B. Osmond and H. Ziegler), Encyclopedia of Plant Physiology, N.S. 12B, Springer, Berlin, pp. 479–547.

Ostfeld, R.S. (1988) Fluctuations and constancy in populations of small rodents. *Am. Nat.* **131**, 445–452.

Owen-Smith, N., Robbins, C.T. and Hagermann, A.E. (1993) Browse and browsers: interactions between woody plants and mammalian herbivores. *Tree* **8**, 158–160.

Packer, C. (1986) The ecology of sociality in felids, in *Ecological Aspects of Social Evolution* (eds D.I. Rubenstein and R.W. Wrangham), Princeton University Press, Princeton, NJ, pp. 429–451.

Pant, H.C. and Roy, A. (1970) Studies on the rumen microbiological activity of buffalo and zebu cattle. *Ind. J.Anim. Sci.* **40**, 600–609.

Parker, G.A. (1974) Assessment strategy and the evolution of fighting behaviour. *J. Theor. Biol.* **47**, 223–243.

Parkins, J.J., Holmes, P.H. and Bremner, K.C. (1973) The pathophysiology of ovine ostertagiosis: some nitrogen balance and digestibility studies. *Res. Vet. Sci.* **14**, 21–28.

Pastoret, P.P., Thiry, E., Brochier, P., Schwers, A., Thomas, I. and Dubuisson, J. (1988) Maladies de la faune sauvage transmissibles aux animaux domestiques. *Revue scientifique et technique: office international des épizooties (Paris)* **7**, 661–704.

Patterson, J.H. (1909) *In the Grip of the Nyika: Further Adventures in British East Africa*, Macmillan, New York.

Pellew, R.A.P. (1983) The impacts of elephant, giraffe and fire upon the *Acacia tortilis* woodlands of the Serengeti. *Afr. J. Ecol.* **21**, 41–74.

Penning, P.D. (1986) Some effects of sward condition on grazing behaviour and intake by sheep, in *Grazing Research at Northern Latitudes* (ed. O. Gudmundson), Plenum Press, New York, pp. 219–226.

Perret, M. (1985) Influence of social factors on seasonal variations in plasma testosterone levels in *Microcebus murinus*. *Z. Tierpsychol.* **69**, 265–280.

Perret, M. (1986) Plasma-testosterone-binding globulin-binding capacity in the male Lesser mouse lemur (*Microcebus murinus*): relationship to seasonal and social factors. *J. Endocrinol.* **110**, 169–175.

Peterson, R.O. (1977) *Wolf Ecology and Prey Relationships on Isle Royal*, US Department of Interior and the National Park Service, Fauna Series 11.

Phillipson, J. (1975) Rainfall, primary production and 'carrying capacity' of Tsavo National Park (East), Kenya. *E. Afr. Wildl. J.* **13**, 171–201.

Pianka, E.R. (1973) The structure of lizzard communities. *Ann. Rev. Ecol. Syst.* **4**, 53–74.

Pienaar, U. de V. (1960) 'N uitbraak van die miltsiekte onder wild in die Nasionale Krugerwildtuin 28.9.59 tot 20.11.59. *Koedoe* **3**, 238–251.

Pienaar, U. de V. (1961) A second outbreak of anthrax amongst game animals in the Kruger National Park. *Koedoe* **4**, 4–17.

Pienaar, U. de V. (1963) The large mammals of the Kruger National Park: their distribution and present day status. *Koedoe* **6**, 1–37.

Pienaar, U. de V. (1967) Epidemiology of anthrax of wild animals and the control of anthrax epizootics in the Kruger National Park, South Africa. *Fed. Proc.* **26**, 1496–1502.

Pienaar, U. de V. (1969a) Observations on developmental biology, growth and some aspects of the population ecology of African buffalo (*Syncerus caffer caffer* Sparrman) in the Kruger National Park. *Koedoe* **12**, 29–53.

Pienaar, U. de V. (1969b) Predator–prey relationships amongst the larger mammals of the Kruger National Park. *Koedoe* **12**, 108–176.

Pienaar, U. de V., Van Wijk, P. and Fairall, N. (1966) An aerial census of elephant and buffalo in the Kruger National Park, and the implications thereof on intended management schemes. *Koedoe* **9**, 40–107.

Plotkin, H. (1994) *The Nature of Knowledge: Concerning Adaptations, Instinct and the Evolution of Intelligence*, Allen Lane/The Penguin Press, London.

Plowright, W. (1982) The effects of rinderpest and rinderpest control on wildlife in Africa. *Symp. Zool. Soc. Lond.* **50**, 1–28.

Pollard, J.H. (1977) *A Handbook of Numerical and Statistical Techniques, with Examples Mainly From the Life Sciences*, Cambridge University Press, Cambridge.

Pottinger, G. (1983) Explanation, rationality and micro-economic theory. *Behav. Sci.* **28**, 109–126.

Potts, G.R., Tapper, S.C. and Hudson, P.J. (1984) Population fluctuations in red grouse: analysis of bag records and a simulation model. *J. Anim. Ecol.* **53**, 21–36.

Powell, G.V.N. (1974) Experimental analysis of the social value of flocking by starlings (*Sturnus vulgaris*) in relation to predation and foraging. *Anim. Behav.* **22**, 501–505.

Pratchet, D., Capper, B.G., Light, D.E., Miller, M.D., Rutherford, A.S., Rennie, T.W., Buck, N.G. and Trail, J.C. (1977) Factors limiting live weight gain of beef cattle on rangeland in Botswana. *J. Range Mgt.* **30**, 442–445.

Pratt, D.J. and Gwynne, M.D. (1977) *Rangeland Management and Ecology in East Africa*, Hodder & Stoughton, London.

Pratt, R.M., Putman, R.J., Ekins, J.R. and Edwards, P.J. (1986) Use of habitat by free-ranging cattle and ponies in the New Forest, Southern England. *J. Appl. Ecol.* 23, 539–557.

Prins, H.H.T. (1987) Nature conservation as an integral part of optimal land use in East Africa: action plan for the Masai ecosystem of northern Tanzania. *Biol. Cons.* **40**, 141–161.

Prins, H.H.T. (1988) Plant phenology patterns in Lake Manyara National Park, Tanzania. *J. Biogeog.* **15**, 465–480.

Prins, H.H.T. (1992). The pastoral road to extinction: competition between wildlife and traditional pastoralism in East Africa. *Env. Cons.* **19**, 117–124.

Prins, H.H.T. and Beekman, J.H. (1989) A balanced diet as a goal for grazing: the food of the Manyara buffalo. *Afr. J. Ecol.* **27**, 241–259.

Prins, H.H.T. and Douglas-Hamilton, I. (1990) Stability in a multi-species assemblage of large herbivores in East Africa. *Oecologia* **83**, 392–400.

Prins, H.H.T. and Loth, P.E. (1988) Rainfall in time and space in the Masai ecosystem of northern Tanzania: the background to plant phenology patterns. *J. Biogeog.* **15**, 451–463.

Prins, H.H.T. and Van der Jeugd, H.P. (1993) Herbivore crashes and woodland structure in East Africa. *J. Ecol.* **81**, 305–414.

Prins, H.H.T. and Weyerhaeuser, F.J. (1987). Epidemics in populations of wild ruminants: anthrax and impala, rinderpest and buffalo in Lake Manyara National Park, Tanzania. *Oikos* **49**, 28–38.

Prins, H.H.T., Ydenberg, R.C. and Drent, R.H. (1980) The interaction of brent geese (*Branta bernicla*) and sea plantain (*Plantago maritima*): field observations and experiments. *Acta Bot. Neerl.* **29**, 585–596.

Prins, H.H.T., Van der Jeugd, H.P. and Beekman, J.H. (1994) Elephant decline in Lake Manyara National Park, Tanzania. *Afr. J. Ecol.* (in press).

Pyke, G.H., Pulliam, H.R. and Charnov, E.L. (1977) Optimal foraging: a selective review of theory and tests. *Q. Rev. Biol.* **52**, 137–155.

Radclyffe Dugmore, A. (1910) *Camera Adventures in the African Wilds*, Doubleday, Page, New York.

Radclyffe Dugmore, A. (1925) *The Wonderland of Big Game: Being an Account of Two Trips through Tanganyika and Kenya*, Arrowsmith, London.

Rapport, D.J. and Turner, J.E. (1977) Economic models in ecology. *Science* **195**, 367–374.

Reanny, D.C. (1982) The evolution of RNA species. *Ann. Rev. Microbiol.* **36**, 47–73.

Rees, W. (1974) Preliminary studies into bush utilization by cattle in Zambia. *J. Appl. Ecol.* 11, 207–214.

Reid, J.T. (1974) Energy metabolism of the whole animal, in *The Control of Metabolism* (ed. J.D. Sink), Pensylvania State University Press, London, pp. 113–151.

Reinhardt, V. (1983a) Movement orders and leadership in a semi-wild cattle herd. *Behaviour* **83**, 251–264.

Reinhardt, V. (1983b) Flehmen, mounting and copulation among members of a semi-wild cattle herd. *Anim. Behav.* **31**, 641–650.

Reinhardt, V. and Reinhardt, A. (1982) Mock fighting in cattle. *Behaviour* **81**, 1–13.

Reiter, J., Stinson, N.L. and LeBoeuf, B.J. (1978) Northern elephant seal development: the transition from weaning to nutritional independence. *Behav. Ecol. Sociobiol.* **3**, 337–367.

Rejmánek, M. (1992) Stability in a multi-species assemblage of large herbivores in East Africa: an alternative interpretation. *Oecologia* **89**, 454–456.

Renouf, D. and Lawson, W.J. (1986) Harbour seal vigilance: watching for predators or mates. *Biol. Behav.* **11**, 44–49.

Rittenhouse, L.R. (1986) The relative efficiency of rangeland use by ruminant and non-ruminant herbivores, in *Grazing Research at Northern Latitudes* (ed. O. Gudmundson), Plenum Press, New York, pp. 179–191.

Robertson, A. (1976) *Handbook on Animal Diseases in the Tropics*, British Veterinary Association, London.

Roseby, F.B. (1973) Effects of *Trichostrongylus colubriformis* (Nematoda) on the nutrition and metabolism of sheep. I. Feed intake, digestion and utilization. *Aust. J. Agric. Res.* **24**, 947–953.

Roseby, F.B. (1974) Effects of *Trichostrongylus colubriformis* (Nematoda) on the nutrition and metabolism of sheep. II. Metabolism of urea. *Aust. J. Agric. Res.* **25**, 363–370.

Roseby, F.B. (1977) Effects of *Trichostrongylus colubriformis* (Nematoda) on the nutrition and metabolism of sheep. III. Digesta flow and fermentation. *Aust. J. Agric. Res.* 28, 155–164.

Rowell, T.E. (1975) Growing up in a monkey group. *Ethos* **3**, 113–128.

Rowell, T.E. (1979) How would we know if social organizations were not adaptive?, in *Primate Ecology and Human Origins: Ecological Influences on Social Organization* (eds I.S. Bernstein and E.O. Smith), Garland STPM Press, New York, pp. 1–22.

Rudnai, J. (1974) The pattern of lion predation in Nairobi National Park. *E. Afr. Wildl. J.* **12**, 213–225.

Sachser, N. and Prîve, E. (1986) Social status and plasma-testosterone-titers in male Guinea pigs (*Cavia aparea f. porcellus*). *Ethology* **71**, 103–114.

Scarce, D. (1986) Selection of plant species by buffalo in Lake Manyara National Park by analysis of faeces, MSc Thesis, University of Groningen/ University of Utrecht.

Schaller, G.B. (1967) *The Deer and the Tiger: A Study of Wildlife in India*, Chicago University Press, Chicago.

Schaller, G.B. (1972) *The Serengeti Lion, A Study of Predator–Prey Relationships*, Chicago University Press, Chicago.

Schein, M.W. and Fohrman, M.H. (1955) Social dominance relationships in a herd of dairy cattle. *Br. J. Anim.Behav.* **3**, 45–55.

Schoener, T.W. (1971) Theory of feeding strategies. *Ann. Rev. Ecol. Syst.* **2**, 369–405.

Scott, G.R. (1981) Rinderpest, in *Infectious Diseases of Wild Mammals* (eds J.W. Davis, L.H. Karlstad and D.O. Trainer), Iowa State Press, Ames, pp. 18–30.

Scrimshaw, N.S., Taylor, C.E. and Gordon, J.E. (1968) *Interactions of Nutrition and Infection* (World Health Organization paper 57), WHO, Geneva.

Short, J. (1986) The effect of pasture availability on food intake, species composition, and grazing behaviour of kangaroos. *J. Appl. Ecol.* 23, 559–571.

Sinclair, A.R.E. (1975) The resource limitation of trophic levels in tropical grassland ecosystems. *J. Anim. Ecol.* **44**, 497–520.

Sinclair, A.R.E. (1977) *The African Buffalo: A Study in Resource Limitations of Populations*, Chicago University Press, Chicago.

Sinclair, A.R.E. (1979) Dynamics of the Serengeti ecosystem: process and pattern, in *Serengeti: Dynamics of an Ecosystem* (eds A.R.E. Sinclair and M. Norton-Griffiths), Chicago University Press, Chicago, pp. 1–30.

Sinclair, A.R.E. (1985) Does interspecific competition or predation shape the African ungulate community? *J. Anim. Ecol.* **54**, 899–918.

Sinclair, A.R.E. and Norton-Griffiths, M. (1982) Does competition or facilitation regulate migrant ungulate populations in the Serengeti? A test of hypotheses. *Oecologia* **53**, 364–369.

Sinclair, A.R.E., Dublin, H. and Borner, M. (1985) Population regulation of Serengeti wildebeest: a test of the food hypothesis. *Oecologia* **65**, 266–268.

Slobodkin, L.B. (1974) Prudent predation does not require group selection. *Am. Nat.* **108**, 665–678.

Slyter, L.L. (1976) Influence of acidiosis on rumen function. *J. Anim. Sci.* **43**, 910–929.

Snaydon, R.W. (1981) The ecology of grazed pastures, in *Grazing Animals* (ed. F.H.W. Morley), Elsevier Scientific, Amsterdam, pp. 13–31.

Sobani, N. (1988) Fishermen herders: subsistence, survival and cultural change in northern Kenya. *J. Afr. Hist.* **29**, 41–56.

Soulsby, E.J.L. (1968) *Helminths, Artropods and Protozoa of Domestic Animals*, Baillière, Tindall & Cassell, London.

Spalding, D.J. and Lesowski, J. (1971) Winter food of the cougar in south-central British Columbia. *J. Wildl. Mgt.* **35**, 378–381.

Spinage, C.A. (1973) The role of photoperiodism in the seasonal breeding of tropical ungulates. *Mammal Rev.* **3**, 71–84.

Stanley, H.M. (1890) *In Darkest Africa or the Quest, Rescue and Retreat of Emin Governor of Equatoria*, Vol. I, Sampson Low, Marston, Searle & Rivington, London.

Stanley Price, M.R. (1974) The feeding ecology of Coke's hartebeest, *Alcelaphus buselaphus cokei* Günther, in Kenya, PhD Thesis, Oxford University.

Stanley Price, M.R. (1978) The nutritional ecology of Coke's hartebeest (*Alcelaphus buselaphus buselaphus*) in Kenya. *J. Appl. Ecol.* 15, 33–49.

Stelfox, J.G., Eden, G.G., Epp, H., Hudson, R.J., Mbugua, S.W., Agatsiva, J.L. and Amunyunzu, C.L. (1985) Herbivore dynamics in southern Narok, Kenya. *J. Wildl. Mgt.* **50**, 339–347.

Stephens, D.W. and Krebs, J.R. (1986) *Foraging Theory*, Princeton University Press, Princeton.

Stephenson, L.S., Crompton, D.W.T., Latham, M.C., Schulpen, T.W.J., Nesheim, McC. and Jansen, A.A.J. (1980) Relationships between *Ascaris* infection and growth of malnourished preschool children in Kenya. *Am. J.Clin. Nutr.* **33**, 1165–1172.

Stevenson-Hamilton, J. (1911) The relation between game and tsetse-flies. *Bull. Entom. Res.* **2**, 113–118.

Stevenson Hamilton, J. (1937, reprinted 1974) *South African Eden: From Sabi Game Reserve to Kruger National Park*, Collins, London

Stewart, D.R.M. and Stewart, J. (1970) Food preference data by faecal analysis for African plain ungulates. *Zool. Afr.* **15**, 115–129.

Stobbs, T.H. (1970) Automatic measurement of grazing time by dairy cows on tropical grass and legume pastures. *Trop. Grassl.* **4**, 237–244.

Stobbs, T.H. (1973a) The effects of plant structure on the intake of tropical pastures. I. Variation in the bite size of grazing cattle. *Austr. J. Agric. Res.* **24**, 809–819.

Stobbs, T.H. (1973b) The effects of plant structure on the intake of tropical pastures. II. Differences in sward structure, nutritive value, and bite size of animals grazing *Setaria anceps* and *Chloris gayana* at various stages of growth. *Austr. J. Agric. Res.* **24**, 821–829.

Stolba, A. (1979) Entscheidungsfindung in Verbänden von *Papio hamadryas*, PhD Thesis, University of Zürich.

Struhsaker, T.T. and Leland, L. (1979) Socioecology of five sympatric monkey species in the Kibale Forest, Uganda. *Adv. Study Behav.* **9**, 159–228.

Sunquist, M.E. and Sunquist, F.C. (1989) Ecological constraints on predation by large felids, in *Carnivore Behavior, Ecology and Evolution* (ed. J.L. Gittleman), Chapman & Hall, London, pp. 283–301.

Sykes, A.R. and Coop, R.L. (1977) Intake and utilization of food by growing sheep with abomasal damage caused by daily dosing with *Ostertagia circumcincta* larvae. *J. Agric. Sci.* **88**, 671–677.

Taylor, C.R. (1968) The minimum water requirements of some East African bovids. *Symp. Zool. Soc. Lond.* **21**,195–206.

Teunissen, W., Spaans, B. and Drent, R.H. (1985) Breeding success in Brent in relation to individual feeding opportunities during spring staging in the Wadden Sea. *Ardea* **73**, 109–119.

Thienpont, D., Rochette, F. and Vanparijs, O.F.J. (1970) *Diagnose van Verminose door Koprologisch Onderzoek,* Janssen Reserarch Foundation, Beerse.

Thomas, A.D. and Reid, N.R. (1944) Rinderpest in game: a description of an outbreak and an attempt at limiting its spread by means of a bush fence. *Onderstepoort J. Vet. Sci. Anim. Ind.* **20**, 7–23.

Tieszen, L.L., Senyimba, M.M., Imbamba, S.K. and Troughton, J.H. (1979) The distribution of C_3 and C_4 grasses and carbon isotope discrimination along an altitudinal and moisture gradient in Kenya. *Oecologia* **37**, 373–350.

Tinbergen, N. (1962) The evolution of animal communication: a critical examination of the methods. *Symp. Zool.Soc. Lond.* **8**, 1–6.

Tinbergen, N. (1963) On aims and methods of ethology. *Z. Tierpsychol.* **20**, 410–433.

Trudell, J. and White, R.G. (1981) The effect of forage structure and availability on food intake, biting rate, bite size and daily eating time of reindeer. *J. Appl. Ecol.* **18**, 63–81.

Trueman, R.E. (1981) *Quantitative Methods for Decision Making in Business,* The Dryden Press/Holt Saunders, Chicago.

Tulloch, D.G. (1969) Home range in feral water buffalo, *Bubalis bubalus* Lydekker. *Aust. J. Zool.* **17**, 143–152.

Tulloch, D.G. (1979) The water buffalo, *Bubalis bubalus,* in Australia: reproductive and parent–offspring behaviour. *Aust. Wildl. Res.* **6**, 265–287.

Tyler, S.J. (1972) The behaviour and social organization of the New Forest ponies. *Anim. Behav. Monogr.* **5**, 85–196.

Tyrell, J.G. and Coe, M.J. (1974) The rainfall regime of Tsavo National Park, Kenya and its potential phenological significance. *J. Biogeog.* **1**, 187–192.

Underwood, R. (1982) Vigilance behaviour in grazing African antelopes. *Behaviour* **79**, 81–108.

Van de Nieuwenhuizen, J. (1979) Het effect van continue infectie met *Cooperia* spp. en *Ostertagia* spp. op het metabolisme van kalveren. Thesis 95, Veehouderij, Wageningen Agricultural University.

Van der Honing, Y., Steg, A. and Van Es, A.J.H. (1977) Feed evaluation for dairy cows: tests on the system proposed in the Netherlands. *Livestock Prod. Sci.* **4**, 57–67.

Van der Post, L. (1965, reprinted 1973) *Flamingo Feather: A Story of Africa,* Penguin Books, Harmondsworth.

Van de Veen, H. E. (1979) Food selection and habitat use in the red deer *(Cervus elaphus* L.), PhD Thesis, University of Groningen.

Van Es, A.J.H. and Boekholt, H.A. (1976) Protein requirements in relation to the lactation cycle, in *Protein Metabolism and Nutrition* (eds D.J.A. Cole, K.N. Boorman, P.J. Buttery, D. Lewis, R.J. Neale and H.Swan), Butterworths, London, pp. 441–455.

Van Kasteren, H.W.J. (1982) A spectrometer to determine soil convenge and biomass *in situ*. Colloques de l'Institut National de la Recherche *Agronomique* **5**, 125–132.

Van Orsdol, K.G. (1984) Foraging behaviour and hunting success of lions in Queen Elizabeth National Park, Uganda. *Afr. J. Ecol.* **22**, 79–99.

Van Roosmalen, M.G.M. (1980) Habitat preferences, diet, feeding studies and social organization of the black spider monkey (*Ateles paniscus paniscus* Linnaeus 1758) in Surinam, PhD Thesis, Agricultural University, Wageningen.

Van Soest, P.J. (1982) *Nutritional Ecology of the Ruminant*, O & B Books, Cornwallis, OR.

Van Wijngaarden, W. (1985) Elephants–Trees–Grass–Grazers: relationships between climate, soils, vegetation and large herbivores in a semi-arid savanna ecosystem (Tsavo, Kenya), PhD Thesis, University of Wageningen. ITC Publication 4, International Institute for Aerospace Survey and Earth Sciences ITC, Enschede.

Van Wirdum, C. (1983) Basic computer program 'SPEDIS 8303C' made for the programmable light reflectance meter. RIN, Leersum (unpublished).

Venkatachalam, P.S. and Patwardhan, V.N. (1953) The role *of Ascaris lumbricoides* in the nutrition of the host: effect of ascariasis on digestion of protein. *Trans. Royal Soc. Trop. Med. Hyg.* **47**, 169–175.

Verlinden, C. and Wiley, R.H. (1989) The constraints of digestive rate: an alternative model of diet selection. *Evol. Ecol.* **3**, 264–273.

Verma, M.L., Singh, N., Sidhu, G.S., Kochar, A.S. and Bathia, I.S. (1970) The 'in vitro' cellulose digestion and VFA production from some of the common Indian feeds using rumen inocula from zebu cattle and buffalo. *Ind. J. Dairy Sci.* **23**, 155–160.

Vesey-FitzGerald, D.F. (1965) Buffalo in Lake Manyara National Park. Unpublished report to the Director of Tanganyika National Parks, Arusha.

Vesey-FitzGerald, D.F. (1969) Utilisation of the habitat by buffalo in Lake Manyara National Park, Tanzania. *E. Afr. Wild. J.* **7**, 131–145.

Vesey-FitzGerald, D.F. (1974) Utilization of the grazing resources by buffaloes in the Arusha National Park. *E. Afr. Wild. J.* **12**, 107–134.

Vine, I. (1971) Risk of visual detection and pursuit by a predator and selective advantage of flocking behaviour. *J. Theor. Biol.* **30**, 405–422.

Vlek, C. and Stallen, P.J. (1981) Judging risks and benefits in the small and in the large. *Org. Behav. Human Perform.* **28**, 273–300.

Von Ostertag, R. and Kulenkampff, G. (1941) *Tierseuchen und Herdenkrankheiten in Afrika*. Band IX. *Afrika: Handbuch der praktischen Kolonialwissenschaften*. Walter de Gruyter, Berlin.

Wagnon, K.A., Loy, R.G., Rollin, W.C. and Carroll, F.D. (1966) Social dominance in a herd of Angus, Hereford and Shorthorn cows. *Anim. Behav.* **14**, 474–479.

Walker, J., Robertson, J.A., Penridge, L.K. and Sharpe, P.J.H. (1986) Herbage response to three thinning in a *Eucalyptus cebra* woodland. *Aust. J. Ecol.* **11**, 135–140.

Wallmo, O.C., Carpenter, L.H., Regelin, W.L., Gill. R.B. and Baker, D.L. (1977) Evaluation of deer habitat on a nutritional basis. *J. Range Mgt.* **30**, 122–127.

Walter, H. (1962) *Die Vegetation der Erde in Ökologischer Bertrachtung. Band I: Die tropischen und subtropischen Zonen*, Gustav Fisher, Jena.

Waltz, E.C. (1987) A test of the information-centre hypothesis in two colonies of colonial terns, *Sterna hirundo*. *Anim. Behav.* **35**, 48–59.

Wangersky, P.J. (1978) Lotka-Volterra population models. *Ann. Rev. Ecol. Syst.* **9**, 189–219.

Ward, P. (1965) Feeding ecology of the black-faced Dioch *Quelea quelea* in Nigeria. *Ibis* **107**, 173–214.

Ward, P. and Zahavi, A. (1973) The importance of certain assemblages of birds as 'information-centres' for food finding. *Ibis* **115**, 517–534.

Warden's Reports (1959–1973) Unpublished monthly reports by respective wardens of the Lake Manyara Game Reserve/Lake Manyara National Park to the Director of Tanganyika/Tanzania National Parks, Arusha.

Watermeyer, A.M. and Elliott, H.F.I. (1943) Lake Manyara. *Tangan. Notes Rec.* **1943**, 58–70.

Watson, L., Dallwitz, M.J. and Johnston, C.R. (1986) Grass genera of the world: 728 detailed descriptions from an automated database. *Aust. J. Bot.* **34**, 223–230 (plus microfiches).

Watson, R.M. (1967) The population biology of the wildebeeste (*Connochaetes taurinus albojubatus* Thomas) in the Serengeti, PhD Thesis, Cambridge University.

Watson, R.M. and Turner, M.I.N. (1965) A count of the large mammals of the Lake Manyara National Park: results and discussion. *E. Afr. Wildl. J.* **3**, 95–98.

Watts, D.P. (1985) Relation between group size and composition and feeding competition in mountain gorilla troops. *Anim. Behav.* **33**, 72–85.

Welch, J.G. and Smith, A.M. (1969) Effect of varying amounts of forage intake on rumination. *J. Anim. Sci.* **28**, 827–832.

Welch, J.G. and Smith, A.M. (1970) Forage quality and rumination time in cattle. *J. Dairy Sci.* **53**, 797–803.

Wenderoth, P. and Johnson, M. (1982a) Visual illusions with acute and obtuse angles: configurational effects and observer strategies. *Percept. Psychophys.* **31**, 243–250.

Wenderoth, P. and Johnson, M. (1982b) Constant errors do not occur in the matched reproduction of angles. *Percept. Psychophys.* **32**, 388–392.

Weyerhaeuser. F.J. (1982) On the ecology of the Lake Manyara elephants, MSc Thesis, Yale School of Forestry and Environmental Studies.

White, T.C.R. (1978) The importance of relative shortage of food in animal ecology. *Oecologia* **33**, 71–86.

Whiteman, P.C. (1980) *Tropical Pasture Science*, Oxford University Press, Oxford.

Wiens, J.A. (1989) *The Ecology of Bird Communities*, Cambridge University Press, Cambridge.

Wilde, J.K.H. (1953) The game animal factor in the control of rinderpest in tropical Africa, in *Proceedings of the International Veterinary Congress*, Stockholm, Vol. 15, 283–287.

Wilkinson, P.F. and Shank, C.C. (1976) Rutting-fight mortality among Musk oxen on Banks Islands, North-westTerritories, Canada. *Anim. Behav.* **24**, 756–758.

Williams, C. (1859, reprinted 1972) *Narratives and Adventures of Travellers in Africa*, Black Heritage Library Collection/Books for Libraries Press, Freeport, NY.

Wonnacott, T.H. and Wonnacott, R.N. (1977) *Introductory Statistics*, 3rd edn, John Wiley, New York.

Wrangham, R.W. (1977) Feeding behaviour of chimpanzees in Gombe National Park, Tanzania, in *Primate Ecology Studies of Feeding and Ranging Behaviour in Lemurs, Monkeys and Apes* (ed. T.H. Clutton-Brock), Academic Press, New York, pp. 503–538.

Wright, B. S. (1960) Predation on big game in East Africa. *J. Wildl. Mgt.* **24**, 1–15.

Wyatt, J.R. and Eltringham, S.K. (1974) The daily activity of the elephant in the Rwenzori National Park, Uganda. *E. Afr. Wildl. J.* **12**, 273–289.

Yakimoff, N. and Mitrani, L. (1979) A concept of perceived orientation. *Percept. Psychophys.* **26**, 327–330.

Ydenberg, R. C. and Dill, L.M. (1986) The economics of fleeing from predators. *Adv. Study Behav.* **16**, 229–249.

Ydenberg, R.C. and Prins, H.H.T. (1981) Spring grazing and the manipulation of food quality by barnacle geese. *J. Appl. Ecol.* **18**, 443–453.

Ydenberg, R.C. and Prins, H.H.T. (1984) Why do birds roost communally?, in *Coastal Waders and Wildfowl in Winter* (eds P.R. Evans, J.D. Goss-Custard and W.G. Hale), Cambridge University Press, Cambridge, pp. 123–139.

Ydenberg, R.C., Prins, H.H.T. and Van Dijk, J. (1983) Post-roost gatherings of wintering barnacle geese: information centres? *Ardea* **71**, 125–131.

Zadoks, J.C. (1972) Epidemieën als populatiebiologisch verschijnsel, in *Populatiebiologie*, PUDOC, Wageningen, pp. 57–79.

Zar, J.H. (1984) *Biostatistical Analysis*, 2nd edn, Prentice-Hall, Englewood Cliffs, NJ.

Index

Note: page numbers in **bold** type refer to **figures**; page numbers in *italic* type refer to *tables*.

Lightning Source UK Ltd.
Milton Keynes UK
UKOW06f0046250315

248425UK00003BA/11/P

9 780412 725203